Flow Batteries

Flow Batteries

From Fundamentals to Applications

Edited by Christina Roth, Jens Noack, and Maria Skyllas-Kazacos

Volume 3

Editors

Prof. Dr. Christina Roth
University Bayreuth
Electrochemical Process Engineering
Universitätsstraße 30
95448 Bayreuth
Germany

Adj. Assoc. Prof. (UNSW) Dr. Jens Noack
Fraunhofer-Institute for Chemical Technology
Applied Electrochemistry
Joseph-von-Fraunhofer-Str. 7
76327 Pfinztal
Germany

University of New South Wales
Mechanical and Manufacturing Engineering
2052 Sydney
Australia

Prof. Dr. Maria Skyllas-Kazacos
University of New South Wales
School of Chemical Engineering
2052 Sydney
Australia

Cover Images: Courtesy of Jens Noack; © Mimadeo/Shutterstock

All books published by **WILEY-VCH** are carefully produced. Nevertheless, authors, editors, and publisher do not warrant the information contained in these books, including this book, to be free of errors. Readers are advised to keep in mind that statements, data, illustrations, procedural details or other items may inadvertently be inaccurate.

Library of Congress Card No.: applied for

British Library Cataloguing-in-Publication Data
A catalogue record for this book is available from the British Library.

Bibliographic information published by the Deutsche Nationalbibliothek
The Deutsche Nationalbibliothek lists this publication in the Deutsche Nationalbibliografie; detailed bibliographic data are available on the Internet at <http://dnb.d-nb.de>.

© 2023 WILEY-VCH GmbH, Boschstraße 12, 69469 Weinheim, Germany

All rights reserved (including those of translation into other languages). No part of this book may be reproduced in any form – by photoprinting, microfilm, or any other means – nor transmitted or translated into a machine language without written permission from the publishers. Registered names, trademarks, etc. used in this book, even when not specifically marked as such, are not to be considered unprotected by law.

Print ISBN: 978-3-527-35201-2
ePDF ISBN: 978-3-527-83278-1
ePub ISBN: 978-3-527-83277-4
oBook ISBN: 978-3-527-83276-7

Typesetting   Straive, Chennai, India
Printing and Binding   CPI Group (UK) Ltd, Croydon, CR0 4YY

C9783527351725_230323

Contents

Volume 1

Foreword xxi
Preface xxiii
About the Editors xxvii

Part I Fundamentals 1

1 **The Need for Stationary Energy Storage** 3
   Anthony Price

2 **History of Flow Batteries** 29
   Jens Noack, Maria Skyllas-Kazacos, Larry Thaller, Gerd Tomazic, Bjorn Jonshagen, and Patrick Morrissey

3 **General Electrochemical Fundamentals of Batteries** 53
   Rudolf Holze

4 **General Aspects and Fundamentals of Flow Batteries** 69
   Luis F. Arenas, Frank C. Walsh, and Carlos Ponce de León

5 **Redox-mediated Processes** 99
   Danick Reynard, Mahdi Moghaddam, Cedrik Wiberg, Silver Sepp, Pekka Peljo, and Hubert H. Girault

6 **Membranes for Flow Batteries** 121
   Giovanni Crivellaro, Chuanyu Sun, Gioele Pagot, Enrico Negro, Keti Vezzù, Francesca Lorandi, and Vito Di Noto

7 **Standards for Flow Batteries** 155
   Jens Noack

8   **Safety Considerations of the Vanadium Flow Battery**   *175*
Adam H. Whitehead

9   **A Student Workshop in Sustainable Energy Technology: The Principles and Practice of a Rechargeable Flow Battery**   *193*
C.T. John Low, Carlos Ponce de León, Richard G.A. Wills, and Frank C. Walsh

Part II   Characterization of Flow Batteries and Materials   *213*

10   **Characterization Methods in Flow Batteries: A General Overview**   *215*
Christina Roth and Marcus Gebhard

11   **Electrochemical Methods**   *229*
Jonathan Schneider, Tim Tichter, and Christina Roth

12   **Radiography and Tomography**   *263*
Roswitha Zeis

13   **Characterization of Carbon Materials**   *281*
Michael Bron, Julia Melke, and Matthias Steimecke

14   **Characterization of Membranes for Flow Batteries**   *307*
Jochen Kerres, Nico Mans, and Henning Krieg

Part III   Modeling and Simulation   *333*

15   **Quantum Mechanical Modeling of Flow Battery Materials**   *335*
Piotr de Silva

16   **Mesoscale Modeling and Simulation for Flow Batteries**   *355*
Jia Yu and Alejandro A. Franco

17   **Continuum Modelling and Simulation of Flow Batteries**   *379*
Jakub K. Włodarczyk, Gaël Mourouga, Roman P. Schärer, and Jürgen O. Schumacher

18	**Pore-scale Modeling of Flow Batteries** 413
	Amadeus Wolf, Susanne Kespe, and Hermann Nirschl

19	**Dynamic Modelling of Vanadium Flow Batteries for System Monitoring and Control** 443
	Jie Bao and Yitao Yan

20	**Techno-economic Modelling and Evaluation of Flow Batteries** 463
	Christine Minke and Thomas Turek

21	**Machine Learning for FB Electrolyte Screening** 487
	Laura-Sophie Berg, Jan Hamaekers, and Astrid Maass

Volume 2

Foreword *xvii*
Preface *xix*
About the Editors *xxiii*

Part IV Vanadium Flow Batteries 507

22	**The History of the UNSW All-Vanadium Flow Battery Development** 509
	Maria Skyllas-Kazacos

23	**Vanadium Electrolytes and Related Electrochemical Reactions** 539
	Nataliya V. Roznyatovskaya, Karsten Pinkwart, and Jens Tübke

24	**Electrodes for Vanadium Flow Batteries (VFBs)** 563
	D.N. Buckley, A. Bourke, N. Dalton, M. Alhajji Safi, D. Oboroceanu, V. Sasikumar, and R.P. Lynch

25	**Membranes for Vanadium Flow Batteries** 589
	Purna Chandra Ghimire, Arjun Bhattaraj, Nyunt Wai, and Tuti Mariana Lim

26	**Advanced Flowfield Architecture for Vanadium Flow Batteries** 607
	Yasser Ashraf Gandomi, D. Aaron, and M.M. Mench

27 **State-of-Charge Monitoring for Vanadium Redox Flow Batteries** *627*
    Yifeng Li

28 **Rebalancing/Regeneration of Vanadium Flow Batteries** *641*
    Nicola Poli, Andrea Trovò, and Massimo Guarnieri

29 **Life Cycle Analysis of Vanadium Flow Batteries** *659*
    Carmen M. Fernández-Marchante, María Millán, and Justo Lobato

30 **Next-Generation Vanadium Flow Batteries** *673*
    Chris Menictas and Maria Skyllas-Kazacos

31 **Asymmetric Vanadium-based Aqueous Flow Batteries** *689*
    Soowhan Kim, Litao Yan, and Wei Wang

    Part V Other Important Inorganic Flow Battery Technologies *709*

32 **Zn/Br Battery – Early Research and Development** *711*
    Gerd Tomazic

33 **Iron–Chromium Flow Battery** *741*
    Huan Zhang and Chuanyu Sun

34 **An Overview of the Polysulfide/Bromine Flow Battery** *765*
    Patrick Morrissey

35 **Fe/Fe Flow Battery** *791*
    Robert F. Savinell, Nicholas Sinclair, Xiaochen Shen, Julia Song, and Jesse S. Wainright

36 **Zinc–Cerium and Related Cerium-Based Flow Batteries: Progress and Challenges** *819*
    Luis F. Arenas, Frank C. Walsh, and Carlos Ponce de León

37 **Undivided Copper–Lead Dioxide Flow Battery Based on Soluble Copper and Lead in Aqueous Methanesulphonic Acid** *837*
    R.C. Tangirala, F.C. Walsh, and C. Ponce de León

38 All-copper Flow Batteries 855
Laura Sanz, Wouter D. Badenhorst, Giampaolo Lacarbonara, Luigi Faggiano, David Lloyd, Pertti Kauranen, Catia Arbizzani, and Lasse Murtomäki

39 Hydrogen-Based Flow Batteries 875
Douglas I. Kushner and Adam Z. Weber

Volume 3

Foreword *xvii*
Preface *xix*
About the Editors *xxiii*

Part VI Organic Flow Batteries 895

40 Aqueous Organic Flow Batteries 897
Yan Jing, Roy G. Gordon, and Michael J. Aziz

40.1 Advantages 898
40.2 Challenges and Opportunities 900
40.2.1 Solubility 900
40.2.2 Viscosity 900
40.2.3 Crossover 901
40.2.4 Lifetime 901
40.2.5 Analytic Methods 903
40.2.6 Molecular Engineering 904
40.2.7 Cost 905
40.2.8 Membrane 906
40.2.9 pH Imbalance 907
40.2.10 Toxicity 907
40.3 Classes of Aqueous Organic Redox Actives 908
40.3.1 Redox Aromatic Carbonyl Compounds (Quinones, Fluorenones) 908
40.3.2 Aza-/Azo-Aromatics (Alloxazines, Phenazines, Azobenzenes, and Viologens) 910
40.3.3 Nitroxide Radicals 910
40.3.4 Metal Coordination Complexes 911
40.4 Properties of Aqueous Organic Redox Actives 912
40.4.1 Range of Redox Potential 912
40.4.2 Range of Solubility 912
40.4.3 Range of Fade Rate 913
40.4.4 General Decomposition Mechanisms 913

40.4.5	Range of Operational pH	*913*
40.4.6	Recomposition of Redox-active Molecules	*915*
40.5	Performance of AOFBs	*915*
40.6	Outlook for AOFBs	*918*
	Acknowledgments	*919*
	References	*919*

41 Metal Coordination Complexes for Flow Batteries *923*
Benjamin D. Silcox, Curt M. Wong, Xiaoliang Wei, Christo Sevov, and Levi T. Thompson

41.1	Introduction	*923*
41.1.1	Background	*923*
41.1.2	Overview of Metal Coordination Complexes in Flow Batteries	*924*
41.1.3	Chapter Overview	*925*
41.2	Aqueous Metal Coordination Complex Based Flow Batteries	*926*
41.2.1	Catholyte Chemistries	*926*
41.2.1.1	Ferrocene Complexes	*926*
41.2.1.2	Ferro/ferricyanide Complexes	*929*
41.2.1.3	Fe Oligo-Aminocarboxylate Complexes	*930*
41.2.1.4	Co-Based Complexes	*931*
41.2.2	Anolyte Chemistries	*931*
41.2.2.1	Fe-Based Complexes	*931*
41.2.2.2	Cr-Based Complexes	*933*
41.3	Non-Aqueous Metal Coordination Complex Based Flow Batteries	*933*
41.3.1	Chalcogen (M–O/S) Chemistries	*933*
41.3.1.1	Acetylacetonate Complexes	*933*
41.3.1.2	Amino-Alcohol Complexes	*937*
41.3.1.3	Tunable Oxo Complexes	*937*
41.3.2	Pyridyl Chemistries	*938*
41.3.3	Metallocene Chemistries	*941*
41.4	Conclusion	*941*
	References	*941*

42 Organic Redox Flow Batteries: Lithium-Ion-based FBs *951*
Feifei Zhang and Qing Wang

42.1	Introduction	*951*
42.2	Semi-solid Electroactive Materials for LFBs	*951*
42.2.1	Electroactive Materials Based on Li^+ Insertion-Extraction Chemistry	*953*
42.2.2	Electroactive Materials Based on Precipitation-Dissolution Chemistry	*955*
42.2.3	Electroactive Materials Based on Multiple Redox Reactions	*959*
42.3	Redox Targeting-based LFBs	*960*
42.3.1	Principles of Redox Targeting-based LFBs	*960*
42.3.2	Development of Redox Targeting-based LFBs	*961*

42.3.3	Redox Targeting-based Lithium–Sulfur Flow Batteries	965
42.3.4	Redox Targeting-based Lithium–Oxygen Flow Batteries	966
42.4	Challenges and Outlook	968
	References	969

43 Nonaqueous Metal-Free Flow Batteries 975
Kathryn Toghill and Craig Armstrong
Preamble 975

43.1	Introduction	976
43.2	Catholytes	979
43.2.1	Nitroxyl Radical	979
43.2.2	Dialkoxybenzene Derivatives	980
43.2.3	Phenothiazine and Phenazine	980
43.2.4	Cyclopropenium Derivatives	981
43.2.5	Amines	982
43.3	Anolytes	982
43.3.1	Phthalimide (N-MP)	982
43.3.2	Benzophenone (BP)	984
43.3.3	Benzothiadiazole (BTZ)	984
43.3.4	Nitrobenzene (NB)	984
43.3.5	Fluorenone (FL)	985
43.3.6	Pyridine Derivatives (Py and BPy)	985
43.3.7	Viologen	986
43.3.8	Quinones and Quinoxaline	987
43.4	Symmetric and Bipolar Redox Materials	987
43.5	Limitations and Challenges	989
43.5.1	Electrolyte Stability	989
43.5.2	Electrolyte Conductivity	992
43.5.3	Membrane Incompatibility	993
43.5.4	Electrolyte Cost	993
43.5.5	Solubility and Energy Density	995
43.5.6	Outlook – Transitional Developments	996
43.5.7	Conclusions	997
	References	998

44 Polymeric Flow Batteries 1007
Oliver Nolte, Martin D. Hager, and Ulrich S. Schubert

44.1	Introduction	1007
44.2	Basic Organic Redox Moieties	1008
44.3	Oligomers and Polymers	1010
44.4	Examples of Polymeric FBs	1014
44.5	Countering the Challenges	1017
44.6	Conclusion and Outlook	1018
	References	1019

Part VII Industrial and Commercialization Aspects of Flow Batteries 1025

45 Inverter Interfacing and Grid Behaviour 1027
John Fletcher and Jiacheng Li
45.1 Introduction 1027
45.2 The Six-Switch, Three-Phase Inverter Circuit 1027
45.2.1 The Inverter DC-Link Model 1031
45.2.2 Three-Phase Inverter and Pulse-Width Modulation 1031
45.2.3 Inverter Control Schemes – Grid Feeding 1033
45.3 Inverter Control Modes for Energy Storage Applications 1035
45.3.1 Fault Response 1037
45.3.2 Single-phase Inverters 1037
45.3.3 Inverters for Flow Battery Energy Storage 1038
45.4 Conclusions 1039
 Reference 1039

46 Flow-Battery System Topologies and Grid Connection 1041
Thomas Lüth, Thorsten Seipp, and David Kienbaum
46.1 Introduction 1041
46.1.1 Power-Conditioning System (PCS) 1041
46.1.2 Shunt Currents 1042
46.1.3 Reliability 1042
46.1.4 SoC Band Limitation 1043
46.1.5 Modularity/Flexibility 1043
46.1.6 System Size 1043
46.2 Topologies 1044
46.2.1 Low-Voltage Parallel Connection (LV-P) 1044
46.2.2 Low-Voltage System with Several Inverters (LV-DC/AC) 1045
46.2.3 Low-Voltage System with Several DC/DC Converters and One Central Inverter (LV-DC/DC) 1046
46.2.4 High-Voltage System with One Tank Pair (HV-1T) 1046
46.2.5 High-Voltage System with Multiple Tank Pairs (HV–MT) 1048
46.2.6 Mixed Parallel-Series High-Voltage System (HV-MIX) 1049
46.3 Evaluation of the Topologies 1050
46.4 Summary 1052
 References 1052

47 Vanadium FBESs installed by Sumitomo Electric Industries, Ltd 1055
Toshio Shigematsu and Toshikazu Shibata
47.1 Historical Overview [1–3] 1055
47.2 Typical Vanadium FBESs Delivered by Sumitomo Electric 1058

47.2.1	1990s *1058*
47.2.2	2000s *1059*
47.2.2.1	A 500 kW×10 hours Vanadium FBES with Underground Tanks Installed at a University [2001–2011] *1059*
47.2.2.2	A 1.5 MW × 1 hours Vanadium FBES with Momentary Voltage Drop Compensation Function in a Factory [2001–2007] *1059*
47.2.2.3	A 4 MW × 1.5 hour Vanadium FBES Installed in a Wind Farm [2003–2007] *1061*
47.2.3	Recent Vanadium FBESs Since 2010 *1065*
47.2.3.1	A Vanadium FBES for Optimal Energy Management in a Factory with Photovoltaic Power Generation (in Operation Since 2012) *1065*
47.2.3.2	Operation Example of Grid Control (2015–2018 Demonstration, 2019 ∼ Practical Operation) *1067*
47.2.3.3	Operation Example of Wholesale Market [2017∼] *1070*
47.2.3.4	Microgrid Operation [2018∼] *1072*
47.2.3.5	Operation in an Off-grid Area [2019∼] *1074*
47.3	Summary *1076*
	References *1076*

48   Industrial Applications of Flow Batteries *1079*
Pavel Mardilovich and Martin Harrer

48.1	Company History: How Funktionswerkstoffe Forschungs und Entwicklung GmbH Became Enerox, and What Is CellCube *1079*
48.2	Elephant in the Room: Can Flow Batteries Deliver the Duration They Promise? *1085*
48.3	Anatomy of A Project: What Does It Take to Put in a MW-Plus Battery in the Field? *1088*
48.4	I'm on a Boat!: Innovation at CellCube *1095*
48.5	Sunny Upside: An Egg Pun or a Realistic Outlook? *1096*
	Acknowledgements *1097*
	Reference *1098*

49   Applications of VFB in Rongke Power *1099*
Huamin Zhang

49.1	Development and Application of Core Materials for VFB *1100*
49.1.1	Electrolyte *1100*
49.1.2	Bipolar Plate *1101*
49.1.3	Ion-Conducting Membrane *1102*
49.1.3.1	Excellent Ion Conductivity *1103*
49.1.3.2	High ion Selectivity *1103*
49.1.3.3	Outstanding Mechanical and Chemical Stability *1103*
49.1.3.4	Low Cost for Large-Scale Commercial Application *1103*

49.1.4	Key Technologies of Stack and Energy Storage System of VFB in RKP–DICP Team *1105*
49.1.5	Applications of VFB Technology *1107*
49.1.5.1	VPower *1107*
49.1.5.2	TPower *1108*
49.1.5.3	ReFlex *1108*
	References *1112*

50 Metal-Free Flow Batteries Based on TEMPO *1115*
Tobias Janoschka and Olaf Conrad

50.1	Introduction *1115*
50.2	Properties, Synthetic Procedures, and Redox Reactions *1115*
50.2.1	TEMPO *1115*
50.2.2	Viologen *1118*
50.3	TEMPO Flow Batteries – Selected Examples *1120*
50.3.1	Fundamental Research *1120*
50.3.2	Scale-up *1122*
50.4	Raw Materials Outlook and Summary *1124*
	References *1124*

51 Commercialization of All-Iron Redox Flow-Battery Systems *1127*
Julia Song

51.1	Introduction *1127*
51.2	Background *1128*
51.3	Key IFB Technology Breakthroughs and IFB System Commercialization *1131*
51.3.1	The Proton Pump *1131*
51.3.2	IFB Power Module *1134*
51.3.3	System Optimization and LCOS *1138*
51.4	Conclusions *1142*
	References *1143*

52 Application of Hydrogen–Bromine Flow Batteries: Technical Paper *1145*
Wiebrand Kout and Yohanes A. Hugo

52.1	Introduction *1145*
52.2	Energy Domain *1145*
52.2.1	Electrolyte *1145*
52.2.2	Hydrogen *1146*
52.3	Power Domain *1147*
52.3.1	Membrane Electrode Assembly *1147*

52.3.1.1	H_2 Electrode	*1147*
52.3.1.2	Membranes	*1147*
52.3.1.3	Liquid-Side Electrode	*1148*
52.3.2	Stack	*1148*
52.3.3	Balance of System	*1149*
52.4	Application	*1150*
52.4.1	Field Tests	*1150*
52.4.2	Permittability	*1150*
52.4.3	Outlook	*1151*
52.5	Conclusion	*1151*
	References	*1152*

53	**Some Notes on Zinc/Bromine Flow Batteries**	*1153*
	Bjorn Hage	
53.1	Modern Large Scale ZB Flow Battery	*1153*
53.2	Energy Storage Cost	*1153*
53.3	The Gould/ERC ZBF Battery	*1155*
53.4	The Gould/ERC ZBF Battery Stack	*1155*
53.5	The Gould/ERC ZBF Battery Stack Electrolyte System	*1157*
53.6	The Gould/ERC ZBF Battery Stack Electrolyte Circulation System	*1157*
53.7	Shunt Currents – Are They for Real?	*1158*
53.8	Can Battery Efficiency be Over 100% ?!	*1161*
53.9	Zinc–Bromine Battery Efficiency	*1163*
53.10	The Solid Bed $ZnBr_2$ Flow Battery	*1166*
53.11	Flow Battery Auxiliary Power	*1167*
53.12	Summary	*1169*

54	**Mobile Applications of the ZBB**	*1171*
	Gerd Tomazic	
54.1	Introduction	*1171*
54.2	Scheme of the Zink–Bromine Battery	*1172*
54.3	EV-Application Issues	*1173*
54.3.1	Performance	*1174*
54.3.2	Operation	*1175*
54.3.2.1	Stripping After Every Cycle (SAEC)	*1175*
54.3.2.2	Random Cycling (RC)	*1175*
54.3.2.3	Standby	*1176*
54.3.2.4	Shunt Current Interruption	*1176*
54.3.2.5	State-of-Charge (SOC) Measurement	*1176*
54.3.2.6	Thermal Management	*1177*
54.3.3	Safety	*1178*
54.3.3.1	Measures Against and in Case of Leakage	*1178*

54.3.3.2 Crash Test *1181*
54.3.3.3 Vibration Test *1183*
54.3.3.4 Electrical Short *1184*
54.3.3.5 Hydrogen Recombination *1184*
54.4 Testing of Electric Vehicles *1185*
54.4.1 Controller *1185*
54.4.2 Basic Design at Project Start *1185*
54.4.3 Colenta Test Vehicle *1186*
54.4.4 City Stromer *1188*
54.4.5 Mini-El City *1190*
54.4.6 Colenta Mini Cab Bus *1191*
54.4.7 Hotzenblitz Commuter Car *1192*
54.4.8 ELIN-VW Bus *1192*
54.4.9 Fiat Panda *1193*
54.4.9.1 12 Electric Hours of Namur 1991 *1196*
54.4.10 DAEWOO Electric Vehicle *1198*
54.4.11 Geo PRISM Sedan: 35 kWh Battery Delivery *1199*
54.5 Summary *1200*

Index *1201*

Foreword

Flow Batteries – in the Beginning
Larry Thaller

The energy crisis in the early 1970s saw the emergence of battery technologies that were previously nonexistent or were not seen to be of any use. However, some within the electric utility industry sensed a useful application which would store off-hour electricity production for later use during times of peak usage. That would be one way of addressing the energy crisis. The term "bulk energy storage" came into use.

Traditional batteries at that time had bothersome morphology problems leading to slumping, shedding, shorting, etc. as well as expensive and time-consuming manufacturing and assembly costs. This resulted in an opening for out-of-the-box ideas for long-life, low-cost batteries that might be of use for storing larger amounts of energy than had been the norm. In response to this situation, NASA headquarters gave permission to their field centers to apply their technologists and scientists to address terrestrial energy-related issues.

Over the past fifty years, ideas have come and gone, been tried again, and carefully examined to see whether they could be developed, tested, and brought into service for a particular application. One such idea originating at the NASA Glenn Research Center was named redox flow batteries. This concept used two liquid redox couples, one for the anode and a second one for the cathode. The anode and cathode liquids were separated by a membrane through which acid ions could move during the charge–discharge process. The majority of the solutions were stored in tanks and pumped through stacks of cells placed hydraulically in parallel and electrically in series. This arrangement allowed the voltage and capacity to be varied separately according to the particular application.

A series of engineering studies and laboratory experiments were conducted starting from scratch to explore how the idea of redox flow batteries could be transformed into working hardware. At the same time materials had to be evaluated for use as electrodes, separators, cell frames, pumps, etc. About six years later a fully functioning system was displayed and explained at an energy-related trade show in Washington DC. As a consequence of these trade studies aimed at increasing the round-trip efficiency of the energy storage process the interrelationship between the cell performance, shunt losses, and pumping power required an in-depth understanding of these three parameters to maximize this all-important feature.

Other concepts and technologies were of course being explored in parallel with the NASA development – each seeking funding for their ideas within other private and government laboratories. In the interim, early versions of several flow battery systems entered service in electric utility applications as well as on customer side of the meter applications. These early applications pointed to the vast potential market size for flow batteries. This accelerated supply chain interest in research and development activities directed toward components for these systems. The book you have before you summarizes the current status of some of the flow battery-related concepts as they are today.

Preface

Dear Readers,

The research and development of flow batteries has experienced considerable growth in the last ten years. Whereas before that time, very few people were familiar with the term, today a multitude of research groups are engaged in the development of flow batteries of all kinds. Companies are emerging for the commercialisation of flow batteries and their components, and an ever-increasing number and size of flow batteries have been installed around the world. Flow batteries have always had the dilemma of being ahead of their time. With the exception of the zinc/bromine system, flow batteries have been under development since the 1940s with the aim of storing energy from fluctuating renewable energy sources. Renewable energy sources were far too expensive after World War II and there were other problems to overcome. It was not until the oil price shock at the beginning of the 1970s and after publication of the study *The Limits to Growth* in 1972 that the world first became aware that this type of energy production was not sustainable, thus motivating NASA's Flow Battery developments. However, the shock was short-lived and hegemony turned again in favour of fossil energy sources. This only changed with the arrival of affordable photovoltaic and wind energy from the mid-2000s. To this day, however, electrical grids, with the exception of isolated solutions, are characterised by a main energy input from fossil energy sources. However, the enormous growth in recent years and the resulting mass production of photovoltaic and wind energy systems have led to a drastic reduction in the price of wind and solar energy, so that today they are by far the cleanest and cheapest energy sources. This lays the foundation for the restructuring of the energy grids, which in the future will be based primarily and to a greater extent on renewable energy sources.

Despite various efforts by different actors, a clear dominance of renewable energy sources over fossil resources in the future seems inevitable and is also to be preferred for a variety of reasons. Along with this, the situation for stationary storage has changed significantly to compensate the power fluctuations and realise the needed energy shifts. More and ever-larger installations of the most diverse technologies have taken place in the last ten years. Today, storage systems in the range of over 100 MWh are state of the art. Due to their many advantages, flow batteries are candidates for storage times of approximately four to six hours for inexpensive, safe, and clean energy storage. Starting from home storage systems of a few kWh, up to sizes of several hundred MWh, they can accommodate the storage needs of most electrical grids. Even as storage for mobile applications in the field of maritime electric-based transport, the use of flow batteries is currently being considered.

Due to their niche existence, flow batteries have previously been largely neglected in public funding for research and development. Today, there are special funding programmes for flow batteries in Germany and other countries in the European Union, China, the USA, and many other countries around the world. This has led to a significant increase in publications, which means that on the one hand, many more aspects of well-known technologies such as the vanadium/vanadium flow battery are being investigated, and on the other hand, completely new types of flow battery storage systems are being developed. These involve above all, a large number of different new active materials and material combinations that might potentially reduce cost and/or increase energy density. In particular, the research activities on organic active materials have increased greatly since 2015 due to the immense possibilities of these compounds. In addition, there are completely new flow battery concepts. These are mediator-based flow batteries or those with solid boosters to increase energy density, as well as photocatalytically active systems and those based on biological materials.

The significant increase in the relevance of flow batteries and the resulting sharp rise in knowledge has spurred us to compile this book. Our aim was to publish a book that is as comprehensive as possible, covering all important aspects of research and development in the field of flow batteries, as well as their context, such as history, standardisation, and industry aspects, so that it can be used as a daily reference book in which all relevant information is included and appropriate references provided. We have arranged the information in a strictly systematic way. To do this, we first considered the most important aspects in the field of flow batteries and then in turn asked some of the important actors in the respective fields for contributions. This resulted in 55 individual contributions, which we divided into seven different parts. There are six academic parts, written by representatives from different universities and research institutes. The first part covers general aspects of flow batteries and is intended to provide a general understanding of the structure and features of flow batteries. The following five parts address different aspects of flow batteries, with each chapter reflecting the current state of research and development. For the seventh part, the most important representatives of commercial actors were asked for contributions. This part includes aspects of the application of flow batteries and is also intended to provide the reader with an overview of various already-commercialised technologies.

The systematics also motivated us to think about the terms 'redox flow battery' and 'flow battery'. In the literature, the term 'redox flow battery' or sometimes 'redox battery' is used for flow batteries. We wanted to avoid mixing the different terms and ensure consistency of the terms throughout the book. Fortunately, since 2020 there has been an international standard, IEC 62932-1:2020, which defines these battery types and recommends a standardised nomenclature. We have therefore decided to adopt these definitions and terms and consistently refer to electrochemical energy converters that use flowing media storage materials as 'flow batteries' and those with a deposition reaction as 'hybrid flow batteries'.

We would like to take this opportunity to thank Larry Thaller in particular for the foreword and the authors of the numerous contributions, without whom this book would not have been possible. In our opinion, all contributions are of outstanding quality and provide even seasoned researchers in flow batteries with a wealth of new information that has not been published anywhere before. We hope you will enjoy this book, find the information you need, and succeed in your research and development activities.

Sincerely,
Christina, Maria, Jens

About the Editors

Christina Roth is a trained Materials Scientist who earned her doctorate in 2002 at the Technical University of Darmstadt in the group of Prof. Hartmut Fueß. In 2003, she joined the team of Prof. Richard Nichols in Liverpool with a Feodor-Lynen fellowship. Christina was awarded a German Junior Professorship at TU Darmstadt in 2004 before becoming a full professor of Applied Physical Chemistry at the Freie Universität Berlin in 2012.

In 2019, she accepted an appointment at the University of Bayreuth, where she has since held the Chair of Electrochemical Process Engineering at the Faculty of Engineering. Her research focuses on fuel cells, redox flow batteries, lithium-ion batteries, and CO_2 electro-reduction with emphasis on operando spectroscopy and dedication toward structuring of 3D porous electrodes.

Christina co-authored more than 150 publications, including 8 book chapters. She has held almost 100 oral presentations at conferences worldwide, many of them upon invitation. She is a member of several scientific committees and advisory boards, including ChemElectroChem, GDCh Fachgruppe Elektrochemie, CENELEST and the Bavarian Center for Battery Technology, and the Center of Energy Technology at the University of Bayreuth.

Jens Noack studied chemical engineering and environmental technology at the Dresden University of Applied Sciences and received his PhD from Karlsruhe Institute of Technology. Since 2007, he has been working at the German Fraunhofer Institute for Chemical Technology in the Department of Applied Electrochemistry, where he has mainly been involved in the development of flow batteries.

From 2009 to 2011, he was acting group leader of the newly formed Redox Flow battery group and subsequently deputy group leader. Since then he has been working at Fraunhofer ICT as project lead, scientist, and senior engineer. Since 2019, he is the Deputy Director of the German–Australian Alliance for Electrochemical Technologies for Storage of Renewable Energy (CENELEST). Since 2020, he is an Adjunct Associate Professor at the University of New South Wales in Sydney, Australia. His research and development focus on stationary energy storage systems for renewable energies and the development of novel energy storage and conversion systems.

Jens is the author of over 100 publications, including 57 peer-reviewed papers and three review articles. He has given over 70 talks at conferences worldwide and has 29 patent applications. He is a member of the scientific committee of the International Flow Battery Forum and several national and international standardization committees. He is chairman of the German National Committee "Flow Batteries" in the German Commission for Electrical Engineering and Information Technology DKE AK 371.0.6, and also a member of the International Electrotechnical Commission IEC in the groups TC 105/21 JWG 7 "Flow Batteries" and JWG 82 "Secondary cells and batteries for renewable energy storage."

After graduating with 1st Class Honors and the University Medal in Industrial Chemistry at UNSW Sydney, **Maria Skyllas-Kazacos** completed her PhD at the same university in 1979 in the area of High-Temperature Molten Salt Electrochemistry under the supervision of Professor Barry Welch. She then spent a year as a Member of the Technical Staff at Bell Telephone Laboratories in Murray Hill, New Jersey, USA, where she worked with Adam Heller and Barry Miller on liquid junction solar cells and with John Broadhead on lead–acid batteries. On returning to Australia, she was awarded a Queen Elizabeth II Postdoctoral Fellowship to continue her work on liquid-junction solar cells with Prof Dan Haneman at UNSW Sydney. In 1982, she was appointed to the academic staff in the School of Chemical Engineering and Industrial Chemistry where she initiated research programs in aluminum electrowinning and began her pioneering work on vanadium flow batteries. After filing the first patent on the vanadium redox flow battery in 1986, Maria expanded her research team's efforts into all areas of the vanadium battery technology, from electrolyte production to stack materials, design and manufacture, sensors, and control system development, while also completing several field trials. Over the next 30 years, her group's work led to more than 40 new patents, which currently form the basis of the vanadium flow battery technology that is being commercialized around the world.

In addition to these patents, Maria has published over 200 refereed papers in international journals, has written 8 book chapters, and has edited or co-edited more than 12 books and conference proceedings. Her pioneering work on the vanadium battery has been recognized through many honors and awards, including the Chemeca,

Wiffen, R.K. Murphy, and Castner Medals. In 1999 she was made a Member of the Order of Australia, and in 2009, was Invested as Grand Lady of the Byzantine Order of St Eugene of Trebizond. Maria is a Fellow of the Australian Academy of Technological Sciences and Engineering, the Royal Australian Chemical Institute, and the Institution of Engineers Australia.

Part VI

Organic Flow Batteries

40

Aqueous Organic Flow Batteries

Yan Jing[1], Roy G. Gordon[1,2], and Michael J. Aziz[2]

[1] Harvard University, Department of Chemistry and Chemical Biology, 12 Oxford Street, Cambridge, MA 02138, USA
[2] Harvard University, Harvard John A. Paulson School of Engineering and Applied Sciences, 29 Oxford Street, Cambridge, MA 02138, USA

An organic flow battery (OFB) is a flow battery that reversibly converts electrical energy into the chemical energy of bond formation and cleavage through redox reactions of organic or metalorganic molecules. The two electrolytes contain redox species with distinct redox potentials: to conveniently distinguish them, hereafter we call the negolyte (negative electrolyte) the electrolyte comprising the lower-potential species and the posolyte (positive electrolyte) the electrolyte comprising the higher-potential species.

Featuring their rich redox chemistry, structural diversity and tunability, and earth-abundance of composing elements, organic molecules are promising as energy-carrier materials. OFBs are therefore the subject of a great deal of R&D. Depending on whether the redox-active compounds are dissolved in aqueous or nonaqueous solvents, OFBs are further categorized into aqueous organic flow batteries (AOFBs) and nonaqueous organic flow batteries (NAOFBs). Organic solvents typically utilized demonstrate broad stability windows of >4 V; and therefore, nonaqueous FBs offer opportunities for higher energy density if organic redox couples with high solubility and extreme redox potentials are developed to leverage the broad voltage window of organic solvents [1]. Aqueous electrolytes are highly ion-conductive and offer higher power density; they are also typically nonflammable and have negligible solvent cost. For these reasons, we focus on AOFBs.

AOFBs have typically utilized quinones, viologens, phenazines, nitroxide radicals, aza-aromatics, and iron coordination complexes [2]. They operate in acid, base, or near-neutral pH, typically with ion-selective membranes designed to transport monatomic cations or anions. Great progress has been made in developing organic redox molecules for AOFBs, particularly for negolytes. In contrast, the development of active species for posolytes has lagged behind. Several review articles have been published recently by different groups with their own perspectives on the development of AOFBs [2–8]. For example, Kwabi, Ji, and Aziz recently collected, analyzed, and compared capacity fade rates from all aqueous organic

Flow Batteries: From Fundamentals to Applications, First Edition.
Edited by Christina Roth, Jens Noack, and Maria Skyllas-Kazacos.
© 2023 WILEY-VCH GmbH. Published 2023 by WILEY-VCH GmbH.

electrolytes that have been utilized in the capacity-limiting side of flow or hybrid flow/nonflow cells; categorized capacity fade rates as being "high" (>1%/day), "moderate" (0.1–1%/day), "low" (0.02–0.1%/day), and "extremely low" (≤0.02%/day), and discussed the degree to which the fade rates have been linked to decomposition mechanisms. Because of the aforementioned reviews, here we aim to provide a brief and general overview.

40.1 Advantages

Advantageous features of aqueous-soluble redox organics are elaborated below.

1. Organic and metalorganic compounds composed of earth-abundant elements such as C, H, O, N, S, P, and Fe are potentially low-cost. In addition, redox organic molecules could possibly be derived from renewable biomass. In particular, depolymerized biomass can be an important feedstock of redox organic molecules [9]. Manufacturing organics or metalorganics on a large scale could be advantageous if cost-effective and readily accessible bio-based feedstock is used.
2. Redox organic molecules are able to undergo multiple electron transfers. For example, quinone and phenazine derivatives have demonstrated two-electron transfer in both aqueous and nonaqueous electrolytes. This doubles the volumetric capacity compared to single-electron transfer molecules of the same solubility, and also decreases the concentration of radicals, which tend to undergo undesirable side reactions.
3. Redox organic and metalorganic compounds have much larger molecular sizes than redox inorganic compounds. Crossover rates of redox organics are thus effectively suppressed, leading in some cases to extremely low permeability [10–12].
4. Molecular engineering is continually being used to improve the performance of AOFBs. Improved molecular design has proven to be powerful in adjusting redox potentials, enhancing molecular solubility, improving molecular stability, and ultimately developing high-performing molecule-based electrolytes with adequate volumetric capacity and extremely long lifetime.
    a. The redox potentials of molecules can be adjusted by introducing functional groups. For instance, the sulfonate group is electron-withdrawing; hence, the redox potential of anthraquinone-2,7-disulfonic acid (AQDS) is 50 mV higher than that of anthraquinone-2-sulfonic acid (AQS) because of the two sulfonate groups in AQDS and one in AQS [13]. Similarly, the deprotonated hydroxyl group is electron-donating; hence the potential of 2,3,6,7-tetrahydroxyanthraquinone is found 130 mV lower than that of 2,6-dihydroxyanthraquinone because of the two extra negative charges from phenoxide groups [14].
    b. Introducing water-solubilizing functional groups, including ionic groups such as SO_3^- [15], O^- [14], PO_3^{2-} [16], COO^- [10], ammonium cations [17]; and nonionic hydrophilic groups, e.g. polyethylene glycol (PEG) [11], to hydrophobic redox molecules via covalent bonds can enhance water solubility

by several orders of magnitude. For instance, anthraquinone itself is virtually insoluble (<1 mM) in water; the solubility of 2,6-dihydroxyanthraquinone (DHAQ) in pH 14 KOH reaches 0.5 M; the solubility of AQDS in pH 0 H_2SO_4 reaches 1.5 M; and 1,8-bis(2-(2-(2-hydroxyethoxy)ethoxy)-ethoxy)anthracene-9,10-dione (AQ-1,8-3E-OH) is completely water-miscible with a molecular concentration of 2.2 M.

c. Significant improvements in molecular stability have been made by different groups through rational molecular design [1, 2, 18], and extremely stable redox molecules, exhibiting capacity fade rates <0.02%/day in AOFBs, have been demonstrated recently [2, 10, 12, 16, 19].

Figure 40.1 shows a schematic of redox organic molecule discovery paradigm. Discoveries of important organic redox molecules for OFBs have often been achieved by a multi-disciplinary team. First, guided mainly by chemical insight, organic chemists come up with a molecule pool filled with numerous desirable redox molecules in mind. Second, they select a synthesizable molecule from the pool with retrosynthetic analysis. Sometimes they have to "redesign" or slightly modify molecular structure given the accessible precursors. They explore a series of synthetic conditions to obtain the desired pure product, whose structure is well-characterized by different techniques. Third, electrochemists evaluate and further downselect those molecules with fast electrochemical reaction kinetics, high aqueous solubility, and high stability. Fourth, flow cell experts run the selected molecule-based aqueous flow cells including cell cycling and polarization curves to evaluate metrics of cell performance such as open-circuit voltage, area-specific resistance, peak galvanic power density, and capacity fade rate. It is essential to perform post-cycling chemical analyses as, in most instances, redox molecules decompose too rapidly to meet a decadal lifetime criterion: learning the decomposition mechanisms is the first step in redesigning molecules for improved performance. After a few rounds of these iterations, there might be a candidate meeting the critical

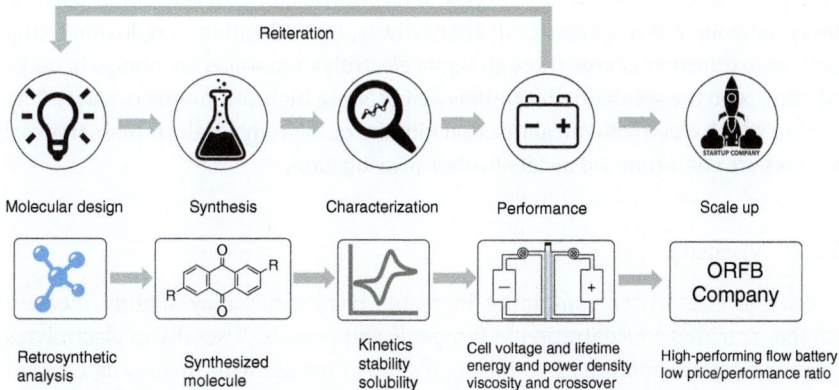

Figure 40.1 Schematic of redox organic molecule discovery paradigm. The top row presents the logical flow chart in general; the bottom row indicates what needs to be implemented and evaluated to move the species toward OFB commercialization.

requirements for commercialization. Once the molecules are determined, organic chemists further optimize the synthetic routes and conditions to make sure the syntheses are environmental friendly and inexpensive at mass-production scale. If development goes well, the intellectual property is commercialized by a startup company or licensed to an established company.

40.2 Challenges and Opportunities

40.2.1 Solubility

The aqueous solubility of redox molecules designed for AOFBs is significantly influenced by the type and number of water-solubilizing groups and the choice of supporting salt. Most of the reported water-soluble redox molecules have demonstrated charge-storing capability of more than one mole electrons per liter of electrolyte. The aqueous solubility of redox molecules can be increased by incorporating multiple water-solubilizing groups via covalent bonds [15]. Introducing molecular asymmetry [20, 21], or mixing redox organic isomers [19] have been found helpful in raising solubility.

Redox molecules designed for NAOFBs are usually screened in different organic solvents to obtain high solubility. Following the "like dissolves like" principle, a redox molecule with polar groups demonstrated good solubility in a polar organic solvent [22].

It has been found that both quantum-mechanical molecular static [23] and molecular dynamic [24] theoretical simulations can reasonably well predict redox potentials of organic molecules in electrolytes; nonetheless, it has been less successful in predicting their solubilities. Unlike the redox potential, the solubility involves a comparison of molecular free energies in solution and in the crystalline form and predicting the crystal structure has proved elusive.

Compared to dissolving redox species in electrolytes, the suspension of particulates comprising redox species [25]; alternatively, the utilization of redox-targeting reactions to transport charge through liquid electrolytes to solids for storage in tanks [26], transcend the solubility limitations and promise high volumetric capacity. The extent to which electrochemical reaction kinetics or hydrodynamic or mass transfer processes are compromised awaits further investigation.

40.2.2 Viscosity

Viscosity of electrolytes containing inorganic compounds may slightly increase with the increased concentration of inorganic compounds. Viscosity of electrolytes containing organic molecules could significantly increase with increasing concentration because of the increased molecular size and the corresponding range of intermolecular interactions. The high energy density of concentrated redox organics may trade-off against slow mass transfer and high pumping losses caused by high viscosity in OFBs in ways that are not apparent from the behavior of inorganic FBs.

Many organic molecules used in OFBs possess aromatic systems, which generate π–π interactions. These interactions can be significantly strengthened in concentrated electrolytes where distances between organic molecules are reduced. The associated molecular interactions lead to higher viscosity.

40.2.3 Crossover

If an OFB is projected to have decades-long operation time, besides the extremely stable redox organic couples, crossover rates of redox molecules in both states through the membrane must be extremely low. Otherwise, the OFB will have a low Coulombic efficiency, a fast capacity fade rate, and a short operational lifetime. In vanadium flow batteries (VFBs), although fast vanadium crossover decreases current and energy efficiencies, it does not cause a chemical contamination problem leading to permanent capacity fade, because the posolyte and negolyte can be remixed and the states of charge rebalanced. In contrast, crossover of redox species in OFBs can result in irreversible capacity fade even if redox molecules remain intact. Increased molecular size in OFBs compared to inorganic redox species can drastically reduce the crossover rate of active species [10]. Nevertheless, when the molecular size is large enough to make crossover rates virtually negligible, e.g. by forming oligomers or polymers, and the concentrations are raised to useful levels, unacceptably large viscosity increases tend to result [27]. Analogously to the operation of VFBs, Winsberg et al. demonstrated a symmetric aqueous OFB [28], in which a low potential molecule (phenazine) and a high potential molecule (2,2,6,6-tetramethylpiperidinyl-N-oxyl) are covalently bound to form a bipolar molecule composing both negolyte and posolyte active species. In an analogous approach to Fe–Cr FBs utilizing porous separators, Luo et al. simply mixed negolyte with posolyte species to minimize the effects of crossover of redox species [29]. In both these cases, however, because roughly half of redox-active material is not utilized, the viscosity and cost of the electrolytes are increased. Efforts based on a simple redox molecule intrinsically possessing two widely-separated redox potentials [30, 31], so that the single molecule can serve as the redox species of both negolyte and posolyte, have met with limited success. Attaching multiple ionic groups to a redox molecule increases Coulomb repulsion (electrostatic repulsion) against membranes [12, 32], thereby reducing crossover of the redox molecule without increasing the viscosity of the electrolyte.

40.2.4 Lifetime

Evaluating the lifetime of a single redox molecule in full cells is difficult because both molecular decomposition and crossover across the membrane contribute to capacity fade. Goulet and Aziz developed the "unbalanced, compositionally symmetric flow cell" method in which the electrolytes are compositionally symmetric but volumetrically unbalanced. One side is the capacity limiting side during both charge and discharge, and the other side is non-capacity limiting side [33]. Diffusion of redox species through the membrane is virtually eliminated because the oxidized

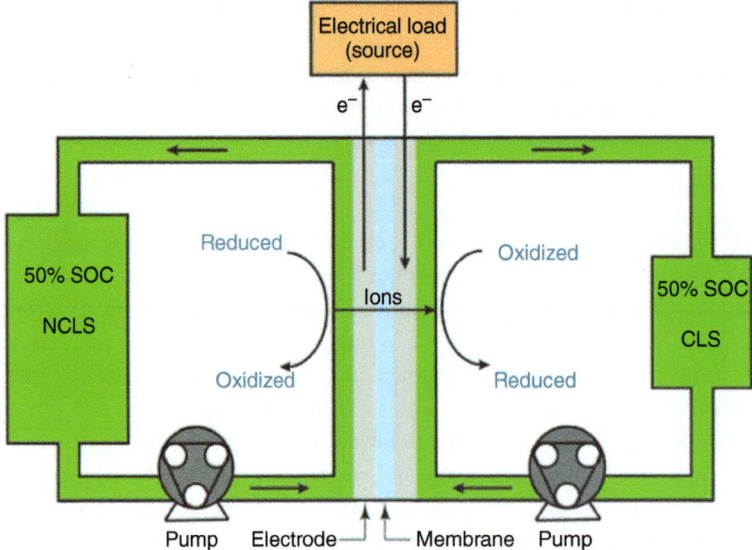

Figure 40.2 Schematic of volumetrically unbalanced compositionally symmetric cell, with identical electrolytes at identical concentrations at 50% state of charge in reservoirs of different volumes. Active species crossover is suppressed, and measured capacity fade may be attributed to the state of the capacity limiting side. Source: Adapted from Goulet and Aziz [33].

and reduced states of the single redox molecule reside on both sides of the membrane with equal durations, suppressing the influence of crossover on capacity fade (Figure 40.2).

When the crossover of redox species in an OFB is negligibly small or eliminated in an unbalanced, compositionally symmetric flow cell and precipitation is avoided, molecular instability, i.e. electrochemical decompositions of redox molecules become the main reasons leading to capacity fade of an OFB. Electrochemical decomposition involves the electrode as an electron donor or acceptor; its trajectory depends on the history of the electrode potential. Typically, this results in the capacity fade rate being proportional to the charge–discharge cycling rate. Chemical decomposition tends to occur homogeneously throughout the electrolyte at a rate that is independent of cycling. Both electrochemical and chemical decomposition can be strongly influenced by aspects of the chemical environment, including concentration, solvent, salt, pH, temperature, and electrolyte-contacting materials. Identified chemical decomposition mechanisms include nucleophilic addition/substitution, disproportionation, tautomerization, and dimerization [2]. Electrochemical decomposition mechanisms including over-reduction or over-oxidation, which can be highly sensitive to the applied potential. When evaluating the capacity fade rate of a cell utilizing a new redox molecule of unknown lifetime and decomposition mechanism, it is important to report both time-denominated and cycle-denominated fade rates (%/day and %/cycle). In the absence of confounding mechanisms such as leakage, the time-denominated fade rate reflects chemical instability and the cycle-denominated fade rate reflects electrochemical instability of the molecule. In virtually all cases exhibiting reasonably long lifetimes reported to date, in which sufficient experimentation has been performed to distinguish between time-denominated and cycle-denominated mechanisms, the capacity fade has been shown to be time-denominated [2].

40.2.5 Analytic Methods

Galvanostatic and/or potentiostatic methods are commonly used to evaluate cycling stability of OFBs. If an OFB fades rapidly (faster than about 1%/day) then purely galvanostatic cycling is adequate for measuring the fade rate with a reasonable degree of uncertainty. If, however, a FB exhibits a low fade rate (less than ~1%/day), artifacts in the purely galvanostatic cycling method render it inaccurate, as galvanostatic cycling is highly sensitive to fluctuations caused by diurnal temperature changes and drifts in membrane resistance. To overcome the inaccuracy, purely potentiostatic cycling, or galvanostatic cycling followed with potentiostatic holds at the end of a charging step and of a subsequent discharging step (also known as CCCV or IU mode charging and discharging), is essential for accurate capacity measurements (Figure 40.3) [5, 33].

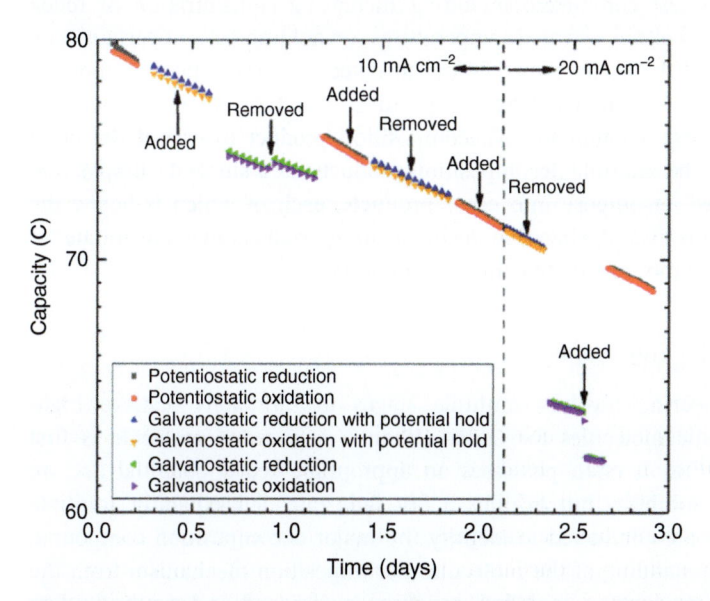

Figure 40.3 Semi-log plot of volumetrically unbalanced compositionally symmetric cell cycling of 2,6-dihydroxyanthraquinone (DHAQ), showing equivalence of potentiostatic and galvanostatic cycling with same potential holds. In both cases cycling was performed by imposing voltage holds at ± 200 mV and switched when current density dropped to 1 mA cm^{-2}. Galvanostatic conditions were started at 10 mA cm^{-2} and increased to 20 mA cm^{-2} after roughly 2.1 days. Vertical arrows indicate times at which a 0.1 Ω resistor was added and removed in series with the cell. The jumps in measured capacity demonstrate the dependence of measured cycling capacity on cell resistance when strictly galvanostatic conditions are used. Source: Adapted from Goulet and Aziz [32].

Because molecular decomposition is a major cause of capacity fades of OFBs, it is indispensable to perform postmortem analyses to characterize decomposition compounds in cycled electrolytes. Reported characterization techniques include cyclic voltammetry (CV), nuclear magnetic resonance (NMR), electron paramagnetic resonance (EPR), Fourier transform infrared spectroscopy (FT–IR), ultraviolet–visible spectrophotometry (UV–Vis), and liquid chromatography–mass spectrometry (LC–MS). Multiple techniques should be employed to examine whether molecular decompositions occur over cycling and to determine decomposition compounds, in particular when decomposition compounds are elusive [34].

Of the aforementioned techniques, CV, EPR, UV–Vis, and FT–IR are usually qualitative, but they can determine concentrations quantitatively if suitably calibrated, e.g. with known concentrations. Capacity measurements during cycling are quantitative but are incapable of providing structural information on decomposition products. NMR coupled with LC–MS has been shown to be effective at identifying and quantifying decomposition compounds [35].

It is worth noting that each technique has its own instrumental sensitivity limit, i.e. a lower limit of detection. It is possible to extend the cycling time [36] or cycle count or to accelerate fade rates to accumulate appreciable decomposition compounds for postmortem analyses. Acceleration of fade rates can be achieved by changing cell test conditions, including increasing concentration of redox molecules, cycling cells at elevated temperatures, or holding cells at certain state of charge (SOC) [10]. In any event, failing to detect decomposition compounds does not mean there is no molecular decomposition even if the amount of observed capacity loss is large enough for a decomposition product to exceed detection limits. There may be multiple decomposition products, or a single decomposition product may itself decompose into other products, each of which is below the detection limit. It is also possible that decomposition products may precipitate, or may polymerize on the electrolyte-contacting materials.

40.2.6 Molecular Engineering

Molecular engineering involves multiple stages to ultimately achieve high-performing organic molecules for OFBs. When a new redox molecule is first developed for OFBs, it often possesses an appropriate redox potential and an adequately high solubility but exhibits a fast fade rate. Subsequently, multiple characterizations are conducted to identify the major decomposition compound. With better understanding of the molecular decomposition mechanism from the identification of products, a modified structure is designed and synthesized to protect against the hypothesized or proven decomposition mechanism. Iteration leads to improved performance in cell operation.

Molecular engineering has been employed for both AOFBs and NAOFBs [10, 37]. For example, the alkaline flow battery using commercially available 2,6-DHAQ in the negolyte shows a fast fade rate of ~5%/day [14, 33]. With significant efforts in structural characterization, the decomposition compounds were determined to be redox-inactive 2,6-dihydroxyanthrone or 2,6-dihydroxyantranol, and a corresponding decomposition mechanism was proposed. Theoretical simulations suggest that the anthraquinone negolyte decomposition rate by this mechanism decreases with increasing molecular redox potential, albeit at the cost of a decreased cell voltage [38]. Raising the redox potential was thus expected to slow the decomposition rate. Indeed, 4,4′-((9,10-anthraquinone-2,6-diyl)dioxy)dibutyrate (2,6-DBEAQ), whose redox potential is 100 mV higher than that of 2,6-DHAQ, was synthesized and demonstrated a temporal fade rate 0.04%/day [10], which is two orders of magnitude lower than that of 2,6-DHAQ. Further experimental studies suggest that 2,6-DBEAQ is susceptible to nucleophilic attack at high pH and elevated temperature, leading to the cleavage of the ether bonds between anthraquinone and the water-solubilizing chains. Carbon–carbon bonds are chemically more resistant against extreme pH or elevated temperature than carbon–oxygen bonds. Thus, 3,3′-(9,10-anthraquinone-diyl)bis(3-methyl-butanoic acid) (DPivOHAQ) and 4,4′-(9,10-anthraquinone-diyl)dibutanoic acid (DBAQ), possessing carbon-linked water-solubilizing groups, indeed demonstrate outstanding thermal stability and extremely low temporal fade rates of <1%/year [19].

As a general strategy, molecular engineering has been applied to different redox cores with improved molecular stability. However, complicated synthesis and expensive chemicals adversely increase the synthetic cost of targeted molecules. Alternatively, restricting the SOC of an AOFB [38], or chemically regenerating redox-active molecules from decomposition compounds [19, 38] have been employed to reduce AOFB fade rates.

40.2.7 Cost

Because organics are composed of earth-abundant elements, organic molecules are potentially inexpensive compared to inorganic compounds; nevertheless, the synthetic cost of a high-performing organic redox molecule must quite low to be commercially viable.

The cost of a molecule decreases with the increase of the production scale and will, at sufficiently large scale, approach the cost of raw materials used for the synthesis; therefore, the raw materials should be sufficiently inexpensive. In addition, mild synthetic conditions, fewer synthetic steps, recyclable solvents and catalysts, higher product yields, and minimal waste are favored to further reduce synthetic cost. Both innovation and optimization in syntheses should be accomplished for a promising low-cost high-performing redox molecule. C–H activations simplify synthetic procedures by eliminating intermediate steps; thus it is expected to dramatically reduce synthetic cost if catalysts used for C–H activations can be easily recovered and reused. Electrosyntheses can replace toxic chemicals with clean electrons and possibly synthesize redox molecules in place; hence it can reduce the cost of waste disposal [36, 39]. Biosynthesis is in its infancy [40]; it may hold the promise of producing desired molecules with high yields under mild conditions assisted by directed evolution of enzyme catalysts [41].

Note that, under certain scenarios, molecular lifetime does not have to be infinite for an organic chemical when the molecule is extremely cheap [5, 42]. The up-front

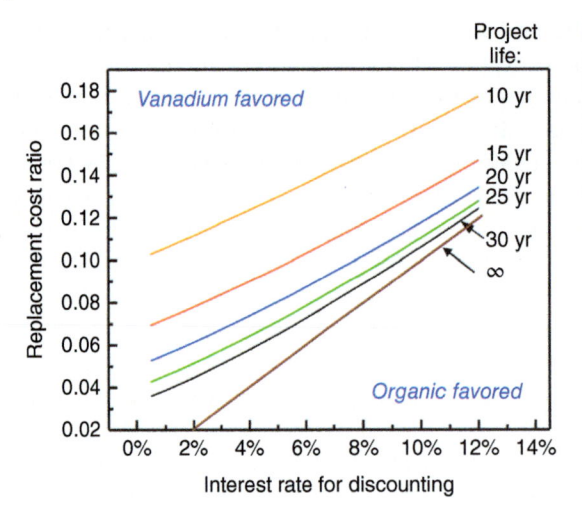

Figure 40.4 Break-even value of replacement cost ratio vs. interest rate for discounting, assuming project lifetimes as indicated. Source: Adapted from Brushett et al. [5], Yang et al. [42].

capital cost savings of using an inexpensive molecule instead of vanadium can be compared with the present value of a series of future replacement costs, recognizing the time value of money. The trade-off is quantified by a replacement cost ratio, defined as the annual replacement cost divided by the up-front capital cost savings. The break-even value of the replacement cost ratio depends on the interest rate for discounting and the project lifetime, as shown in Figure 40.4. When the actual value of the replacement cost ratio is less than this break-even value, organic molecules are potentially lower cost than vanadium.

40.2.8 Membrane

Ion-selective membranes include cation-exchange membranes (CEMs) and anion-exchange membranes (AEMs). Negatively charged sulfonate groups are attached covalently as pendants in typical CEMs, positively charged ammonium groups are attached as pendants in typical AEMs. A CEM selectively allows cations to permeate because of Coulomb attraction (charge attraction) between cations and the sulfonate groups but constrains negatively charged redox molecules from permeation due to Coulomb repulsion between anions and the sulfonate groups. Increasing the degree of sulfonation of a CEM increases the population of sulfonate groups and Coulomb repulsion between membrane and negatively-charged redox molecules, thus reducing the crossover rates of redox molecules.

Size-selective membranes are capable of blocking large redox molecules and conducting ionic or neutral species that are smaller than the pore size of membrane [43]. Reducing the pore size of membranes can reduce crossover rates of redox molecules. Because different sterically bulky groups can be incorporated into the repeating unit of a polymer, the pore size of final size-selective polymer membrane is thus adjustable, selectively conducting ions through the polymer membrane. The permeability as a function of molecular size tends to have a somewhat broad transition from high to low with increasing size.

It is extremely challenging to develop a durable membrane maintaining high ionic conductivity and selectivity in a varying electrical field, at extreme pH, and at fluctuating temperatures for decades. Commercially available Nafion® membranes exhibit reasonable stability, moderate ion selectivity, and ionic conductivity. However, the cost of Nafion is still too high because of its per-fluorination [44]. Developing low-cost, long-lasting membranes with high ionic conductivity and selectivity is as important as developing high-performing redox organic molecules [45].

40.2.9 pH Imbalance

pH imbalance results from side reactions during AOFB cycling including oxygen evolution reactions (OERs), oxygen reduction reactions (ORRs), and hydrogen evolution reactions (HERs). For example, when a charged (reduced) negolyte containing reduced organic molecules (Q^{2-}) is exposed to oxygen, the reduced molecules will become their oxidized forms (Q) and meanwhile, oxygen will be reduced to hydroxide, increasing the pH of the negolyte.

$$O_2 + 2Q^{2-} + 2H_2O \rightarrow 2Q + 4OH^-$$

The pH gradient produces concentration overpotential, which penalizes energy efficiency. Furthermore, pH imbalance induced by side reactions penalizes Coulombic efficiency and ultimately results in system imbalance. The pH drift may also bring the electrolyte into a regime in which the molecule is unstable or precipitates.

When one electrolyte involves proton-coupled electron transfer (PCET) whereas the other one does not [11], cycling will induce temporary pH imbalance, which will periodically intensify/relieve during the charge/discharge process.

$$\text{Charge: } Q + 2H_2O + 2e^- \rightarrow QH_2 + 2OH^-;$$

$$Q + H_2O + 2e^- \rightarrow QH^- + OH^-$$

$$\text{Discharge: } QH_2 + 2OH^- - 2e^- \rightarrow Q + 2H_2O;$$

$$QH^- + OH^- - 2e^- \rightarrow Q + H_2O$$

This cyclic pH imbalance can lead to Coulombic efficiency loss due to proton or hydroxide crossover, which may require engineering to control the losses. Sometimes it can be avoided by increasing the pH of electrolytes to above the pKa values of the protonated form of the reduced molecule (for example QH_2), so that the AOFB has proton-decoupled electron transfer on both the negolyte and posolyte sides [10].

40.2.10 Toxicity

When a FB is considered commercially viable, toxicology studies of the redox species need to be conducted. Although VFBs are being deployed, it is worth noting that vanadate (V^{5+}) and vanadyl (V^{4+}) are reproductive and developmental toxicants in mammals [46].

Polyaromatic hydrocarbons (PAHs) are frequently related to cancers, cardiovascular disease, and poor fetal development [47]. As one of the typical PAHs, anthracene is the precursor of anthraquinone (AQ), which is the starting material of anthraquinone-based redox molecules. Because of the similarity in structure between anthraquinone and anthracene, concern about the toxicity and carcinogenicity of AQ derivatives used for large-scale electricity storage is sensible. Earlier studies seem to suggest that substituent groups and their positions on AQ can have different effects [48]. For example, 1-(methylamino)-anthraquinone appears to alleviate toxicity; emodin and 1,8-dihydroxyanthraquinone are two phenolic anthraquinones; both are noncarcinogenic and used as cathartic laxatives [49]. In contrast, 1-nitro-anthraquinone consistently produced tumors in experimental animals [48]. Therefore, a conclusive statement on redox-active molecular toxicity should be cautious and it is essential to investigate the toxicity of the redox species before an OFB is deployed on a commercial scale.

It is worth noting that methyl viologen is proven notoriously toxic to mammals [50]. Toxicology studies should be performed for other viologen derivatives before they are considered for practical AOFB applications.

40.3 Classes of Aqueous Organic Redox Actives

40.3.1 Redox Aromatic Carbonyl Compounds (Quinones, Fluorenones)

Carbonyl groups when connected in a conjugated manner, such as quinones or fluorenones, can reversibly accept and donate electrons, thus being redox active.

Quinones, including benzoquinones, naphthoquinones, and anthraquinones, are the most studied redox molecular family due to their structural richness and their facile two-electron transfer. As the simplest quinone, benzoquinone derivatives were first studied as posolyte active species because of their high-redox potential. Yang et al. investigated 1,2-benzoquinone-3,5-disulfonic acid (BQDS) as the posolyte active species that has a potential of 0.85 V vs. SHE at pH 0 [51]. Subsequent characterization done by the same group demonstrate that BQDS is susceptible to nucleophilic addition, forming 1,2,4-trihydroxybenzene-3,5-sulfonic acid (THBS), which is still redox active. THBS can be further oxidized during a charging process and attacked by water, ultimately becoming 1,2,4,6-tetrahydroxybenzene-3,5-disulfonic acid with a much lower potential due to the electron-donating hydroxyl groups [52]. To avoid the susceptibility of the molecule to nucleophilic addition, the group hypothesized that it is essential to minimize the number of unsubstituted positions on the benzene; therefore, they synthesized 3,6-dihydroxy-2,4-dimethylbenzenesulfonic acid (DHDMBS), which indeed does not undergo any more nucleophilic additions. Postmortem analysis of the cycled electrolytes indicates that it is the crossover of DHDMBS causing the capacity fade [18].

2,3,5,6-tetrakis((dimethylamino)methyl)hydroquinone (FQH_2) is a fully substituted hydroquinone bearing four (dimethylamino)methyl groups that will become water-soluble ammonium cations in strong acid. The full substitution improves the stability of FQH_2 and the multiple charges and increased molecular size reduce the crossover of FQH_2 [32]. Similarly, 2,3,5,6-tetrakis(propylsulfanyl-2'-sulfonate)-1,4-hydroquinone tetrasodium salt bearing four longer chains and four negative charges, shows improved molecular stability and inappreciable crossover of redox molecules [53].

Naphthoquinone (NQ) has been less studied because of its intermediate redox potential, which is higher than that of anthraquinone but lower than that of benzoquinone [54]. It can exhibit desired potentials when the structure is properly modified, for example, bislawsone is a dimerized 2-hydroxyl-naphthoquinone with a potential of −0.551 V vs. SHE at pH 14, which was used as a negolyte active species with four-electron transfers per molecule. Postmortem analysis of the cycled electrolyte indicates that tautomerization of the half-reduced bislawsone is the major reason leading to the capacity fade [55].

The redox potentials of AQ derivatives are close to but slightly higher than the potential of the hydrogen evolution reaction over a broad pH range; therefore, anthraquinones with water solubilizing groups are ideal negolytes. In general, the more aromatic rings a quinone is fused with, the lower redox potential it will have due to the extended π-electron delocalization. The extended aromatics also stabilize structures over redox reactions; hence, extremely stable anthraquinone negolytes have been reported with projected decades-long lifetimes. Compared to BQ and NQ, AQ has the largest conjugation and the highest molecular weight; the bare AQ core is thus the least water-soluble. Significantly improved solubility of AQ in water has been achieved by introducing water-solubilizing groups to AQ.

Redox-active fluorenone derivatives have been studied as negolyte active species for AOFBs [56]. For example, 9-fluorenone-2-carboxylic acid shows a solubility of 0.8 M in 1 M KOH with a redox potential of −0.69 V vs. SHE. When paired with the ferro-/ferricyanide redox couple, an open-cell potential of 1.18 V was achieved. However, significant capacity loss was observed within the first few cycles. Given that fluorenone is derived from coal tar [57], its derivatives are potentially low cost; water-soluble fluorenone derivatives deserve more investigations in the future. Recently, Feng et al. synthesized a series of water-soluble fluorenones and reported 4-carboxylic-7-sulfonate fluorenone (4C7SFL) with two-electron transfer and a claimed temporal fade rate of 0.0209%/day [58]. The 4C7SFL molecule demonstrates significantly improved cell performance and extended lifetime over other fluorenone derivatives. Interestingly, the capacity utilization of 4C7SFL is highly dependent on current density and concentration of redox molecule. For example, at room temperature, 0.5 M 4C7SFL demonstrates 77.5% of theoretical discharge capacity utilization at $2\,mA\,cm^{-2}$, but only 31% at $20\,mA\,cm^{-2}$; at the same current density of $20\,mA\,cm^{-2}$, 0.5 M 4C7SFL shows a 31% of theoretical discharge capacity utilization, 1.36 M 4C7SFL shows a 72.2% of theoretical capacity utilization. This atypical electrochemical behavior deserves further investigation.

40.3.2 Aza-/Azo-Aromatics (Alloxazines, Phenazines, Azobenzenes, and Viologens)

Another redox organic family is aza-/azo-aromatics. They are composed of aromatic rings and multiple nitrogen atoms. The nitrogen atoms are the redox-active sites that accept/donate electrons. The nitrogen atoms are incorporated into aromatics in two different manners: (i) nitrogen atoms replace carbon atoms in aromatics and form aza-aromatics, such as (iso)alloxazine [59, 60], phenazine [20, 21], and viologen [12, 17]; (ii) nitrogen atoms are connected with aromatics through conjugation and form azo-aromatics such as azobenzene [61].

Redox in water-soluble (iso)alloxazine, phenazine, and azobenzene derivatives proceeds by two-electron transfer. Phenazine derivatives undergo PCET even in strong alkaline conditions as the pKa values of 5,10-dihydrophenazine are higher than 14 [20].

Viologens $(C_5H_4NR)_2^{n+}$, derived from 4,4'-bipyridyl, undergo stepwise two-electron transfer at near-neutral pH, with single-electron reduction potentials separated by ~400 mV [12]. Nevertheless, the 1st electron is mostly utilized in aqueous flow batteries as the 2nd electron transfer is less reversible. Extremely stable viologen molecules have been developed with the utilization of only the 1st electron transfer [12, 17].

Recently, through extending π-conjugation of viologens, two-electron transfer in one step has been demonstrated. For example, Luo et al. introduced a planar thiazolo[5,4-d]thiazole in between the two pyridiniums, which undergoes two-electron transfer at −0.44 V vs. SHE [62]. Hu et al. reported a phenylene-bridged bispyridinium with two-electron transfer at −0.763 V vs. SHE; however, chain cleavage leads to rapid capacity fade [63].

40.3.3 Nitroxide Radicals

Featuring the high-redox potentials, fast kinetics, and ease of structural modifications, (2,2,6,6-tetramethylpiperidin-1-yl)oxyl, (TEMPO) is a relatively stable radical. It has been substituted with water-solubilizing groups at its 4 position via carbon–heteroatom bonds that then were used as posolyte active species. Water-soluble TEMPO derivatives are pH sensitive as the oxidized TEMPO derivative tends to react with either proton or hydroxide [64], and corresponding reversibility and stability are demonstrated mainly at neutral pH [65].

TEMPO itself is inexpensive for use on a laboratory scale; however, its price is prohibitive at an industrial scale. Instead, 4-hydroxy-TEMPO (TEMPOL), produced from acetone and ammonia via triacetone amine, is less expensive than TEMPO. The hydroxyl group in TEMPOL can be further modified and functionalized with water-solubilizing groups [65]. Given that developing high-performing organic posolyte active species has been challenging, TEMPOL derivatives deserve further investigation.

40.3.4 Metal Coordination Complexes

A typical metal complex used for AOFBs consists of a redox-active transition metal and a few ligands. For example, ferrocene comprises an iron atom and two cyclopentadienyl ligands; ferro-/ferricyanides are composed of one iron atom and six cyanide ligands; tri(bipyridyl)iron complexes are composed of one iron atom and three bipyridyl ligands. Compared to bare Fe^{2+}/Fe^{3+} ions, iron complexes demonstrate faster reaction kinetics and several orders of magnitude lower crossover rates due to the larger complex sizes. The thermodynamic stability of complexes can be affected by their coordination number and the binding constant between the metal center and the donor atoms of the ligands. Side reactions can happen between protons and ligands or between hydroxides and metal centers; neutral-pH is often used for metal complexes to avoid unwanted side reactions.

Developing redox species used in posolytes is fundamentally more challenging because high-potential molecules have lower electron density around molecular skeletons, and their oxidized forms become energized electron-deficient electrophiles vulnerable to nucleophilic attack by hydroxide or even water. Ferro-/ferricyanide solutions, $([Fe(CN)_6]^{4-/3-})$, have been the most studied posolyte active species in alkaline or near-neutral media. The solubility of $K_4Fe(CN)_6$ is 0.76 M in water and the solubility of $Na_4Fe(CN)_6$ is 0.56 M in water. A total concentration of $Fe(CN)_6^{4-}$ is up to 1.5 M when sodium ferrocyanide and potassium ferrocyanide are mixed with 1:1 ratio [66]; the solubility of $(NH_4)_4Fe(CN)_6$ in water could be up to 1.6 M (Figure 40.7a, vide infra) [29]. Recently, Gregory et al. estimated that the mass production cost of $(NH_4)_4Fe(CN)_6$, $K_4Fe(CN)_6$, $Na_4Fe(CN)_6$ are 16, 25, 21 \$/kAh, respectively [67].

Beh et al. introduced (3-trimethylammonio)propyl dichloride to the 1, 1′ positions of ferrocene and synthesized bis((3-trimethylammonio)propyl)-ferrocene dichloride (BTMAP-Fc) [17]. The solubility of BTMAP-Fc in water is approximately 2 M. More important, BTMAP-Fc demonstrates a low temporal fade rate of 0.031%/day.

Because the redox potential (0.77 V vs. SHE) of bare Fe^{2+}/Fe^{3+} is close to that of the oxygen evolution reaction, iron coordination complexes are developed as posolyte active species [17, 68, 69]. In contrast, the redox potential of Cr^{2+}/Cr^{3+} is −0.41 V vs. SHE; chromium (Cr) coordination complexes have been studied as negolyte active species. Ethylenediaminetetraacetic acid (EDTA) is a commonly used chelating agent that has been found to coordinate with Cr in conjunction with water to form $Cr(EDTA)(H_2O)$, which has a redox potential of approximately −1.10 V vs. SHE, which is lower than the thermodynamic potential of the HER. Consequently, the ligand water molecule can be reduced by the Cr, releasing hydrogen gas. 1,3-propylenediaminetertaacetic acid (PDTA) allows the formation of a CrPDTA complex instead of being ligated to one water molecule in $Cr(EDTA)(H_2O)$. PDTA provides six donor atoms that fully surround the Cr center, serving as a molecular barrier to prevent the reaction between Cr^{2+} and water. The difference between EDTA and PDTA is the number of methylene groups ($-CH_2-$)

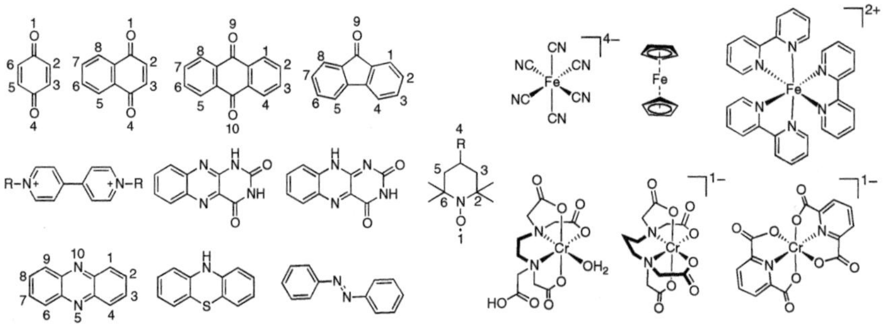

Scheme 40.1 Redox organics and metalorganics used for aqueous organic flow batteries.

in between two adjacent nitrogen atoms (Scheme 40.1): there are two methylene groups in EDTA and three in PDTA. The extra methylene group in PDTA extends the chain length and thus enables full coordination of Cr metal center [70]. In a separate work, two deprotonated dipicolinic acid (DPA) coordinate with Cr to form $Cr(DPA)_2$, whose redox potential (−0.7 V vs. SHE) is 300 mV lower than that of Cr^{2+}/Cr^{3+}, but 300 mV higher than that of CrPDTA or $Cr(EDTA)(H_2O)$. Due to the sluggish kinetics of the HER in the absence of a catalyst, the HER side reaction in a $Cr(DPA)_2$ flow battery is negligible, as reflected by its Coulombic efficiency of >99.5% [71].

40.4 Properties of Aqueous Organic Redox Actives

40.4.1 Range of Redox Potential

Generally, the potentials of negolyte active species are close to but slightly higher than that of HER; the potentials of posolyte active species are close to but slightly lower than that of OER. As such, an AOFB can deliver an open circuit voltage of ~1.0 V. Sometimes, the potentials of redox species are either lower than that of the HER or higher than that of the OER; overall cell voltages of AOFBs can thus be slightly higher than the water stability window of 1.23 V due to the overpotentials of the OER and the HER. For example, $Cr(PDTA)^{2-/1-}$ is reported as a negolyte active species and its redox potential is −1.10 V vs. SHE, 582 mV lower than that of the HER (−0.528 V vs. SHE) at pH 9; and when ferro-/ferricyanide is used as the posolyte active species whose potential is 0.45 V vs. SHE, the full cell thus demonstrates an open-circuit voltage of 1.55 V, which is 320 mV higher than 1.23 V [70].

40.4.2 Range of Solubility

Because of the incorporation of water-solubilizing groups, redox organics, or metalorganics can store electrons ranging from ~1 M up to 5 M; corresponding volumetric capacities range from below 20 to above 100 Ah l^{-1} [11]. Mixing counter ions for a redox species can be helpful to increase its solubility. For instance, the solubility of potassium ferrocyanide and sodium ferrocyanide in water is 0.5 and 0.3 M respectively, whereas the solubility of potassium and sodium ferrocyanide with 1 : 1 molar ratio can reach up to 1.5 M at pH 7. That is because the mixed cations frustrate recrystallizations of ferrocyanide [66].

40.4.3 Range of Fade Rate

A redox core can be introduced with different functional groups at different positions; consequently, the temporal fade rates of the modified redox molecules could vary by order of magnitude. For example, the temporal fade rates of 2,6-DHAQ [14], 1,8-PEGAQ [11], 2,6-DBEAQ [10], and 2,6(7)-DPivOHAQ [19] are ~5%/day, 0.5%/day, 0.04%/day, and 0.0018%/day, respectively. Similarly, order of magnitude improvements in lifetime has been achieved for redox aza-aromatic compounds: in particular, viologen and phenazine derivatives have demonstrated extremely low fade rates (<0.02%/day) after judicious molecular modifications [2, 12, 17, 20, 21, 72, 73].

40.4.4 General Decomposition Mechanisms

Generally, but not always [10], the "energized" state of a redox molecule is susceptible to chemical or electrochemical decompositions. "Energized" states are the reduced form of the redox molecule in the negolyte and the oxidized form of the redox molecule in the posolyte. Reported chemical decompositions include nucleophilic addition, nucleophilic substitution, disproportionation, dimerization, and tautomerization [2]. Electrochemically induced decompositions can also happen during the charge–discharge process [30, 34, 35].

40.4.5 Range of Operational pH

Reported AOFBs operate with a broad pH range, from below 0 to above 14. An appropriate operational pH should be chosen because pH influences solubility, potential, reversibility, and stability of a redox molecule. For instance, the reduced viologen derivatives [17] and the oxidized TEMPO derivatives [65] are susceptible to both acid- and base-induced decompositions, thus (near) neutral operation pH (7 ± 2) should be used to minimize the decomposition rate.

In general, a mild or near-neutral pH is advantageous because electrolytes with extreme pH tend to be highly corrosive, requiring costly acid- or base-resistant electrolyte-contacting materials; corrosive electrolyte leakage may raise safety concerns. Therefore, when a redox molecule exhibits similar properties within a broad pH range, a milder pH would be chosen for cell operations. For example, PEGAQ demonstrates similar electrochemical reversibility, solubility, and redox potentials from pH 7–14, and pH 7 is chosen for the cell cycling [11]. Similarly, DBEAQ exhibits the same reversibility and redox potential, and similar solubility at pH 12 and 14, hence pH 12 is preferred [10].

Solubility of redox species is often dependent on pH, and sometimes active species of negolyte and posolyte may reach their maximum solubilities at different pH values. Because a pH differential across the membrane is not viable over long time periods, negolyte vs. posolyte solubility may need to be traded-off via pH selection for maximizing the energy density.

When the hydroxide ion is involved in decomposition of a redox molecule, the concentration of hydroxide ions affects the decomposition rate. As a consequence, operational pH influences the fade rate of the organic electrolyte. For example, two reduced DPivOHAQ molecules can reversibly disproportionate to one DPivOHAQ in its oxidized state and one anthrone derivative, producing four hydroxide ions (Figure 40.5a). Because the hydroxide ions are a product of this reversible reaction, increasing the hydroxide concentration, i.e. increasing the pH of electrolyte, suppresses the disproportionation reaction. Consequently, the temporal fade rate of DPivOHAQ electrolyte decreases from 0.014%/day at pH 12 to 0.0018%/day at pH 14 (Figure 40.5b) [19].

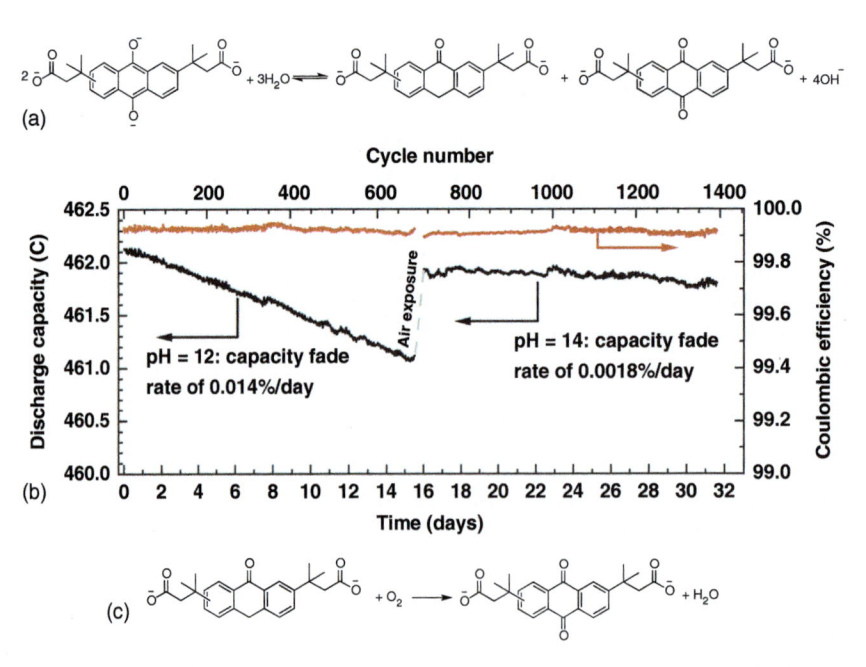

Figure 40.5 Influence of operational pH on molecular decomposition rate and temporal fade rate of DPivOHAQ electrolyte and possible capacity recovery via aeration. (a) The reversible disproportionation of DPivOHAQ in alkaline conditions. (b) DPivOHAQ cell cycling at pH 12 and 14. The negolyte was exposed to air (thereby reversing much of the decomposition) and the pH of both negolyte and posolyte were raised to 14 (thereby suppressing the decomposition rate) by adding KOH pellets. Source: Wu et al. [19]/with permission of Elsevier. (c) Chemical oxidation of the redox-inactive anthrone derivative to the redox-active DPivOHAQ.

40.4.6 Recomposition of Redox-active Molecules

With better understanding of decomposition mechanisms, one can recompose redox-active molecules from their decomposition products in situ. For example, anthrone derivatives are known to be the products of the major decomposition reactions in some anthraquinone-based electrolytes. Anthrone is over reduced and redox inactive compared to the corresponding anthrahydroquinone. O_2 is capable of chemically oxidizing anthrones back to anthraquinones in high yields. As shown in Figure 40.5b, most of the lost capacity of DPivOHAQ electrolyte was recovered after air exposure on day 16 [19].

Another well-studied example is DHAQ, which is characterized by a fast temporal fade rate of ~5%/day. This is caused by the instability of the reduced DHAQ ($DHAQ^{4-}$), which disproportionates to form redox-inactive 2,6-dihydroxyanthrone (DHA) [38]. Instead of increasing the operational pH of the electrolyte as shown in Figure 40.5b, Goulet and Tong et al. reduced the fade rate by restricting the negolyte SOC, thereby limiting the percentage of $DHAQ^{4-}$ in the electrolyte, thereby reducing the rate of $DHAQ^{4-}$ disproportionation. Additionally, a significant fraction of the redox-inactive DHA in the cycled DHAQ electrolyte was converted back to DHAQ via aeration, extending the overall lifetime of the battery.

Redox-active anthraquinones can even be regenerated *electrochemically* from their decomposition products. Zhao et al. demonstrated that $DHAQ^{4-}$ can be electrochemically reduced to DHA^{2-} [35]. Soon after, we found that DHA^{2-} can be electrochemically oxidized to the dimer $(DHA)_2^{4-}$ by one-electron transfer, and the dimer can be electrochemically oxidized to $DHAQ^{2-}$ by three-electron transfer (Figure 40.6a) [74]. Furthermore, electrochemical regeneration rejuvenates not only DHAQ, but also the positive electrolyte, thereby rebalancing the states of charge of both electrolytes without introducing extra ions. We demonstrated repeated capacity recovery with a DHAQ | potassium ferro-/ferricyanide flow battery in basic conditions (Figure 40.6b). The average ratio of capacity recovered to capacity lost in the preceding 50 cycles was 94.7%. The average recovery in Figure 40.6b leads to an overall fade rate of 0.38%/day, which is more than an order of magnitude improvement over the initial instantaneous fade rate of 6.45%/day in this experiment. In addition, preliminary studies with AQDS show the feasibility of electrochemical regeneration in acidic electrolytes [74]. Similar strategies may also be applicable for other redox-active organic molecules.

40.5 Performance of AOFBs

A high-performing AOFB is expected to have high energy and power density, to have decadal operational lifetime, and low-active materials cost. So far, the investigations of AOFBs have been focused primarily at the single electrolyte level, with different cell components and electrolyte compositions and volumes

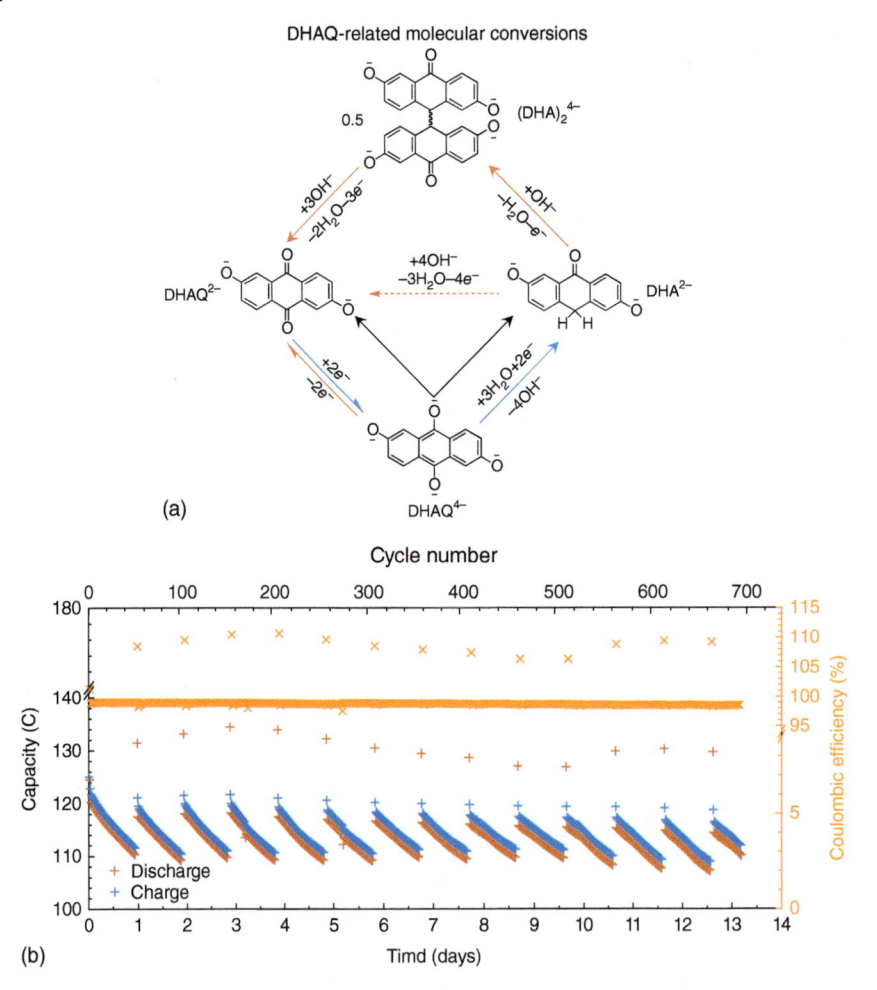

Figure 40.6 Electrochemically rejuvenating a DHAQ | Fe(CN)$_6$ flow battery. (a) Potential-driven DHAQ-related molecular conversions at pH 14. The reduction of DHAQ^{4-} to DHA^{2-} proceeds either chemically via a "disproportionation" of DHAQ^{4-} (2DHAQ^{4-} + 3H$_2$O → DHAQ^{2-} + DHA^{2-} + 4OH$^-$) or electrochemically. The red arrows indicate electrochemical oxidations; the blue arrows indicate electrochemical reductions; the black arrows indicate chemical reactions. The equilibrium arrows indicate reversible reactions. Source: Adapted from Jing et al. [74] (b) Long-term cycling of DHAQ$^{2-/4-}$ | [Fe(CN)$_6$]$^{3-/4-}$ flow battery with repeated daily capacity recovery by electrochemical regeneration of the negolyte. Source: Adapted from Jing et al. [74].

designed to investigate different aspects of the redox organic species. For example, different membranes might be used when studying molecular decomposition and cell power density; different electrolyte volumes might be used when studying the properties of a single molecule as the capacity-limiting side, and when studying the performance of a capacity-balanced cell. These studies have facilitated the development of redox molecules with proper redox potentials, high solubility, high stability, and potential low-cost production. Studies on cell-level performance are not yet systematic and simultaneously reporting all important performance parameters for a system-level optimized cell or stack, with a unique cell build (e.g. membrane choice) and a unique choice of posolyte and negolyte composition. Cell voltage has been reported as high as 2.13 V (Figure 40.7b) [70]; power densities have been reported as high as 1.0 W cm^{-2} (Figure 40.7c) [75]; energy densities have been reported as high as 25–27 Wh l^{-1} (Figure 40.7d) [11, 76]; capacity fade rates have been reported as low as 0.66–0.88%/year (Figure 40.5b) [19, 72]; and combined active species costs have been reported as low as $41.25/kWh [67]; however, these attributes have not been approached simultaneously in any single system. Simply

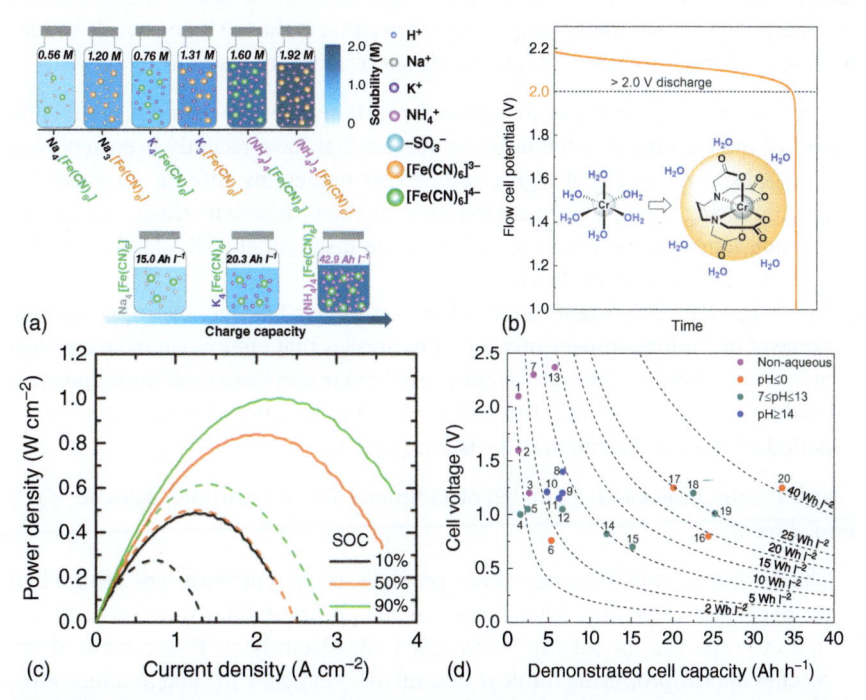

Figure 40.7 Selected measures of performance of redox species used for aqueous organic flow batteries. (a) Aqueous solubilities of ferro-/ferricyanide with different counter cations, and volumetric capacity of $Na_4Fe(CN)_6$, $K_4Fe(CN)_6$, and $(NH_4)_4Fe(CN)_6$. Source: Luo et al. [29]/with permission of Elsevier. (b) CrPDTA | Br_2 cell discharge voltage at 10 mA cm^{-2}. Source: Robb et al. [70]/with permission of Elsevier. Copyright 2019 (graphical abstract) (c) AQDS | Br_2 in 1 M H_2SO_4 modified cell performance at 10%, 50%, and 90% SOC, power density vs. current density. For comparison, the power density curves in the base case are plotted as the dashed lines. Source: Adapted from Chen et al. [75]. Copyright 2016 (Fig. 5, A5013) (d) Energy density contours on cell voltage vs. volumetric capacity plot of different flow battery systems. The pink dots (No. 1, 2, 3, 7, and 13) represent nonaqueous organic redox batteries. The red dots (No. 17 and 20) represent all-vanadium flow batteries with different supporting electrolytes (No. 17: H_2SO_4; No. 20: H_2SO_4/HCl mixture). The red dots (No. 6 and 16) represent aqueous acidic organic flow batteries. The dark green dots (No. 4, 5, 12, 14, 15, 18, and 19) represent aqueous near-neutral organic flow batteries. The blue dots (No. 8, 9, 10, and 11) represent aqueous alkaline organic flow batteries. Source: Republished permission of American Chemical Society from Jin et al. [11], permission conveyed through Copyright Clearance Center, Inc.

collecting those parameters from the reported AOFBs and comparing different AOFB systems would not be particularly informative at this stage. Optimization of system-level performance will be easier once-promising redox molecules are well investigated.

40.6 Outlook for AOFBs

Commercialization of AOFBs requires long lifetime of the battery and all its components, low-mass production cost of active species, and competitive power density and energy density. Significant improvement in either of the following two directions or some combination thereof, might enable successful commercialization:

1. Innovations in molecular syntheses are important to reduce the mass production cost of the extremely stable negolyte species that have recently been reported, or tweaks to molecular design enabling cost reductions without unacceptably compromising other properties. Optimizations on synthetic methods, conditions, reaction temperature, pressure, yields, waste disposal, etc. should also be conducted to further reduce the cost.
2. Innovations are important to extend the molecular lifetime, over the expected range of operation temperatures, of redox species that are known to be low-cost at mass-production scale, or to recompose them in situ from their decomposition products, so that the total active cost – including the present value of the cost of periodic future replacements [5] – is competitive.

Additionally, progress in a number of additional directions would facilitate AOFB development:

1. The development of extremely stable posolyte species currently lags behind that for negolytes. Currently, ferri-/ferrocyanide appears to offer the best combination of properties for weakly to strongly basic electrolytes. Functionalized ferrocenes appear promising for near-neutral pH. For acidic pH and as alternatives at any pH, inorganic redox couples such as bromide/tribromide and may be utilized in a hybrid organic/inorganic FB [15], if issues such as crossover can be managed.
2. It has been proven that potentials and water solubility of redox molecules can be readily adjusted by molecular engineering, improving molecular stability requires longer time, and more efforts. Multiple characterization techniques should be employed to determine decomposition compounds and reveal decomposition mechanisms, thereby informing molecular design strategies.
3. Ion-exchange membranes should possess high ionic conductivity and selectivity, as well as outstanding stability in the relevant (possibly extreme) pH environment. An important direction particular to AOFBs is to develop low-cost, robust ion-selective membranes with a high ratio of monatomic ion conductivity to bulky organic permeability. A combination of charge exclusion and size exclusion may be exploited to this end.

Acknowledgments

We gratefully acknowledge the extraordinary inspiration, skill, and dedication of our students, postdocs, and collaborators in flow battery research: Kiana Amini; Alán Aspuru-Guzik; Meisam Bahari; Eugene Beh; Fikile Brushett; Qing Chen; Xudong Chen; Jaephil Cho; Frank Crespilho; Alessandra D'Epifanio; Ali Davoodi; Diana De Porcellinis; Dian Ding; Louise Eisenach; Süleyman Er; Eric Fell; Taina Gadotti; Cooper Galvin; Thomas George; Michael Gerhardt; Rafael Gómez-Bombarelli; Marc-Antoni Goulet; Rebecca Gracia; Sergio Granados-Focil; Clare Grey; David Hardee; Lauren Hartle; William Hogan; Junling Huang; Brian Huskinson; Yunlong Ji; Shijian Jin; Erlendur Jónsson; Emily Kerr; Sang Bok Kim; David Kwabi; Eugene Kwan; Yuanyuan Li; Silvia Licoccia; Kaixiang Lin; Yahua Liu; Yazhi Liu; Michael Marshak; Barbara Mecheri; Luis Martin Mejia-Mendoza; P. Winston Michalak; Sujit Mondal; Saraf Nawar; the late Susan Odom; Minjoon Park; Daniel Pollack; Kara Rodby; Shmuel Rubinstein; Jason Rugolo; Jaechan Ryu; Mauricio Salles; Graziela Sedenho; Changwon Suh; Daniel Tabor; the late Zhijiang Tang; Liuchuan Tong; Tatsuhiro Tsukamoto; Alvaro Valle; Lucia Vina-Lopez; Baoguo Wang; Andrew Wong; Liang Wu; Min Wu; Kay Xia; Tongwen Xu; Ziang Xu; Zhengjin Yang; Evan Wenbo Zhao. R.G.G. and M.J.A. have significant financial interests in Quino Energy, Inc., which may profit from some of the results reviewed here.

References

1 Yan, Y., Robinson, S.G., Sigman, M.S., and Sanford, M.S. (2019). *Journal of the American Chemical Society* 141: 15301–15306.
2 Kwabi, D.G., Ji, Y., and Aziz, M.J. (2020). *Chemical Reviews* 120: 6467–6489.
3 Winsberg, J., Hagemann, T., Janoschka, T. et al. (2017). *Angewandte Chemie International Edition* 56: 686–711.
4 Liu, Y., Chen, Q., Sun, P. et al. (2021). *Materials Today Energy* 20: 100634.
5 Brushett, F.R., Aziz, M.J., and Rodby, K.E. (2020). *ACS Energy Letters* 5: 879–884.
6 Yao, Y., Lei, J., Shi, Y. et al. (2021). *Nature Energy* 6: 582–588.
7 Nolte, O., Volodin, I.A., Stolze, C. et al. (2021). *Materials Horizons* https://doi.org/10.1039/d0mh01632b.
8 Narayan, S.R., Nirmalchandar, A., Murali, A. et al. (2019). *Current Opinion in Electrochemistry* 18: 72–80.
9 Schlemmer, W., Nothdurft, P., Petzold, A. et al. (2020). *Angewandte Chemie International Edition* 59: 22943–22946.
10 Kwabi, D.G., Lin, K., Ji, Y. et al. (2018). *Joule* 2: 1907–1908.
11 Jin, S., Jing, Y., Kwabi, D.G. et al. (2019). *ACS Energy Letters* 4: 1342–1348.
12 Jin, S., Fell, E.M., Vina-Lopez, L. et al. (2020). *Advanced Energy Materials* 10: 2000100.
13 Gerhardt, M.R., Tong, L., Gomez-Bombarelli, R. et al. (2017). *Advanced Energy Materials* 7: 1601488.
14 Lin, K., Chen, Q., Gerhardt, M.R. et al. (2015). *Science* 349: 1529–1532.

15 Huskinson, B.T., Marshak, M.P., Suh, C. et al. (2014). *Nature* 505: 195–198.
16 Ji, Y., Goulet, M.-A., Pollack, D.A. et al. (2019). *Advanced Energy Materials* 9: 1900039.
17 Beh, E.S., De Porcellinis, D., Gracia, R.L. et al. (2017). *ACS Energy Letters* 2: 639–644.
18 Hoober-Burkhardt, L., Krishnamoorthy, S., Yang, B. et al. (2017). *Journal of the Electrochemical Society* 164: A600–A607.
19 Wu, M., Jing, Y., Wong, A.A. et al. (2020). *Chem* 6: 1432–1442.
20 Hollas, A., Wei, X., Murugesan, V. et al. (2018). *Nature Energy* 3: 508–514.
21 Wang, C., Li, X., Yu, B. et al. (2020). *ACS Energy Letters* 5: 411–417.
22 Wang, W., Xu, W., Cosimbescu, L. et al. (2012). *Chemical Communications (Cambridge, England)* 48: 6669–6671.
23 Er, S., Suh, C., Marshak, M.P., and Aspuru-Guzik, A. (2015). *Chemical Science* 6: 885–893.
24 Yu, J., Zhao, T.-S., and Pan, D. (2020). *Journal of Physical Chemistry Letters* 11: 10433–10438.
25 Duduta, M., Ho, B., Wood, V.C. et al. (2011). *Advanced Energy Materials* 1: 511–516.
26 Fan, L., Jia, C., Zhu, Y., and Wang, Q. (2017). *ACS Energy Letters* 2: 615–621.
27 Janoschka, T., Martin, N., Martin, U. et al. (2015). *Nature* 527: 78–81.
28 Winsberg, J., Stolze, C., Muench, S. et al. (2016). *ACS Energy Letters* 1: 976–980.
29 Luo, J., Hu, B., Debruler, C. et al. (2019). *Joule* 3: 149–163.
30 Tong, L., Jing, Y., Gordon, R.G., and Aziz, M.J. (2019). *ACS Applied Energy Materials* 2: 4016–4021.
31 Potash, R.A., McKone, J.R., Conte, S., and Abruña, H.D. (2015). *Journal of the Electrochemical Society* 163: A338–A344.
32 Park, M., Beh, E.S., Fell, E.M. et al. (2019). *Advanced Energy Materials* 9: 1900694.
33 Goulet, M.-A. and Aziz, M.J. (2018). *Journal of the Electrochemical Society* 165: A1466–A1477.
34 Zhao, E.W., Jónsson, E., Jethwa, R.B. et al. (2021). *Journal of the American Chemical Society* 143: 1885–1895.
35 Zhao, E.W., Liu, T., Jónsson, E. et al. (2020). *Nature* 579: 224–228.
36 Jing, Y., Wu, M., Wong, A.A. et al. (2020). *Green Chemistry* 22: 6084–6092.
37 Sevov, C.S., Brooner, R.E., Chenard, E. et al. (2015). *Journal of the American Chemical Society* 137: 14465–14472.
38 Goulet, M.-A., Tong, L., Pollack, D.A. et al. (2019). *Journal of the American Chemical Society* 141: 8014–8019.
39 Gerken, J.B., Stamoulis, A., Suh, S.-E. et al. (2020). *Chemical Communications (Cambridge, England)* 56: 1199–1202.
40 Price-Whelan, A., Dietrich, L.E., and Newman, D.K. (2006). *Nature Chemical Biology* 2: 71–78.
41 Kuchner, O. and Arnold, F.H. (1997). *Trends in Biotechnology* 15: 523–530.
42 Yang, Z., Tong, L., Tabor, D.P. et al. (2017). *Advanced Energy Materials* 8: 1702056.

43 Tan, R., Wang, A., Malpass-Evans, R. et al. (2019). *Nature Materials* 19: 195–202.
44 Yuan, Z., Duan, Y., Zhang, H. et al. (2016). *Energy & Environmental Science* 9: 441–447.
45 Kusoglu, A. and Weber, A.Z. (2017). *Chemical Reviews* 117: 987–1104.
46 Domingo, J.L. (1996). *Reproductive Toxicology* 10: 175–182.
47 Bostrom, C.-E., Gerde, P., Hanberg, A. et al. (2002). *Environmental Health Perspectives* 110: 451–488.
48 Sendelbach, L.E. (1989). *Toxicology* 57: 227–240.
49 Li, Z., Smith, K.C., Dong, Y. et al. (2013). *Physical Chemistry Chemical Physics* 15: 15833–15839.
50 Haley, T.J. (1979). *Clinical Toxicology* 14: 1–46.
51 Yang, B., Hoober-Burkhardt, L., Wang, F. et al. (2014). *Journal of the Electrochemical Society* 161: A1371–A1380.
52 Yang, B., Hoober-Burkhardt, L., Krishnamoorthy, S. et al. (2016). *Journal of the Electrochemical Society* 163: A1442–A1449.
53 Gerken, J.B., Anson, C.W., Preger, Y. et al. (2020). *Advanced Energy Materials* 10: 2000340.
54 Wang, C., Yang, Z., Wang, Y. et al. (2018). *ACS Energy Letters* 3: 2404–2409.
55 Tong, L., Goulet, M.-A., Tabor, D.P. et al. (2019). *ACS Energy Letters* 4: 1880–1887.
56 Rodriguez, J. Jr., Niemet, C., and Pozzo, L.D. (2019). *ECS Transactions* 89: 49–59.
57 Granda, M., Blanco, C., Alvarez, P. et al. (2014). *Chemical Reviews* 114: 1608–1636.
58 Feng, R., Zhang, X., Murugesan, V. et al. (2021). *Science* 372: 836–840.
59 Lin, K., Gomez-Bombarelli, R., Beh, E.S. et al. (2016). *Nature Energy* 1: 16102.
60 Orita, A., Verde, M.G., Sakai, M., and Meng, Y.S. (2016). *Nature Communications* 7: 13230.
61 Zu, X., Zhang, L., Qian, Y. et al. (2020). *Angewandte Chemie International Edition* 59: 22163–22170.
62 Luo, J., Hu, B., Debruler, C., and Liu, T.L. (2018). *Angewandte Chemie International Edition* 57: 231–235.
63 Hu, S., Li, T., Huang, M. et al. (2021). *Advanced Materials* 33: e2005839.
64 Orita, A., Verde, M.G., Sakai, M., and Meng, Y.S. (2016). *Journal of Power Sources* 321: 126–134.
65 Liu, Y., Goulet, M.-A., Tong, L. et al. (2019). *Chem* 5: 1861–1870.
66 Esswein, A.J., Goeltz, J., and Amadeo, D. (2014). US Patent, Application 2014/0051003 A1.
67 Gregory, T.D., Perry, M.L., and Albertus, P. (2021). *Journal of Power Sources* 499: 229965.
68 Páez, T., Martínez-Cuezva, A., Palma, J., and Ventosa, E. (2020). *Journal of Power Sources* 471: 228453.
69 Ruan, W., Mao, J., Yang, S., and Chen, Q. (2020). *Journal of the Electrochemical Society* 167: 100543.
70 Robb, B.H., Farrell, J.M., and Marshak, M.P. (2019). *Joule* 3: 2503–2512.

71 Ruan, W., Mao, J., Yang, S. et al. (2020). *Chemical Communications (Cambridge, England)* 56: 3171–3174.

72 Pang, S., Wang, X., Wang, P., and Ji, Y. (2021). *Angewandte Chemie International Edition* 60: 5289–5298.

73 Xu, J., Pang, S., Wang, X. et al. (2021). *Joule* https://doi.org/10.1016/j.joule.2021.06.019.

74 Jing, Y., Zhao, E.W., Goulet, M.-A. et al. (2022). *Nat. Chem.* https://doi.org/10.33774/chemrxiv-2021-x05x1.

75 Chen, Q., Gerhardt, M.R., Hartle, L., and Aziz, M.J. (2016). *Journal of the Electrochemical Society* 163: A5010–A5013.

76 Janoschka, T., Martin, N., Hager, M.D., and Schubert, U.S. (2016). *Angewandte Chemie International Edition* 55: 14427–14430.

41

Metal Coordination Complexes for Flow Batteries

Benjamin D. Silcox[1], Curt M. Wong[2], Xiaoliang Wei[3], Christo Sevov[2], and Levi T. Thompson[1]

[1] *University of Delaware, Department of Chemical & Biomolecular Engineering, 150 Academy Street, Colburn Laboratory, Newark, DE 19716, USA*
[2] *The Ohio State University, Department of Chemistry and Biochemistry, 151 W. Woodruff Ave., CBEC Building, Columbus, OH 43210, USA*
[3] *Indiana University–Purdue University Indianapolis, Department of Mechanical & Energy Engineering, 723 West Michigan Street, Indianapolis, IN 46202-5195, USA*

41.1 Introduction

41.1.1 Background

Metals are excellent candidates for use as active species in flow batteries (FBs) because of their rich redox chemistries and, in some cases, low cost. The first modern FB, developed by NASA in the 1970s, relied on the Fe^{2+}/Fe^{3+} and Cr^{2+}/Cr^{3+} couples from the corresponding halide salts in the catholyte (electron-donating electrolyte) and anolyte (electron-withdrawing electrolyte), respectively. This battery demonstrated the promise of FBs but was plagued with poor redox kinetics, elevated operating temperatures (~60 °C), and parasitic hydrogen evolution [1]. This investigation also identified $V^{2+}/V^{3+}/V^{4+}/V^{5+}$ and Br^-/Br_3^- as possible redox-active materials. In the 1980s, Skyllas-Kazacos and coworkers at the University of South Wales [2] described the vanadium redox flow battery (VRFB). This battery is based on the V^{2+}/V^{3+} (anolyte) and V^{4+}/V^{5+} (catholyte) couples from vanadium salts dissolved in aqueous sulfuric acid. The kinetics for this system were promising and the use of a common metal eliminated electrolyte cross-contamination challenges. The VRFB is the most widely installed FB system demonstrating lifetimes up to 20 000 cycles and round-trip efficiencies up to 85%.

While the VRFB has been commercially deployed, it is limited by low-energy densities of less than $25\,Wh\,l^{-1}$, which stems from the limited cell voltage (1.25 V) and low V ion solubilities (1.7 M in sulfuric acid) [3]. Moreover, the irreversible conversion of V^{5+} to V_2O_5(s) at elevated temperatures and the high-freezing point of vanadium electrolytes limit the system to a narrow range of operating temperatures (0–40 °C), creating challenges when considering some geographic locations [4]. Finally, because of its low-energy density, the vanadium raw material (as V_2O_5)

Flow Batteries: From Fundamentals to Applications, First Edition.
Edited by Christina Roth, Jens Noack, and Maria Skyllas-Kazacos.
© 2023 WILEY-VCH GmbH. Published 2023 by WILEY-VCH GmbH.

costs result in system costs of ~US$ 500/kWh for a four-hour system [5]. Zn–Br hybrid flow batteries are the second most common system and can achieve energy densities of up to 65 Wh l^{-1}. However, commercial systems have only demonstrated lifetimes of 2 000 cycles. Despite less expensive redox-active materials compared to VRFBs, system costs are still high (~US$ 500/kWh for a four-hour system) due to the need for expensive complexing agents in the electrolyte and lower-current densities [6, 7]. To compete with lithium-ion batteries (~700 Wh l^{-1} and ~US$ 350/kWh for a four-hour system) [8, 9] at the grid scale, flow batteries with higher energy densities and lower costs need to be demonstrated.

41.1.2 Overview of Metal Coordination Complexes in Flow Batteries

The use of metal coordination complexes (MCC) offers the possibility of increased solubilities and the use of non-aqueous solvents with wider voltage windows than water. Together these features could result in higher energy densities and lower costs. MCCs consist of a central metal atom bonded to ligands that contribute to the chemical and electrochemical behaviors.

The first reported MCC-based FB included a Ru complex in a non-aqueous solvent that yielded a 2.6 V cell [10]. Since then, a variety of metals, including Mg, Mn, Fe, Co, Ni, Zn, V, Cr, and TI have been investigated, although electrolytes based on first-row transition metals are preferred due to their low cost. From a systems perspective, the metal center is the cost driver and creates the most significant environmental, health, and safety issues.

Ligands are generally characterized by their redox (non-)innocence and denticity. Innocent ligands have a single and defined oxidation state, while non-innocent ligands can have multiple oxidation states in an MCC. This means that for MCCs incorporating innocent ligands, the metal center is responsible for all of the redox activity. When non-innocent ligands are incorporated, both the ligands and metal contribute to redox activity [11, 12]. Denticity refers to the number of coordination bonds between the ligand and metal center. Many of the ligands used in metal-ligand MCCs are bidentate (e.g. acetylacetonates, bipyridines, and dithiolates), with a few tridentate (e.g. terpyridines and metallocenes) options.

For MCCs, a number of FB relevant properties, including the redox potentials, solubility, number of redox couples, stability, and cyclability can be varied by changing the MCC structure. This flexibility can lead to higher energy densities and more stable operations. In addition, the presence of ligands increases molecular size, which reduces crossover through some membranes. These materials are not without challenges. Ligand shedding is a common decay mechanism for most MCCs [13, 14]. Recent systematic investigations report that the stability of MCCs at negative potentials is a function of ligand substitution rates at divalent metal centers [13, 14].

A common ligand modification to increase solubility in polar, non-aqueous solvents is to add polyethylene glycol (PEG) ether chains. This strategy has been used with a variety of MCC and organic redox-active materials and has only a small impact on molecular weight [14, 15]. Figure 41.1 illustrates the effect of PEG addition to bipyridylimino isoindoline (BPI) ligands for Ni-based MCCs in acetonitrile (MeCN)

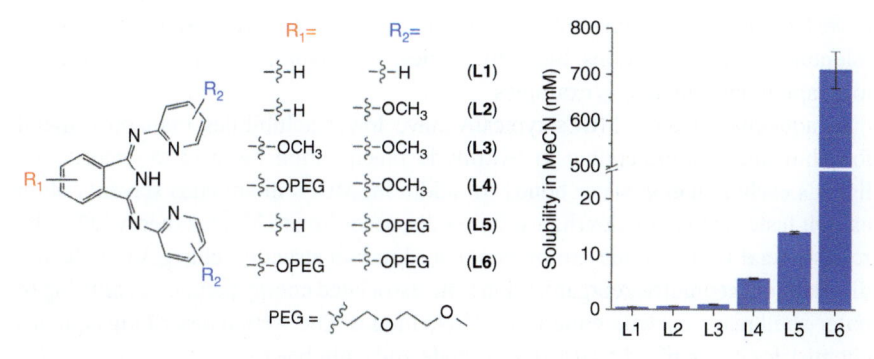

Figure 41.1 Structure modification for Ni BPI complexes by polyethylene glycol addition and its effect on solubility. Structures L1 through L6 have progressively more polyethylene glycol chains, increasing their solubility. Source: Fisher 2017 [16].

[16]. For metallocenes, a common modification to increase solubility in aqueous solvents is tetraalkylammonium ($-NR_4^+$) addition [13–15].

The first MCC-based FB was reported by Matsumura-Inoue and coworkers in 1988 [10] and employed tris(2,2′-bipyridine)ruthenium(II) tetrafluoroborate [Ru(bpy)$_3$](BF$_4$)$_2$ in MeCN for both the anolyte and catholyte. This system had four possible electron transfers across 2.6 V, but the redox-active material solubility was limited to 0.2 M and the Coulombic efficiency at the anolyte was only 40%. Results were also presented for [Fe(bpy)$_3$]$^{2+}$ and Ru(acac)$_3$; both were less stable than [Ru(bpy)$_3$]$^{2+}$ and Ru(acac)$_3$ had a low cell voltage of 1.5 V. Some years later Roberts and coworkers investigated Ru(acac)$_3$ and [Fe(bpy)$_3$](ClO$_4$)$_2$ [17]. These two compounds offer 1.75 and 2.4 V cell voltages, respectively, but Ru(acac)$_3$ could only be charged to 9.5% state of charge (SOC) due to side reactions, and Fe(bpy)$_3$ could only reach an energy efficiency of 6%.

Despite the frequent use of ferrocene as a reference electrode for non-aqueous systems, ferrocene and other metallocenes were only first investigated for use in FBs in 2014. Yu and coworkers and Wang and coworkers reported the use of ferrocene and cobaltocene, respectively, as catholytes in hybrid FB systems with a lithium metal anode in aprotic solvents [18, 19]. The system reached 250 cycles with a power density of 120 W kg^{-1} and specific capacity of 130 mAh g^{-1}. The first use of ferrocene in an aqueous FB was reported by Sun Catalytix Corp [20, 21], which patented a ferrocene catholyte, titanium metal-ligand anolyte system. The MCC solubilities exceeded 1 M and theoretical energy densities up to 35 Wh l^{-1} were achieved, depending on the titanium complex ligand.

41.1.3 Chapter Overview

This chapter will review the use of MCC chemistries in both aqueous and non-aqueous electrolytes for FB applications. Aqueous electrolytes are usually favored for FBs because of their high ionic conductivities, wide pH ranges to accommodate diverse electrochemical reactions, availability of ion-selective membranes, nonflammability, minimal environmental impact, and low cost. These advantages have facilitated electrochemical evaluations and FB system designs, leading to tolerable end-use conditions, high efficiencies, long stabilities and cyclabilities, and low capital investments, as examples.

In aqueous solvents, MCCs typically have lower solubilities than pure metal ions but are less susceptible to hydrolytic precipitation because the coordinate ligands exclude metal–water bonds. In addition, MCCs in aqueous solvents often exhibit faster redox kinetics than simple salts, according to Marcus theory [22]. The robust metal-ligand bonds can prevent coordination structure changes (e.g. ligand dissociation, geometric reorganization) and associated energy penalties, resulting in outer-shell reactions with enhanced redox kinetics. The section describing aqueous chemistries is organized by use (e.g. anolyte and catholyte).

Compared to aqueous systems, non-aqueous systems have the potential to significantly increase cell voltage allowing for multiple electron transfers, although there are other challenges. Reports of non-aqueous MCC systems have focused on pyridyl complexes and chalcogenides with acetylacetonate (acac) ligands. Some metallocenes have also been investigated in non-aqueous hybrid flow batteries with a lithium metal anode. The section on non-aqueous chemistries will be organized by ligand type.

41.2 Aqueous Metal Coordination Complex Based Flow Batteries

41.2.1 Catholyte Chemistries

Iron is among the least expensive metals and has received significant attention for use in electrochemical applications. The ferrous/ferric ($Fe^{2+/3+}$) redox reaction occurs at a suitable redox potential in aqueous media (~0.77 V vs. SHE) and exhibits high stability. The redox potential depends on several parameters, including the complexation constants for the Fe^{2+}-ligand and Fe^{3+}-ligand complexes. For example, a higher constant for the latter than the former negatively shifts the redox potential of the Fe^{2+}/Fe^{3+} couple. This feature provides an effective approach to tailor the redox potential of MCCs, thus expanding the design space and materials pool. Effects of the ligand in determining the redox potential are illustrated in Figure 41.2 [23]. The solubility and chemical stability can also be tailored via careful design of the ligand structure.

41.2.1.1 Ferrocene Complexes

Ferrocene is the best-known metallocene and has a broad range of applications [24]. This MCC consists of two cyclopentadienyl (Cp) rings bound on opposite sides to a Fe ion. The Fe center is responsible for the conversion of ferrocene and ferrocenium. The Cp ligands negatively shift the redox potential of the Fe^{2+}/Fe^{3+} couple from 0.77 to 0.40 V vs. SHE [25]. Because Cp is hydrophobic, structural modification is required to produce a ferrocene with sufficient water solubility. In addition, the ferrocene/ferrocenium redox couple is only chemically stable in pH-neutral electrolytes. In acidic electrolytes, ferrocene is converted to ferrocenonium via formation

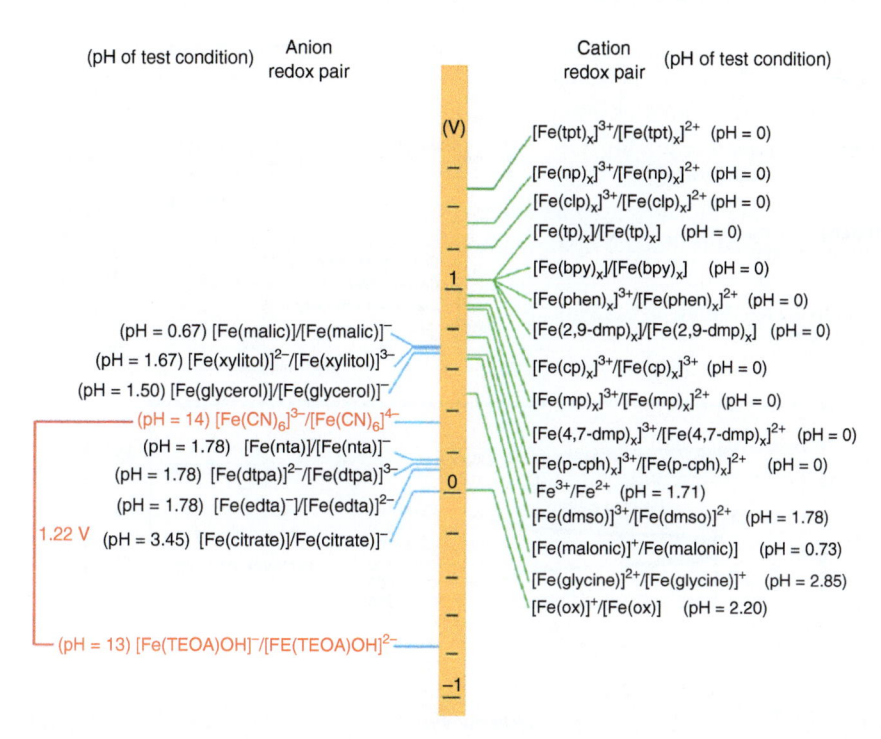

Figure 41.2 Redox potentials of reported Fe-based MCCs vs. the standard hydrogen electrode. Source: Gong et al. 2016 [23]. Further permissions for use of this material should be directed to ACS.

of a Fe—H bonded species [26]. In alkaline electrolytes, ferrocenium reacts with OH^- leading to demetallation and release of Cp radicals and $Fe(OH)_2/Fe(OH)_3$ as precipitates [27].

Liu et al. synthesized a water-soluble ferrocene derivative, ferrocenylmethyl trimethylammonium chloride (FcNCl, Figure 41.3a), and evaluated its key properties [29]. Compared to pristine ferrocene, FcNCl had an enhanced solubility of 3.8 M because of the hydrophilic cationic tetraalkylammonium (NR_4^+) group. Due to its electron-withdrawing nature, the redox potential was positively shifted by ~0.2 V, an attractive characteristic for achieving higher cell voltages. An FB consisting of FcNCl catholyte and methyl viologen (MV) anolyte yielded a cell voltage of 1.06 V. The cationic structures of both FcNCl and MV minimized crossover through an anion-exchange membrane, as indicated by a Coulombic efficiency >99% in flow cell tests. However, the flow cells exhibited noticeable capacity fading over several hundred galvanic cycles (Figure 41.3b), possibly caused by decomposition of the redox materials. The close proximity of NR_4^+ to the aromatic ring makes trimethylamine a good leaving group, resulting in its detachment via nucleophilic attack by water [30].

To overcome this challenge, Aziz et al. designed a similar structure with a longer propylene spacer ($-CH_2CH_2CH_2-$), i.e. bis((3-trimethylammonio)propyl)ferrocene (BTMAP-Fc) [28]. The NR_4^+ enabled a high water solubility of 1.9 M and the

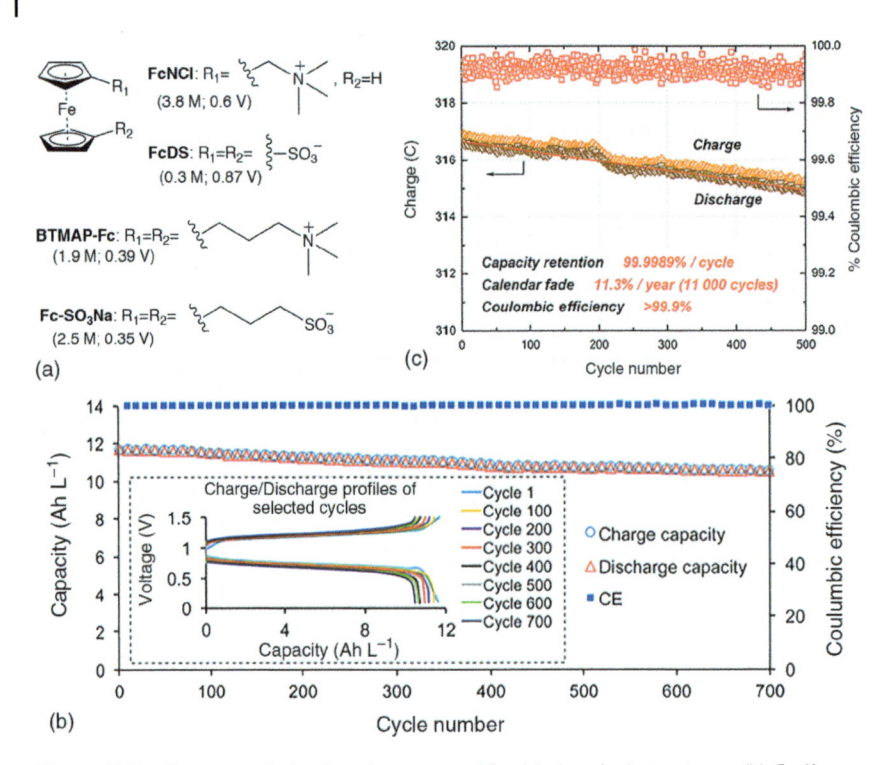

Figure 41.3 Ferrocene derivatives in aqueous FBs: (a) chemical structures; (b) Cycling performance of 0.5M FcNCl-MV flow cell; (c) Cycling performance of BTMAP-Fc/BTMAP-viologen flow cell. Source: Republished with permission of American Chemical Society, from (a, c) Beh et al. 2017 [28]. (b) Hu et al. 2017 [29].

longer spacers reduced the inductive effect of NR_4^+ on the ferrocenyl core, resulting in a redox potential that was nearly identical to the parent ferrocene. More importantly, the propylene spacers afforded greatly improved chemical stability against hydrolysis-induced deamination. Flow cells using BTMAP-Fc catholyte and BTMAP-viologen anolyte demonstrated an extremely low capacity-fade rate of 0.0011%/cycle or 0.033%/day (Figure 41.3c). These excellent properties suggest the promise of BTMAP-Fc in achieving energy-dense, stable aqueous FBs.

Xu and coworkers continued to study the effects of the length and number of $-NR_4^+$ on the stability [31]. Ferrocene bearing two $-NR_4^+$ substituents with longer alkyl spacers afforded improved cycling stability. Recently, Schubert and coworkers designed and evaluated a water-soluble, temperature-stable catholyte copolymer containing both redox-active ferrocene and $-NR_4^+$ pendants, both spaced by at least two carbons from the polymer backbone, in aqueous FBs [32]. When coupled with a BTMAP-viologen anolyte, the flow cell demonstrated near-zero capacity fade over 100 cycles at an elevated temperature (60 °C).

Ferrocene derivatives bearing anionic substituents have been evaluated for pairing with diverse anolytes. Boika and coworkers investigated ferrocene-1,1′-disulfonate (FcDS); the two electron-withdrawing sulfonates afforded an enhanced redox potential of 0.87 V vs. SHE [33]. However, the solubility was low (0.3 M), presumably due to the adjoining rings sterically hindering solvation, and the chemical stability was poor. Wang and coworkers incorporated a longer propylene spacer between ferrocene and the sulfonate (Fc-SO$_3$Na) resulting in greatly increased solubility (2.5 M) and stability [34]. This new ferrocene was coupled with a Zn anode to yield a cell voltage of 1.2 V. The flow cells demonstrated no appreciable capacity fade at both low (0.02 M) and high (1.5 M) concentrations, indicating the high chemical stability of this anionic ferrocene.

41.2.1.2 Ferro/ferricyanide Complexes

The ferrocyanide/ferricyanide (Fe(CN)$_6^{4-}$/Fe(CN)$_6^{3-}$) couple is the second class of Fe-based MCC that has been extensively investigated for aqueous FBs. The complexes consist of low-spin Fe^{2+} or Fe^{3+} centers bound to six CN$^-$ ligands. Unlike other metal cyanide complexes that are highly toxic, ferro/ferricyanide does not release free CN$^-$ due to the strong Fe—CN bond and even has been used as food additives [35]. The Fe(CN)$_6^{4-}$/Fe(CN)$_6^{3-}$ couple has a standard potential of 0.36 V vs. SHE [25], a moderately high value for catholyte materials. Its bulky size and polyanionic structure lead to minimal crossover loss when a cation-exchange membrane is used, which has led to its wide use as a catholyte in aqueous FBs. However, their solubilities, e.g. 1.0 M for K$_3$Fe(CN)$_6$ and 0.7 M for K$_4$Fe(CN)$_6$ in water, may pose a barrier for achieving energy-dense FBs. By using mixed Na$^+$ and K$^+$ cations at a 1 : 1 ratio, Esswein and coworkers achieved a significant increase in the solubility of Fe(CN)$_6^{4-}$ to 1.5 M in water [36]. Kwon and coworkers observed a similar solubility increase [37]. Wang and coworkers developed a sodium polysulfide–potassium ferro/ferricyanide FB with a Nafion membrane, in which the cross-mixed Na$^+$ and K$^+$ allowed a test concentration of 1 M [38].

Recently, Liu et al. used NH$_4^+$ instead of alkali metal ions to increase the solubilities of Fe(CN)$_6^{4-}$ to 1.6 M and Fe(CN)$_6^{3-}$ to 1.9 M; this was ascribed to strong solvation interactions with NH$_4^+$ via H-bonding [39]. Symmetric Fe(CN)$_6^{4-}$/Fe(CN)$_6^{3-}$ and Fe(CN)$_6^{4-}$/(SPr)$_2$V flow cells using 1 : 1 mixed-reactant electrolyte exhibited near-zero capacity fading during long-term operations (Figure 41.4a). These high solubilities and cycling stability represent a significant advancement toward achieving energy-dense, durable, cost-effective FBs.

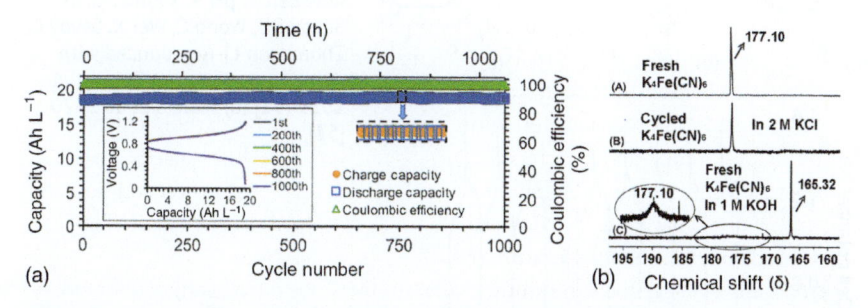

Figure 41.4 (a) Cycling stability of 0.9 M (NH$_4$)$_2$Fe(CN)$_6$/(SPr)$_2$V symmetric flow cell; (b) ^{13}C NMR of fresh and cycled K$_4$Fe(CN)$_6$ electrolytes (chemical shifts in ppm). Source: Reproduced with permissions from: (a) Luo et al. 2019 [39]. (b). Luo et al. 2017 [40].

Interestingly, the $Fe(CN)_6^{4-}/Fe(CN)_6^{3-}$ couple is also used in alkaline FBs, such as zinc-ferricyanide system [41], and with organic anolyte materials [42]. However, the long-term stability of $Fe(CN)_6^{4-}/Fe(CN)_6^{3-}$ in strongly alkaline electrolytes remains a concern. Liu and coworkers demonstrated that the symmetric $Fe(CN)_6^{4-}/Fe(CN)_6^{3-}$ flow cells exhibited pH-dependent cycling stability, with non-negligible capacity fading in pH 14 electrolyte [40]. ^{13}C NMR of the cycled electrolytes revealed the formation of new species with δ 165 ppm peak that was assigned to free CN^- (Figure 41.4b), perhaps due to the release of CN^- by OH^- attack. In a separate report, Ventosa and coworkers observed similar phenomena but claimed that the capacity fade originated from the oxygen evolution reaction as its redox potential overlaps with $Fe(CN)_6^{4-}/Fe(CN)_6^{3-}$ at pH 14 [43]. The δ 165 ppm peak in ^{13}C NMR was assigned to the K_2CO_3 impurity in KOH (85% purity). Nevertheless, both proposed side reactions are slow. Therefore, the $Fe(CN)_6^{4-}/Fe(CN)_6^{3-}$ catholyte has been widely used, often in excess as a non-limiting side, to evaluate new organic anolyte materials, including derivatives of anthraquinone [44], alloxazine [45, 46], phenazine [47], azobenzene [48], and 9-fluorenone [49]. Stable capacity profiles have been generally obtained during these FB tests, reflecting the high stability of $Fe(CN)_6^{4-}/Fe(CN)_6^{3-}$.

41.2.1.3 Fe Oligo-Aminocarboxylate Complexes

Oligo-aminocarboxylate (OAC) ligands such as ethylenediaminetetraacetic acid (EDTA) are well-known for forming Fe-OAC chelates that have high complexation constants [50]. However, Fe-EDTA is not suitable for FB use. Its hexadentate heptacoordinate structure includes the 4 carboxylates and 2 N atoms of EDTA, plus a H_2O molecule, bonded to Fe (Figure 41.5a) [53–56]. This chelate structure (i) inhibits the approach of solvating water leading to a low solubility (~0.2 M); (ii) makes Fe^{2+}-EDTA vulnerable to oxidation by O_2 via initial substitution of the coordinate water by O_2 [57]; and (iii) raises an instability concern due

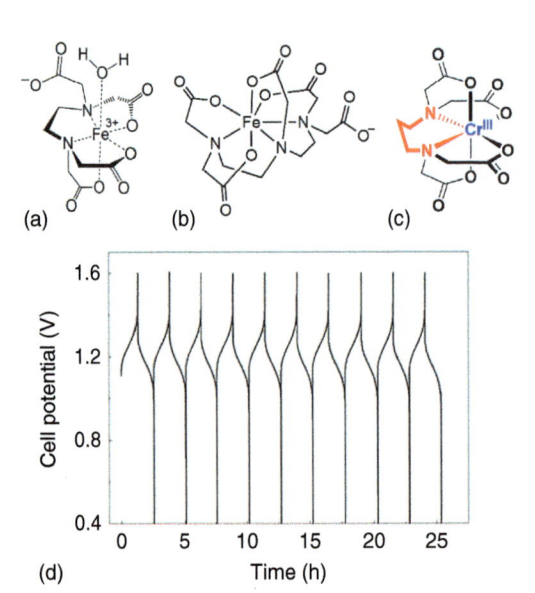

Figure 41.5 Metal-OAC complex structures: (a) Fe-EDTA; (b) Fe-DTPA; (c) Cr-PDTA; (d) voltage curves of 0.5M Fe-DTPA/Cr-PDTA flow cell at pH 9. Source: (a, b) Silcox BD, Wong C, Wei X, Sevov C, Thompson LT Reproduced with permissions from: (c) Robb et al. 2019 [51] (d) Waters et al. 2020 [52].

to $Fe(OH)_2/Fe(OH)_3$ formation if deprotonated. In contrast, Fe chelates with polydentate OAC ligands such as diethylenetriamine pentaacetic acid (DTPA) have improved properties.

Marshak and coworkers investigated Fe-DPTA chelate as a cost-effective catholyte material and correlated its solution structure with aqueous FB-relevant properties [52]. Fe-DTPA exhibits a heptadentate heptacoordinate structure in solution, with an unbonded carboxylate but without directly coordinated water (Figure 41.5b) [58]. The pendant carboxylate offers a solvation site to afford a high solubility (1 M). The absence of coordinate water eliminates the possibility of $Fe(OH)_2/Fe(OH)_3$ formation and blocks the oxidation pathway, likely leading to improved chemical stability. Despite a relatively low-redox potential of 0.043 V vs. SHE, Fe-DTPA was coupled with a chromium chelate anolyte material to achieve a voltage of 1.24 V. The flow cell demonstrated energy efficiencies of 92%, 78%, and 56% at 20, 50, and 100 mA cm^{-2}, respectively. The relatively stable capacities for 40 galvanic cycles (Figure 41.5d) indicate the chemical stability of Fe-DTPA, although long-term stability testing is recommended to further validate the design.

41.2.1.4 Co-Based Complexes

Cobalt is the other important base metal for MCC applications. Bard and coworkers successively prepared a chemically stable, soluble catholyte coordination compound of Co^{2+}/Co^{3+} with 1-[bis(2-hydroxyethyl)amino]-2-propanol (mTEOA) and paired it with a Fe^{2+}/Fe^{3+} anolyte complex with triethanolaime (TEOA) in an alkaline FB (Figure 41.6a) [59]. This Co complex existed as $[Co(mTEOA)(H_2O)]^-/[Co(mTEOA)(H_2O)]^0$ in 5 M NaOH, with Co^{2+}/Co^{3+} bonded to the deprotonated alcoholic ends of mTEOA. The Co-mTEOA compound had a solubility of 0.7 M and a redox potential of 0.077 V vs. SHE. Compared with the ligand-free $Co^{2+/3+}$ couple, the complex exhibited faster redox kinetics, as illustrated by the smaller peak separation in cyclic voltammograms (Figure 41.6b). The flow cell test with 0.5 M complexes yielded a 10% capacity fade over 30 cycles (~20 hours), which was attributed to a combination of oxidation by leaked O_2, complex crossover and electrolyte volume change. To mitigate the risk of $Co(OH)_2/Co(OH)_3$ formation in the strongly alkaline electrolyte, Kwon and coworkers investigated the use of a bulky triisopropanolamine (TiPA) ligand to complex with Co^{2+}/Co^{3+} [60]. Compared to TEOA, the three methyl groups in TiPA may provide a steric hindrance to prevent the approaching of HO^-, reducing the tendency for $Co(OH)_2/Co(OH)_3$ formation.

41.2.2 Anolyte Chemistries

41.2.2.1 Fe-Based Complexes

Compared to catholytes, only a few Fe- and Cr-based MCCs have been reported for use in anolytes. For the Fe^{2+}/Fe^{3+} redox couple, it is difficult to obtain sufficiently low-redox potentials for anolyte use merely via ligand tailoring. Often, the electrolyte is engineered to alkaline pH to introduce a hydroxide (OH^-) in the coordination structure. The stronger association of $Fe^{3+}-OH^-$ than $Fe^{2+}-OH^-$ [61] shifts the Fe^{2+}/Fe^{3+} redox potential negatively to make the MCC suitable for use as an anolyte.

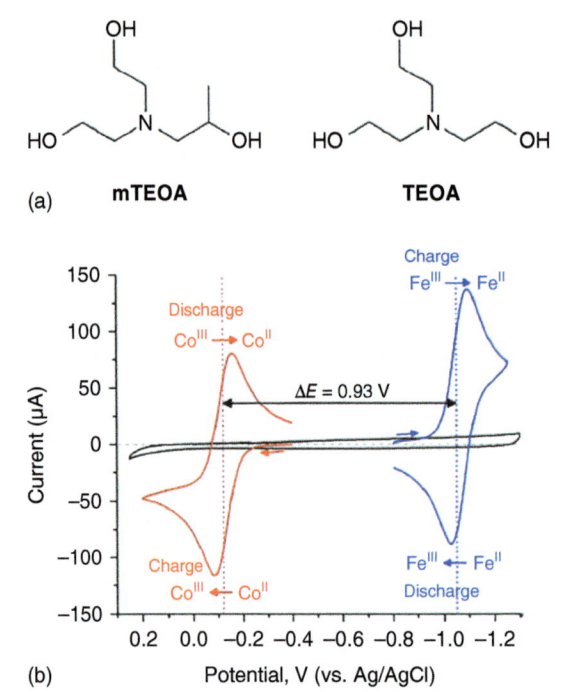

Figure 41.6 (a) Structures of TEOA and mTEOA ligands; (b) CV curves of Fe-TEOA and Co-mTEOA in 5M NaOH. Source: Reproduced with permission from: Arroyo-Currás et al. 2014 [59].

Yang and coworkers developed a Fe-triethanolamine (TEOA) anolyte complex and observed that a high NaOH concentration (3–5 M) was needed to stabilize the Fe-TEOA complexes [62]. Later, Bard and coworkers determined the chemical formula for the complexes in concentrated NaOH as [Fe(TEOA)(OH)]$^-$/[Fe(TEOA)(OH)]$^{2-}$, with deprotonated alcoholic ends in TEOA and a hydroxide group coordinating to the Fe^{2+}/Fe^{3+} ions [59]. Molar excesses of both TEOA and HO$^-$ were always needed to maintain the formation of the complexes, especially at high concentrations. The solubility of Fe-TEOA was 0.8 M and the redox potential was remarkably as low as −0.86 V vs. SHE. Since its discovery, this anolyte complex couple has been coupled with bromine and Co-mTEOA catholyte materials, respectively, in aqueous FBs. More recently, Yan and coworkers developed an alkaline all-soluble, all-iron FB by pairing [Fe(TEOA)(OH)]$^-$ anolyte with ferrocyanide catholyte [23]. Unstable capacity profiles and low CEs were commonly observed for the [Fe(TEOA)(OH)]$^-$ flow cell cycling, which have been ascribed mainly to the crossover of free TEOA used in excess on the anolyte side. ^{1}H NMR confirmed that the crossed TEOA was oxidized by ferricyanide leading to loss of the latter. The electrochemical deposition of Fe(s) is also possible due to the very low-redox potential. To address this issue, Kwon and coworkers investigated the use of 3-[bis(2-hydroxyethyl)amino]-2-hydroxyl-propanesulfonic acid (DIPSO) chelate ligand [60, 63]. The stronger coordination bonds formed by this ligand with Fe ions stabilized the Fe complexes, resulting in a reduced Fe deposition peak in its cyclic voltammograms.

41.2.2.2 Cr-Based Complexes

Due to its low-redox potential (−0.41 V vs. SHE), parasitic H_2 evolution is common for Cr^{2+}/Cr^{3+} based aqueous FBs [64]. The coordinated water in the close proximity to the highly reducing Cr^{2+} may be one of the main reasons for H_2 production. The presence of organic ligands is expected to reduce the coordination of water and increase electrolyte stability. Marshak and coworkers demonstrated that the Cr-(1,3-propylenediaminetetraacetic acid [PDTA]) chelate formed a hexadentate hexacoordinate structure in an aqueous solution (Figure 41.5c), which greatly suppressed H_2 evolution because of the absence of coordinate water [51]. This feature enabled Cr-PDTA to be used beyond the H_2 evolution potential, offering a "molecular SEI" characteristic for enhanced protection. The Cr-PDTA had a low-redox potential of −1.1 V vs. SHE in a pH 9 buffer, and impressively high cell voltages of 2.13 and 1.62 V were achieved when coupled with bromine and ferrocyanide catholytes, respectively. When the SOC was controlled to 80% during flow cell cycling, the Coulombic efficiency remained near-quantitative and the H_2 level remained nondetectable, indicating the promise of the design.

41.3 Non-Aqueous Metal Coordination Complex Based Flow Batteries

41.3.1 Chalcogen (M−O/S) Chemistries

41.3.1.1 Acetylacetonate Complexes

Extensive efforts in the early 2000s were dedicated to translating aqueous redox chemistries into non-aqueous systems to allow for operating potentials that exceed the limits of aqueous media. Many of the resulting complexes, illustrated in Table 41.1, rely on chalcogenide ligands to solubilize and stabilize metals in non-aqueous media like acetonitrile. These ligands are particularly effective for coordination to inexpensive, early-row transition metals. The literature is largely dominated by oxygen-containing ligands, with some examples of sulfide-based MCCs that show promising redox chemistry but are generally unstable to bulk cycling (Table 41.1, entry 1) [11, 65]. Acetylacetonate complexes of vanadium (Table 41.1, entry 2) were among the first chalcogenide MCCs to be evaluated as non-aqueous flow battery (NFB) electrolytes [66]. $V(acac)_3$ was found to have modest solubility (0.4 M) in acetonitrile but exhibited two quasi-reversible redox waves at −1.8 V and +0.4 V (vs. Ag/Ag^+). As a result, an NFB that utilizes $V(acac)_3$ as both anolyte and catholyte can achieve a theoretical cell voltage of 2.2 V, compared to the aqueous analog of 1.26 V.

Despite the reversibility implied by cyclic voltammetry analyses, charge-discharge cycling of $V(acac)_3$ results in significant capacity losses. Detailed mechanistic studies revealed that this capacity fade was the result of decomposition that forms vanadyl acetylacetonate, $V(O)(acac)_2$, which could form from adventitious water or by degradation of acac at the oxidized V^V center [67]. The long-term instability of $V(acac)_3$ led to the evaluation of other metal acac derivatives. Acac analogs of

Table 41.1 Performance of non-aqueous MCCs with chalcogenide ligands.

Entry	MCC	Solubility (M)	Cycled anodic/cathodic couples (V vs. Fc/Fc⁺)	Concentration of MCC during cycling (M)	Performance (%capacity decay per cycle)
1	[V(S₂C₂(CN)₂)₃]²⁻ (NEt₄⁺)₂	0.9 M in MeCN	−2.032, −0.849, +0.230	0.02 M in MeCN	1.3% per cycle
2	M = V, R = H	0.65 M in MeCN	−1.85, +0.35	0.1 M in MeCN	9% per cycle
3	M = Ru, R = H	—	−1.21, +0.58	0.1 M in MeCN	—
4	M = Mn, R = H	0.5 M in MeCN	−0.50, +0.60	0.05 M in MeCN	—
5	M = Cr, R = H	0.4 M in MeCN	−0.50, +0.60	0.05 M in MeCN	—
6	M = Cr R = (CH₂)₂CO₂(CH₂)₂OMe	1.8 M in MeCN	−2.51, −2.29, +0.87, +1.24	0.05 M in MeCN	—
	M = V R = (CH₂)₂CO₂(CH₂)₂OMe	1.32 M in MeCN	−2.06, +0.05	0.05 M in MeCN	—
7	Co(acacen)	—	−2.25, −0.25	0.01 M in MeCN	—
8	M = Fe R₁ = HO(CH₂)₂NH₂, X = OTf⁻	—	+0.318	0.1 M in PC	4.34% per cycle
	M = Fe R₂ = HO, X = OTf⁻	—	+0.344, +1.018	0.1 M in PC	2.71% per cycle
	M = Cu R₁ = HO(CH₂)₂NH₂, X = BF₄⁻	—	−1.00	0.1 M in PC	—
9	[V(...)]²⁻ Ca²⁺	—	−0.34	0.01 M in DMSO	0.625% per cycle

#	Compound	Solubility	Redox potentials (V)	Concentration	Capacity fade
10	[Ce(NO₃)₆·(py)₆]³⁺(Tf₂N)₃ (structure shown)	>2.5 M in MeCN	+0.58	0.05 M in MeCN	11.43% per cycle
	(NH₄)₂[Ce(NO₃)₆]	0.6 M in MeCN	+1.10	0.05 M in MeCN	0.56% per cycle
11	TBA₄H₃SiV₃W₉O₄₀	—	−2.75, +1.05	0.02 M in PC	—
12	Li₃PMo₁₂O₄₀	0.8 M in MeCN	−0.67, −0.31	0.01 M in MeCN	—
13	[V$^{IV}_4$V$^{V}_2$O₇(OCH₃)₁₂]	0.202 M in MeCN	−0.82, −0.32, +0.20, +0.75	0.01 M in MeCN	—
14	[V₆O₇(OC₂H₅)₉(OCH₂)₃CCH₂OC₂H₄OCH₃]	0.586 M in MeCN	−0.89, −0.34, +0.21, +0.74	0.5 M in MeCN	—
15	[Ti^{IV}V$^{IV}_5$O₆(OCH₃)₁₃]⁻	0.513 M in MeCN	−2.17, +0.13	0.005 M in MeCN	—
	[Ti$^{IV}_2$V$^{IV}_4$O₅(OCH₃)₁₄]	0.193 M in MeCN	−2.20, −1.68, +0.05, +0.54	0.005 M in MeCN	—

Ru [13, 17], Mn [68], and Cr [69] were evaluated based on promising data from CV analysis (Table 41.1, entries 3-6). Redox couples and cell cycling performances for these complexes are reported in Table 41.1. While these acac MCCs all exhibit significant capacity decay during cycling, it should be noted that these studies utilize the MCC as both the catholyte and anolyte in symmetric NFBs. As a result, the poor performance is generally the result of instability at only one of the two redox couples that undergo cycling in the single-component FB. In the case of acac-based MCCs, redox at the positive couple is generally less reversible than at the negative couple. Asymmetric FBs that employ MCCs as dedicated anolytes or catholytes generally exhibit better cycling performance, as will be discussed in Sections 41.3.2 and 41.3.3.

Complementing work to produce stable MCCs is the development of MCCs that are highly-soluble in non-aqueous media. Synthetic chemistry has led to the design and synthesis of new ligands that increased the solubilities of V and Cr analogs by more than four orders of magnitude over the parent acac derivatives (Table 41.1, entry 6) [12]. These materials represent a significant advance in theoretical energy density relative to the state-of-the-art for aqueous chemistries. In addition to the improved solubility, the new MCCs retain their high-potential redox chemistry because the redox-active core is largely unchanged.

Collectively, these empirical investigations guided theoretical investigations by Thompson and coworkers that yielded predictive models for metal(acac) performance [70]. Summarized in Figure 41.7, a training set of 16 complexes from 5 different metals with 11 different ligand substituents were used to correlate structural and electronic properties with cycling stability. A predictive equation based on easily-computed solvation energies and dipole moments successfully modeled the experimentally-determined solubilities. In addition, a correlation between structure and cycling performance was determined using parameterization approaches established by Sigman and Sanford for organic electrolytes [71]. Structure-performance correlations were created by evaluating the number of times an MCC could be cycled before 20% of the capacity was lost. These studies reveal that the measured stability correlates with the percentage of the highest occupied

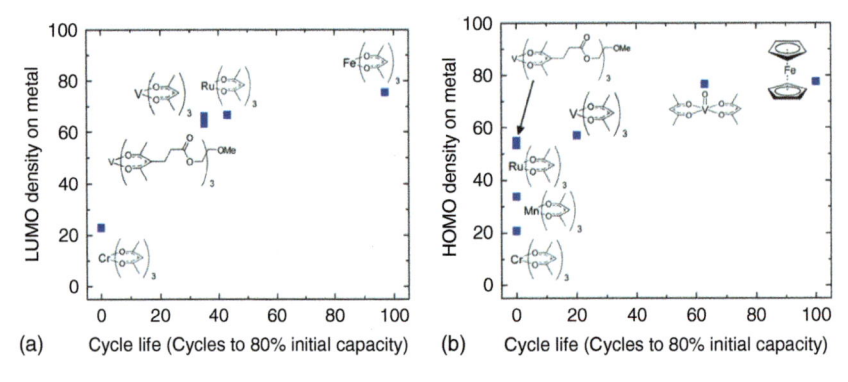

Figure 41.7 Correlation of LUMO or HOMO densities of M(acac)$_3$ complexes with their cycling stability. Source: Reproduced with permission from: Kucharyson et al. 2017 [70].

(HOMO) or lowest unoccupied molecular orbital (LUMO) on the metal center of the complex. This percentage is influenced by the degree of ligand innocence. As a result, acac complexes where redox activity is isolated to the metal center are most stable to cycling. While these results are not general to all MCCs – and actually contrast stability trends for pyridyl MCCs – they provide critical insights into the design of MCCs based on acac ligands.

41.3.1.2 Amino-Alcohol Complexes

Chelating ligands based on oxygen and nitrogen coordination to metals have been evaluated for MCCs with a variety of metals. Tetradentate Co complexes of bis(acetylacetone)ethylenediamine (Table 41.1, entry 7) exhibit two quasi-reversible couples that can be cycled to afford a 2 V cell, albeit with poor capacity retention [72]. Anderson and coworkers developed a series of Fe, Mn, and Cu MCCs bearing simple alkyl amino–alcohols that are liquids at room temperature (Table 41.1, entry 8) [73–75]. Despite the promise of high storage capacities with liquid MCCs, these complexes have not been successfully cycled. Moreover, Zhang and coworkers have reported on limits to solvation in non-aqueous media that may limit NFBs to concentrations well below that which an ionic liquid can reach [76].

Some of the most promising MCCs bearing N/O chelating ligands have been inspired by nature. Cappillino and coworkers reported a dicarboxyl N-oxide chelator of V that is biosynthesized by Amanita mushrooms (Table 41.1 entry 9) [77]. The resulting vanadium complex is robust and exhibits excellent chemical stability in MeCN although the corresponding electrolyte loses capacity during cycling by pathways that do not directly involve degradation of the MCC [78]. The N-oxy motif was found to stabilize other redox-active metal salts and provide high solubility (>2.5 M). In particular, an octa-pyridyl-N-oxide complex of Ce (Table 41.1, entry 10) was cycled against Zn to yield a cell potential of 2.3 V [79]. While the unmodified Ce system exhibited capacity decay (0.5% loss per cycle), the work demonstrated the application of simple coordination complexes for NFB applications.

41.3.1.3 Tunable Oxo Complexes

The success of mono-metallic MCCs inspired research into multi-metallic complexes that were hypothesized to provide improved stability by charge distribution over multiple metal centers (Table 41.1, entries 11–15). One common limitation to such polyoxometallates (POMs) is their poor solubility in non-aqueous solvents [80]. Independent efforts by Barteau and coworker [81] and Mattson and coworkers [82–84] have demonstrated that POMs with high solubilities can be easily prepared and that both the oxy bridging fragments and the metals themselves can be selectively modified. As an example, polyoxovanadate complexes have evolved from methoxy-bridged clusters, to more soluble alkoxy analogs with reduced symmetry, and finally to systems with desirable redox potentials following a substitution of V with Ti at key sites. The Ti-doped vanadium POMs exhibited high cycling stability in symmetric NFBs while achieving cell voltages that are nearly 1 V greater than the all-vanadium POM (Table 41.1, entries 14 vs. 15) [83].

41.3.2 Pyridyl Chemistries

MCCs of pyridyl-based ligands are distinguished from acac-based MCCs because the pyridyl ligands participate in redox chemistry. In particular, pyridyl-based complexes have LUMOs localized more on the ligand than on the metal, which is opposite to that for acac-based MCCs. This charge delocalization generally provides greater stabilization, and ligands serve as electron reservoirs for multi-electron storage. Matsumura-Inoue and coworkers [10] demonstrated that the dicationic [Ru(bpy)$_3$]$^{2+}$ (Table 41.2, entry 1) undergoes 4 quasi-reversible redox events (3 reductions and 1 oxidation) but cycling was only performed at the first reduction (1.7 V vs. Fc/Fc$^+$) and the first oxidation (+0.9 V vs. Fc/Fc$^+$) to yield a cell potential of 2.6 V [85].

The high cell potentials achieved by this first example inspired the design of a wide range of other pyridyl MCCs [86]. Notably, the half-wave potential ($E_{1/2}$) for reduction of the metal bpy complexes (−1.8 V) was independent of metal, which further supports electron storage in a ligand-based orbital. Doo and coworkers at Samsung substituted Ru with inexpensive metals and demonstrated an NFB based on a Ni(bpy)$_3$$^{2+}$ anolyte and Fe(bpy)$_3$$^{2+}$ catholyte with modest cycling stability (Table 41.2, entries 2 and 3) [87]. Since this time, new bpy derivatives based on inexpensive metals have been reported with improved potentials [88, 89] and greater solubilities [90, 91].

Ligand shedding from tris-ligated bpy complexes has been proposed as a major decomposition pathway for these MCCs during charge-discharge cycling. To mitigate this degradation, tri- and tetra-dentate pyridyl analogs were evaluated as MCCs. However, ligands with increased denticity are generally rigid and poorly soluble. As examples, Ni complexes of tetradentate cyclam exhibited solubilities below 400 mM, while bis terpyridine complexes had solubilities below 100 mM (Table 41.2, entry 5) [92]. Moreover, highly-chelating ligands can be difficult to synthesize and modify. As a result, most polydentate ligands employed for high-stability MCCs are those that can be assembled through modular syntheses. As examples, MCCs based on easily-prepared picolinamide [93], pyridyldiimine (PDI) [94–96], and BPI [14] ligands have been extensively investigated (Table 41.2, entries 7 and 8). Among these systems, the Ni(BPI)$_2$ complexes developed by Sanford et al. remain among the best performing due to their ability to store multiple high-voltage electrons for days at high concentrations without decomposition [14]. Surprisingly, the cycling stabilities of other M(BPI)$_2$ complexes were highly variable despite all complexes exhibiting quasi-reversible redox couples at identical potentials (−1.8 and −2.0 V). Mechanistic investigations on a series of M(BPI)$_2$ complexes revealed that ligand exchange impacted long-term cycling stability. In particular, Ni complexes were most stable because ligand shedding occurs by dissociation of a tridentate BPI ligand, which is highly unfavorable. In contrast, Mn(BPI)$_2$ undergoes associative ligand substitution with assistance from the MeCN solvent and undergoes rapid capacity loss during cycling. Thus, while the ligand-based LUMO dictates the potential of the anolyte, the metal controls the overall chemical stability by binding the ligands.

Table 41.2 Non-aqueous MCCs of pyridyl-based ligands.

Entry	MCC	Solubility (M)	Cycled anodic/cathodic couples (V vs. Fc/Fc$^+$)	Concentration of MCC during cycling (M)	Performance (%capacity decay per cycle)
1	M = Ru, R = H	—	−2.1, −1.9, −1.7, +0.9	0.02 M in MeCN	—
2	M = Ni, R = H; M = Fe, R = H	—	Ni: −1.85 (2 e$^−$); Fe: −2.39, −2.10, −1.86, +0.55	Ni: 0.02 M in PC; Fe: 0.04 M in PC	Ni/Fe asymmetric cell 7.78% per cycle
3	M = Fe, R$_1$ = CF$_3$, R$_2$ = H, R$_3$ = OMe	—	R$_1$: −0.73, +1.55; R$_2$: −1.22, +1.15; R$_3$: −1.37, +0.84	—	3.33% per cycle; 2.61% per cycle; —
4	M = Cr, R = CO$_2$(CH$_2$CH$_2$O)$_2$CH$_3$	0.54 M in MeCN	−2.10, −1.92, −1.73, −1.22, −0.68, −0.27	0.2 M in MeCN; 0.01 M in MeCN	—
5	[Ni(TFSI^D)$_2$]$^{2+}$	0.8 M in PC/EC (1:1)	−1.80, +0.75	0.01 M in PC/EC (1:1)	<0.01% over 50 cycles
6	[Co(phen)$_3$]$^{2+}$[PF$_6^D$]$_2$	0.41 M in MeCN	−2.09, −1.43, +0.02	0.01 M in MeCN	—
7	O(CH$_2$CH$_2$O)$_2$CH$_3$–Ni–O(CH$_2$CH$_2$O)$_2$CH$_3$	0.71 M in MeCN	−2.00, −1.80, +0.6	0.002 M in MeCN	<0.025% per cycle
8	Fe complex; Ar = p-^{t}BuPh; Ar = p-OMePh; Ar = Ph; Ar = p-FPh; Ar = p-BrPh; Ar = p-CNPh	0.13 M in MeCN; 0.26 M in MeCN; —; —; —; —	−1.59, −1.32, 0.90; −1.60, −1.30, 0.86; −1.55, −1.27, 0.97; −1.49, −1.20, 1.03; −1.43, −1.15, 1.05; −1.26, −1.01, 1.16	0.002 M in MeCN; 0.002 M in MeCN; 0.0015 M in MeCN; 0.0015 M in MeCN; 0.0015 M in MeCN; 0.0015 M in MeCN	<0.01% over 25 cycles; 1.78% per cycle; <0.01% over 25 cycles; 3.85% per cycle; 3.69% per cycle; 2.59% per cycle

Table 41.3 Metallocene-based MCCs for NFB systems.

Entry	MCC	Solubility (M)	Cycled anodic/cathodic couples (V vs. Fc/Fc⁺)	Concentration of MCC during cycling (M)	Performance (%capacity decay per cycle)
1	$FeCp_2$	—	0.0	0.1 M in DMF	0.4% per cycle
2	$FeCp_2$	—	0.0	0.01 M in MeCN	—
	$CoCp_2$	—	−1.331	0.01 M in MeCN	—
	Br-CoCp*$_2$	—	+0.178	0.01 M in MeCN	—
3	$CoCp*_2$	—	−1.871	0.01 M in MeCN	—
	$FeCp_2$	—	0.0	0.1 M in DMF	0.67% per cycle
	$CoCp_2$	—	−1.65	0.05 M in DOL	0.67% per cycle
	$CoCp*_2$	—	−2.28	0.05 M in DOL	—
4	(Fe with ether chain)	0.53 M in MeCN	−0.06, +0.12	0.05 M in MeCN	0.42% per cycle
5	(Fe–N⁺ TFSI⁻)	1.7 M in EC/PC/EMC	+0.23	0.1 M EC/PC/EMC	0.05% per cycle
6	(Fe)	—	−0.18	0.05 M in EC/DEC	0.02% per cycle
	(Fe)	—	−0.18	Neat (3 M LiClO₄)	—
7	(Fe–N⁺ BF₄⁻)	—	+0.24 (supp. elec. free)	0.2 M in MeCN	1.76% per cycle

41.3.3 Metallocene Chemistries

For metallocenes, the η^5 binding of Cp inhibits ligand dissociation, and these MCCs are very stable during their redox reaction. Because of their rapid redox kinetics, metallocenes are employed in hybrid solid–liquid batteries as redox pairs for solid-state electrodes or as electron shuttles in redox-targeting systems [18, 19, 97, 98] (Table 41.3).

The redox potentials of metallocenes are generally dictated by the metal. As a result, a ferrocene catholyte can be paired against its Co analog, cobaltocene, to form a 2 V NFB [99, 100]. The parent ferrocene has low solubility in MeCN (<100 mM), leading to the development of more soluble derivatives with multiple ferrocene cores [101, 102], solubilizing ammonium fragments [103], and ionic liquids [104]. Ferrocenyl derivatives with alkylammonium substituents were evaluated as both the redox-active material and the supporting electrolyte in NFBs [105]. The operation of a flow cell in the absence of supporting salt by utilizing ionic redox-active species provides a significant cost reduction. Solutions of ionic ferrocenes in the absence of added salts were found to cycle with energy efficiencies (>75%) that are comparable to state-of-the-art NFBs that employ high concentrations of supporting salt.

41.4 Conclusion

MCCs provide a multi-faceted platform to improve flow battery active materials with the ability to change both the metal-center and ligands. This flexibility combines the known cyclability of metal redox with the tunability of organic molecules, which enables use in both aqueous and non-aqueous electrolytes. While there are currently no chemistries ready to replace traditional metal flow batteries, many chemistries have shown promise and are worthy of further study in the search for a low-cost flow battery. The structure-property understandings acquired so far can provide useful clues for the design and development of viable MCC structures.

References

1. National Aeronautics and Space Administration (1977). *Redox Flow Cell Development and Demonstration Project*. Cleveland, OH.
2. Skyllas-Kazacos, M., Rychcik, M., Robins, R.G. et al. (1986). New all-vanadium redox flow cell. *Journal of the Electrochemical Society* 133 (5): 1057–1058. https://iopscience.iop.org/article/10.1149/1.2108706.
3. Wang, W., Luo, Q., Li, B. et al. (2013). Recent progress in redox flow battery research and development. *Advanced Functional Materials* 23 (8): 970–986. https://onlinelibrary.wiley.com/doi/full/10.1002/adfm.201200694.
4. Li, L., Kim, S., Wang, W. et al. (2011). A stable vanadium redox-flow battery with high energy density for large-scale energy storage. *Advanced Energy Materials* 1 (3): 394–400. http://doi.wiley.com/10.1002/aenm.201100008.

5 Mongird, K., Viswanathan, V., Alam, J. et al. (2020). *2020 Grid Energy Storage Technology Cost and Performance Assessment*. Richland, WA.
6 Baxter, R. (2019). *2018 Energy Storage Pricing Survey*. Albuquerque, NM: https://www.osti.gov/servlets/purl/1592892.
7 Khor, A., Leung, P., Mohamed, M.R. et al. (2018). Review of zinc-based hybrid flow batteries: from fundamentals to applications. *Mater Today Energy* 8: 80–108.
8 Panasonic (2017). Specifications for NCR18650GA.
9 Cole, W., Frazier, A., and Augustine, C. (2021). *Cost Projections for Utility-Scale Battery Storage: 2021 Update*. Golden, CO: https://www.osti.gov/servlets/purl/1786976/.
10 Matsuda, Y., Tanaka, K., Okada, M. et al. (1988). A rechargeable redox battery utilizing ruthenium complexes with non-aqueous organic electrolyte. *Journal of Applied Electrochemistry* 18 (6): 909–914.
11 Cappillino, P.J., Pratt, H.D., Hudak, N.S. et al. (2014). Application of redox non-innocent ligands to non-aqueous flow battery electrolytes. *Advanced Energy Materials* 4 (1): 1300566. http://doi.wiley.com/10.1002/aenm.201300566.
12 Suttil, J.A., Kucharyson, J.F., Escalante-Garcia, I.L. et al. (2015). Metal acetylacetonate complexes for high energy density non-aqueous redox flow batteries. *Journal of Materials Chemistry A* 3 (15): 7929–7938. http://xlink.rsc.org/?DOI=C4TA06622G.
13 Kucharyson, J.F., Gaudet, J.R., Wyvratt, B.M., and Thompson, L.T. (2016). Characterization of structural and electronic transitions during reduction and oxidation of Ru(acac)3 flow battery electrolytes by using X-ray absorption spectroscopy. *ChemElectroChem* 3 (11): 1875–1883. https://chemistry-europe.onlinelibrary.wiley.com/doi/full/10.1002/celc.201600360.
14 Sevov, C.S., Fisher, S.L., Thompson, L.T., and Sanford, M.S. (2016). Mechanism-based development of a low-potential, soluble, and cyclable multi-electron anolyte for nonaqueous redox flow batteries. *Journal of the American Chemical Society* 138 (47): 15378–15384. http://pubs.acs.org/doi/abs/10.1021/jacs.6b07638.
15 Huang, J., Cheng, L., Assary, R.S. et al. (2015). Liquid catholyte molecules for nonaqueous redox flow batteries. *Advanced Energy Materials* 5 (6): 1401782. http://doi.wiley.com/10.1002/aenm.201401782.
16 Fisher, S.L. (2017). *Non-Aqueous Redox Flow Batteries : Active Species Stability and Cost Saving Design Concepts*. University of Michigan.
17 Chakrabarti, M.H., Dryfe, R.A.W., and Roberts, E.P.L. (2007). Evaluation of electrolytes for redox flow battery applications. *Electrochimica Acta* 52 (5): 2189–2195.
18 Zhao, Y., Ding, Y., Song, J. et al. (2014). Sustainable electrical energy storage through the ferrocene/ferrocenium redox reaction in aprotic electrolyte. *Angewandte Chemie* 126 (41): 11216–11220.
19 Pan, F., Yang, J., Huang, Q. et al. (2014). Redox targeting of anatase TiO_2 for redox flow lithium–ion batteries. *Advanced Energy Materials* 4 (15): 1400567.

20 Esswein, A.J., Goeltz, J., Reece, S.Y. et al. (2014). Aqueous redox flow batteries comprising metal ligand coordination compounds. US Patent 8, 691, 413.

21 Esswein, A.J., Goeltz, J., King, E.R. et al. (2014). Aqueous redox flow batteries featuring improved cell design characteristics. US Patent 8, 753, 761.

22 Marcus, R.A. (1956). On the theory of oxidation-reduction reactions involving electron transfer. I. *The Journal of Chemical Physics* 24 (5): 966. https://aip.scitation.org/doi/abs/10.1063/1.1742723.

23 Gong, K., Xu, F., Grunewald, J.B. et al. (2016). All-soluble all-iron aqueous redox-flow battery. *ACS Energy Letters* 1 (1): 89–93. https://pubs.acs.org/doi/abs/10.1021/acsenergylett.6b00049.

24 Heinze, K. and Lang, H. (2013). Ferrocene—beauty and function. *Organometallics* 32 (20): 5623–5625. https://pubs.acs.org/doi/full/10.1021/om400962w.

25 Vanýsek, P. (2014). Electrochemical series. In: *CRC Handbook of Chemistry and Physics*, 9e (ed. W.M. Haynes). Boca Raton, FL: CRC Press/Taylor and Francis. https://hbcp-chemnetbase-com.proxy.lib.umich.edu/faces/documents/05_22/05_22_0001.xhtml.

26 Mueller-Westerhoff, U.T., Haas, T.J., Swiegers, G.F., and Leipert, T.K. (1994). The protonation mechanism of metallocenes and [1.1]metallocenophanes. *Journal of Organometallic Chemistry* 472 (1–2): 229–246.

27 Prins, R., Korswagen, A.R., and Kortbeek, A.G.T.G. (1972). Decomposition of the ferricenium cation by nucleophilic reagents. *Journal of Organometallic Chemistry* 39 (2): 335–344.

28 Beh, E.S., De Porcellinis, D., Gracia, R.L. et al. (2017). A neutral pH aqueous organic–organometallic redox flow battery with extremely high capacity retention. *ACS Energy Letters* 2 (3): 639–644. https://pubs.acs.org/doi/abs/10.1021/acsenergylett.7b00019.

29 Hu, B., DeBruler, C., Rhodes, Z., and Liu, T.L. (2017). Long-cycling aqueous organic redox flow battery (AORFB) toward sustainable and safe energy storage. *Journal of the American Chemical Society* 139 (3): 1207–1214. https://pubs.acs.org/doi/abs/10.1021/jacs.6b10984.

30 Gasser, G., Fischmann, A.J., Forsyth, C.M., and Spiccia, L. (2007). Products of hydrolysis of (ferrocenylmethyl)trimethylammonium iodide: synthesis of hydroxymethylferrocene and bis(ferrocenylmethyl) ether. *Journal of Organometallic Chemistry* 692 (17): 3835–3840.

31 Chen, Q., Li, Y., Liu, Y. et al. (2021). Designer ferrocene catholyte for aqueous organic flow batteries. *ChemSusChem* 14 (5): 1295–1301. https://chemistry-europe.onlinelibrary.wiley.com/doi/full/10.1002/cssc.202002467.

32 Borchers, P.S., Strumpf, M., Friebe, C. et al. (2020). Aqueous redox flow battery suitable for high temperature applications based on a tailor-made ferrocene copolymer. *Advanced Energy Materials* 10 (41): 2001825. https://onlinelibrary.wiley.com/doi/full/10.1002/aenm.202001825 .

33 Zhao, Z., Zhang, B., Schrage, B.R. et al. (2020). Investigations into aqueous redox flow batteries based on ferrocene bisulfonate. *ACS Applied Energy Materials* 3 (10): 10270–10277. https://pubs.acs.org/doi/abs/10.1021/acsaem.0c02259.

34 Yu, J., Salla, M., Zhang, H. et al. (2020). A robust anionic sulfonated ferrocene derivative for pH-neutral aqueous flow battery. *Energy Storage Materials* 29: 216–222.

35 Younes, M., Aggett, P., Aguilar, F. et al. (2018). Re-evaluation of sodium ferrocyanide (E 535), potassium ferrocyanide (E 536) and calcium ferrocyanide (E 538) as food additives. *EFSA Journal* 16 (7): e05374.

36 Esswein, A.J., Goeltz, J., and Amadeo, D. (2017). High solubility iron hexacyanides. *Electrochimica Acta* 61: 10374248 B2.

37 Lee, W., Permatasari, A., and Kwon, Y. (2020). Neutral pH aqueous redox flow batteries using an anthraquinone-ferrocyanide redox couple. *Journal of Materials Chemistry C* 8 (17): 5727–5731. https://pubs.rsc.org/en/content/articlehtml/2020/tc/d0tc00640h.

38 Wei, X., Xia, G.-G., Kirby, B. et al. (2015). An aqueous redox flow battery based on neutral alkali metal ferri/ferrocyanide and polysulfide electrolytes. *Journal of the Electrochemical Society* 163 (1): A5150. https://iopscience.iop.org/article/10.1149/2.0221601jes.

39 Luo, J., Hu, B., Debruler, C. et al. (2019). Unprecedented capacity and stability of ammonium ferrocyanide catholyte in pH neutral aqueous redox flow batteries. *Joule* 3 (1): 149–163.

40 Luo, J., Sam, A., Hu, B. et al. (2017). Unraveling pH dependent cycling stability of ferricyanide/ferrocyanide in redox flow batteries. *Nano Energy* 42: 215–221.

41 Yuan, Z., Duan, Y., Liu, T. et al. (2018). Toward a low-cost alkaline zinc–iron flow battery with a polybenzimidazole custom membrane for stationary energy storage. *iScience* 3: 40–49.

42 Luo, J., Hu, B., Hu, M. et al. (2019). Status and prospects of organic redox flow batteries toward sustainable energy storage. *ACS Energy Letters* 4 (9): 2220–2240.

43 Páez, T., Martínez-Cuezva, A., Palma, J., and Ventosa, E. (2020). Revisiting the cycling stability of ferrocyanide in alkaline media for redox flow batteries. *Journal of Power Sources* 471: 228453.

44 Lin, K., Chen, Q., Gerhardt, M.R. et al. (2015). Alkaline quinone flow battery. *Science* 349 (6255): 1529–1532. https://science.sciencemag.org/content/349/6255/1529.

45 Lin, K., Gómez-Bombarelli, R., Beh, E.S. et al. (2016). A redox-flow battery with an alloxazine-based organic electrolyte. *Nature Energy* 1 (9): 1–8. https://www.nature.com/articles/nenergy2016102.

46 Orita, A., Verde, M.G., Sakai, M., and Meng, Y.S. (2016). A biomimetic redox flow battery based on flavin mononucleotide. *Nature Communications* 7 (1): 1–8. https://www.nature.com/articles/ncomms13230.

47 Hollas, A., Wei, X., Murugesan, V. et al. (2018). A biomimetic high-capacity phenazine-based anolyte for aqueous organic redox flow batteries. *Nature Energy* 3 (6): 508–514. https://www.nature.com/articles/s41560-018-0167-3.

48 Zu, X., Zhang, L., Qian, Y. et al. (2020). Molecular engineering of azobenzene-based anolytes towards high-capacity aqueous redox flow batteries. *Angewandte*

Chemie International Edition 59 (49): 22163–22170. https://onlinelibrary.wiley.com/doi/full/10.1002/anie.202009279.

49 Feng, R., Zhang, X., Murugesan, V. et al. (2021). Reversible ketone hydrogenation and dehydrogenation for aqueous organic redox flow batteries. *Science* 372 (6544): 836–840. https://science.sciencemag.org/content/372/6544/836.

50 Martell, A.E., Motekaitis, R.J., Chen, D. et al. (2011). Selection of new Fe(III)/Fe(II) chelating agents as catalysts for the oxidation of hydrogen sulfide to sulfur by air. *Canadian Journal of Chemistry* 74 (10): 1872–1879. https://doi.org/10.1139/v96-210.

51 Robb, B.H., Farrell, J.M., and Marshak, M.P. (2019). Chelated chromium electrolyte enabling high-voltage aqueous flow batteries. *Joule* 3 (10): 2503–2512.

52 Waters, S.E., Robb, B.H., and Marshak, M.P. (2020). Effect of chelation on iron–chromium redox flow batteries. *ACS Energy Letters* 5 (6): 1758–1762. https://pubs.acs.org/doi/abs/10.1021/acsenergylett.0c00761.

53 Maigut, J., Meier, R., Zahl, A., and van Eldik, R. (2007). Elucidation of the solution structure and water-exchange mechanism of paramagnetic [Fe^{II}(edta)(H_2O)]$^{2-}$. *Inorganic Chemistry* 46 (13): 5361–5371. https://pubs.acs.org/doi/abs/10.1021/ic700472q.

54 Kan, K., Hidekazu, D., Naoya, U. et al. (2006). The Raman spectral study on the solution structure of iron(III)-edta complexes. *Bulletin of the Chemical Society of Japan* 63 (5): 1447–1454. https://doi.org/10.1246/bcsj.63.1447.

55 Sawyer, D.T. and Tackett, J.E. (1963). Properties and infrared spectra of ethylenediaminetetraacetic acid complexes. V. Bonding and structure of several metal chelates in solution (1960) (2) (a). *Journal of the American Chemical Society* 85: 2390–2394.

56 Oakes, J. and Smith, E.G. (1983). Nuclear magnetic resonance studies of transition-metal complexes of ethylenediaminetetra-acetic acid (EDTA) in aqueous solution. *Journal of the Chemical Society, Faraday Transactions 1: Physical Chemistry in Condensed Phases* 79 (2): 543–552. https://pubs.rsc.org/en/content/articlelanding/1983/f1/f19837900543.

57 Seibig, S. and van Eldik, R. (1997). Kinetics of [feII(edta)] oxidation by molecular oxygen revisited. New evidence for a multistep mechanism. *Inorganic Chemistry* 36 (18): 4115–4120. https://pubs.acs.org/doi/abs/10.1021/ic970158t.

58 Finnen, D.C., Pinkerton, A.A., Dunham, W.R. et al. (1991). Structures and spectroscopic characteristics of iron(III) diethylenetriaminepentaacetic acid complexes. A non-heme iron(III) complex with relevance to the iron environment in lipoxygenases. *Inorganic Chemistry* 30 (20): 3960–3964. https://pubs.acs.org/doi/abs/10.1021/ic00020a034.

59 Arroyo-Currás, N., Hall, J.W., Dick, J.E. et al. (2014). An alkaline flow battery based on the coordination chemistry of iron and cobalt. *Journal of the Electrochemical Society* 162 (3): A378. https://iopscience.iop.org/article/10.1149/2.0461503jes.

60 Noh, C., Chung, Y., and Kwon, Y. (2021). Optimization of iron and cobalt based organometallic redox couples for long-term stable operation of aqueous organometallic redox flow batteries. *Journal of Power Sources* 495: 229799.

61 Stumm, W. and Lee, G.F. (1961). Oxygenation of ferrous iron. *Industrial and Engineering Chemistry* 53 (2): 143–146.

62 Wen, Y.H., Zhang, H.M., Qian, P. et al. (2006). A study of the Fe(III)/Fe(II)–triethanolamine complex redox couple for redox flow battery application. *Electrochimica Acta* 51 (18): 3769–3775.

63 Shin, M., Noh, C., Chung, Y., and Kwon, Y. (2020). All iron aqueous redox flow batteries using organometallic complexes consisting of iron and 3-[bis(2-hydroxyethyl)amino]-2-hydroxypropanesulfonic acid ligand and ferrocyanide as redox couple. *Chemical Engineering Journal* 398: 125631.

64 Lopez-Atalaya, M., Codina, G., Perez, J.R. et al. (1992). Optimization studies on a Fe/Cr redox flow battery. *Journal of Power Sources* 39 (2): 147–154.

65 Eisenberg, R. and Gray, H.B. (2011). Noninnocence in metal complexes: a dithiolene dawn. *Inorganic Chemistry* 50 (20): 9741–9751. https://pubs.acs.org/doi/abs/10.1021/ic2011748.

66 Liu, Q., Sleightholme, A.E.S., Shinkle, A.A. et al. (2009). Non-aqueous vanadium acetylacetonate electrolyte for redox flow batteries. *Electrochemistry Communications* 11 (12): 2312–2315.

67 Escalante-García, I.L., Wainright, J.S., Thompson, L.T., and Savinell, R.F. (2014). Performance of a non-aqueous vanadium acetylacetonate prototype redox flow battery: examination of separators and capacity decay. *Journal of the Electrochemical Society* 162 (3): A363. https://iopscience.iop.org/article/10.1149/2.0471503jes.

68 Sleightholme, A.E.S., Shinkle, A.A., Liu, Q. et al. (2011). Non-aqueous manganese acetylacetonate electrolyte for redox flow batteries. *Journal of Power Sources* 196 (13): 5742–5745. https://www.sciencedirect.com/science/article/pii/S0378775311003740.

69 Liu, Q., Shinkle, A.A., Li, Y. et al. (2010). Non-aqueous chromium acetylacetonate electrolyte for redox flow batteries. *Electrochemistry Communications* 12 (11): 1634–1637.

70 Kucharyson, J.F., Cheng, L., Tung, S.O. et al. (2017). Predicting the potentials, solubilities and stabilities of metal-acetylacetonates for non-aqueous redox flow batteries using density functional theory calculations. *Journal of Materials Chemistry A* 5 (26): 13700–13709. http://xlink.rsc.org/?DOI=C7TA01285C.

71 Sevov, C.S., Hickey, D.P., Cook, M.E. et al. (2017). Physical organic approach to persistent, cyclable, low-potential electrolytes for flow battery applications. *Journal of the American Chemical Society* 139 (8): 2924–2927. https://pubs.acs.org/doi/abs/10.1021/jacs.7b00147.

72 Zhang, D., Lan, H., and Li, Y. (2012). The application of a non-aqueous bis(acetylacetone)ethylenediamine cobalt electrolyte in redox flow battery. *Journal of Power Sources* 217: 199–203. https://www.sciencedirect.com/science/article/pii/S0378775312010385.

73 Small, L.J., Pratt, H.D. III, Staiger, C.L., and Anderson, T.M. (2017). MetILs[3]: A strategy for high density energy storage using redox-active ionic liquids. *Advanced Sustainable Systems* 1 (9): 1700066. https://onlinelibrary.wiley.com/doi/full/10.1002/adsu.201700066.

74 Pratt, H.D. III, Ingersoll, D., Hudak, N.S. et al. (2013). Copper ionic liquids: tunable ligand and anion chemistries to control electrochemistry and deposition morphology. *Journal of Electroanalytical Chemistry* 704: 153–158.

75 Pratt, H.D. III, Rose, A.J., Staiger, C.L. et al. (2011). Synthesis and characterization of ionic liquids containing copper, manganese, or zinc coordination cations. *Dalton Transactions* 40 (43): 11396–11401. https://pubs.rsc.org/en/content/articlehtml/2011/dt/c1dt10973a.

76 Zhang, J., Corman, R.E., Schuh, J.K. et al. (2018). Solution properties and practical limits of concentrated electrolytes for nonaqueous redox flow batteries. *Journal of Physical Chemistry C* 122 (15): 8159–8172. http://pubs.acs.org/doi/10.1021/acs.jpcc.8b02009.

77 Huang, H., Howland, R., Agar, E. et al. (2017). Bioinspired, high-stability, non-aqueous redox flow battery electrolytes. *Journal of Materials Chemistry A* 5 (23): 11586–11591. https://pubs.rsc.org/en/content/articlehtml/2017/ta/c7ta00365j.

78 Gokoglan, T.C., Pahari, S.K., Hamel, A. et al. (2019). Operando spectroelectrochemical characterization of a highly stable bioinspired redox flow battery active material. *Journal of the Electrochemical Society* 166 (10): A1745. https://iopscience.iop.org/article/10.1149/2.0271910jes.

79 Li, Y., Geysens, P., Zhang, X. et al. (2020). Cerium-containing complexes for low-cost, non-aqueous redox flow batteries (RFBs). *Journal of Power Sources* 450: 227634.

80 Pratt, H.D., Hudak, N.S., Fang, X., and Anderson, T.M. (2013). A polyoxometalate flow battery. *Journal of Power Sources* 236: 259–264.

81 Chen, J.J.J. and Barteau, M.A. (2017). Molybdenum polyoxometalates as active species for energy storage in non-aqueous media. *Journal of Energy Storage* 13: 255–261.

82 Vangelder, L.E., Kosswattaarachchi, A.M., Forrestel, P.L. et al. (2018). Polyoxovanadate-alkoxide clusters as multi-electron charge carriers for symmetric non-aqueous redox flow batteries. *Chemical Science* 9 (6): 1692–1699.

83 VanGelder, L.E. and Matson, E.M. (2018). Heterometal functionalization yields improved energy density for charge carriers in nonaqueous redox flow batteries. *Journal of Materials Chemistry A* 6 (28): 13874–13882. https://pubs.rsc.org/en/content/articlehtml/2018/ta/c8ta03312a.

84 VanGelder, L.E., Petel, B.E., Nachtigall, O. et al. (2018). Organic functionalization of polyoxovanadate–alkoxide clusters: improving the solubility of multimetallic charge carriers for nonaqueous redox flow batteries. *ChemSusChem* 11 (23): 4139–4149. https://chemistry-europe.onlinelibrary.wiley.com/doi/full/10.1002/cssc.201802029.

85 Morita, M., Tanaka, Y., Tanaka, K. et al. (1988). Electrochemical oxidation of ruthenium and iron complexes at rotating disk electrode in acetonitrile solution. *Bulletin of the Chemical Society of Japan* 61 (8): 2711–2714.

86 Kim, J.H., Kim, K.J., Park, M.S. et al. (2011). Development of metal-based electrodes for non-aqueous redox flow batteries. *Electrochemistry Communications* 13 (9): 997–1000.

87 Mun, J., Lee, M.J., Park, J.W. et al. (2012). Non-aqueous redox flow batteries with nickel and iron tris(2,2'-bipyridine) complex electrolyte. *Electrochemical and Solid-State Letters* 15 (6): 80–83.

88 Cammack, C.X., Pratt, H.D., Small, L.J., and Anderson, T.M. (2021). A higher voltage Fe(ii) bipyridine complex for non-aqueous redox flow batteries. *Dalton Transactions* 50 (3): 858–868. https://pubs.rsc.org/en/content/articlehtml/2021/dt/d0dt03927f.

89 Xing, X., Zhang, D., and Li, Y. (2015). A non-aqueous all-cobalt redox flow battery using 1,10-phenanthrolinecobalt(II) hexafluorophosphate as active species. *Journal of Power Sources* 279: 205–209.

90 Cabrera, P.J., Yang, X., Suttil, J.A. et al. (2015). Complexes containing redox noninnocent ligands for symmetric, multielectron transfer nonaqueous redox flow batteries. *Journal of Physical Chemistry C* 119 (28): 15882–15889. https://pubs.acs.org/doi/abs/10.1021/acs.jpcc.5b03582.

91 Cabrera, P.J., Yang, X., Suttil, J.A. et al. (2015). Evaluation of tris-bipyridine chromium complexes for flow battery applications: impact of bipyridine ligand structure on solubility and electrochemistry. *Inorganic Chemistry* 54 (21): 10214–10223. http://pubs.acs.org/doi/10.1021/acs.inorgchem.5b01328.

92 Kim, H., Hwang, S., Mun, J. et al. (2019). Counter anion effects on the energy density of Ni(II)-chelated tetradentate azamacrocyclic complex cation as single redox couple for non-aqueous flow batteries. *Electrochimica Acta* 308: 227–230. https://www.sciencedirect.com/science/article/pii/S0013468619307029.

93 Andrade, G.A., Popov, I.A., Federico, C.R. et al. (2020). Expanding the potential of redox carriers for flow battery applications. *Journal of Materials Chemistry A* 8 (34): 17808–17816.

94 Sharma, S., Andrade, G.A., Maurya, S. et al. (2021). Iron-iminopyridine complexes as charge carriers for non-aqueous redox flow battery applications. *Energy Storage Materials* 37: 576–586.

95 Duarte, G.M., Braun, J.D., Giesbrecht, P.K., and Herbert, D.E. (2017). Redox non-innocent bis(2,6-diimine-pyridine) ligand–iron complexes as anolytes for flow battery applications. *Dalton Transactions* 46 (47): 16439–16445. https://pubs.rsc.org/en/content/articlehtml/2017/dt/c7dt03915h.

96 Braun, J.D., Gray, P.A., Sidhu, B.K. et al. (2020). Zn-Templated synthesis of substituted (2,6-diimine)pyridine proligands and evaluation of their iron complexes as anolytes for flow battery applications. *Dalton Transactions* 49 (45): 16175–16183. https://pubs.rsc.org/en/content/articlehtml/2020/dt/d0dt00543f.

97 Zhou, M., Huang, Q., Pham Truong, T.N. et al. (2017). Nernstian-potential-driven redox-targeting reactions of battery materials. *Chem* 3 (6): 1036–1049.

98 Jia, C., Pan, F., Zhu, Y.G. et al. (2015). High–energy density nonaqueous all redox flow lithium battery enabled with a polymeric membrane. *Science Advances* 1 (10): e1500886. https://advances.sciencemag.org/content/1/10/e1500886.

99 Hwang, B., Park, M.S., and Kim, K. (2015). Ferrocene and cobaltocene derivatives for non-aqueous redox flow batteries. *ChemSusChem* 8 (2): 310–314.

100 Ding, Y., Zhao, Y., Li, Y. et al. (2017). A high-performance all-metallocene-based, non-aqueous redox flow battery. *Energy & Environmental Science* 10 (2): 491–497. http://xlink.rsc.org/?DOI=C6EE02057G.

101 Chen, H., Niu, Z., Ye, J. et al. (2021). Multicore ferrocene derivative as a highly soluble cathode material for nonaqueous redox flow batteries. *ACS Applied Energy Materials* 4 (1): 855–861.

102 Xie, C., Xu, W., Zhang, H. et al. (2018). A multi-electron transfer ferrocene derivative positive redox moiety with improved solubility and potential. *Chemical Communications* 54 (60): 8419–8422. https://pubs.rsc.org/en/content/articlehtml/2018/cc/c8cc04099k.

103 Wei, X., Cosimbescu, L., Xu, W. et al. (2015). Towards high-performance nonaqueous redox flow electrolyte via ionic modification of active species. *Advanced Energy Materials* 5 (1): 1400678.

104 Cong, G., Zhou, Y., Li, Z., and Lu, Y.C. (2017). A highly concentrated catholyte enabled by a low-melting-point ferrocene derivative. *ACS Energy Letters* 2 (4): 869–875.

105 Milshtein, J.D., Fisher, S.L., Breault, T.M. et al. (2017). Feasibility of a supporting-salt-free nonaqueous redox flow battery utilizing ionic active materials. *ChemSusChem* 10 (9): 2080–2088. http://doi.wiley.com/10.1002/cssc.201700028.

42

Organic Redox Flow Batteries: Lithium-Ion-based FBs

Feifei Zhang and Qing Wang

National University of Singapore, Department of Materials Science and Engineering, 117574, Singapore

42.1 Introduction

Lithium-ion batteries (LIBs), because of their high capacity of charge storage and high cell voltage, are currently the state-of-the-art battery technology dominating the market from consumer electronics to electric vehicles. The lithium-ion-based flow battery (LFB), which has the advantages of great system flexibility of FBs and high-energy density of LIB, provides an intriguing approach for large-scale electricity storage. LFB operates upon the redox reactions of solid active materials *directly* or *indirectly* on the electrode. In both cases, the redox-active materials are stored in external energy storage tanks, while in the former, they are dispersed in electrolyte along with conducting additives and form a semi-solid slurry circulating between the tanks and electrode compartments upon operation [1]; in the latter, they are kept statically in the tank while the redox reactions of the materials proceed with the "redox-targeting" process of redox species dissolved in the electrolyte circulating through the system [2]. Thus, the amount of stored energy is determined by the quantity of active materials stored in the energy tanks, while the power is largely dictated by the cell stack as the FBs. Upon operation, the electrical balance is achieved by the transport of Li^+ ions in the electrolytes migrating between the two electrode compartments. Since the first study of LFB, many lithium-active materials have been investigated in the two different flow battery configurations.

42.2 Semi-solid Electroactive Materials for LFBs

The concept of semi-solid flow cell (SSFC) was first proposed by Yet-Ming Chiang and coworkers in 2011 [1]. In SSFC, electroactive solid particles suspended in a Li^+-containing organic electrolyte instead of the dissolved electroactive redox

Flow Batteries: From Fundamentals to Applications, First Edition.
Edited by Christina Roth, Jens Noack, and Maria Skyllas-Kazacos.
© 2023 WILEY-VCH GmbH. Published 2023 by WILEY-VCH GmbH.

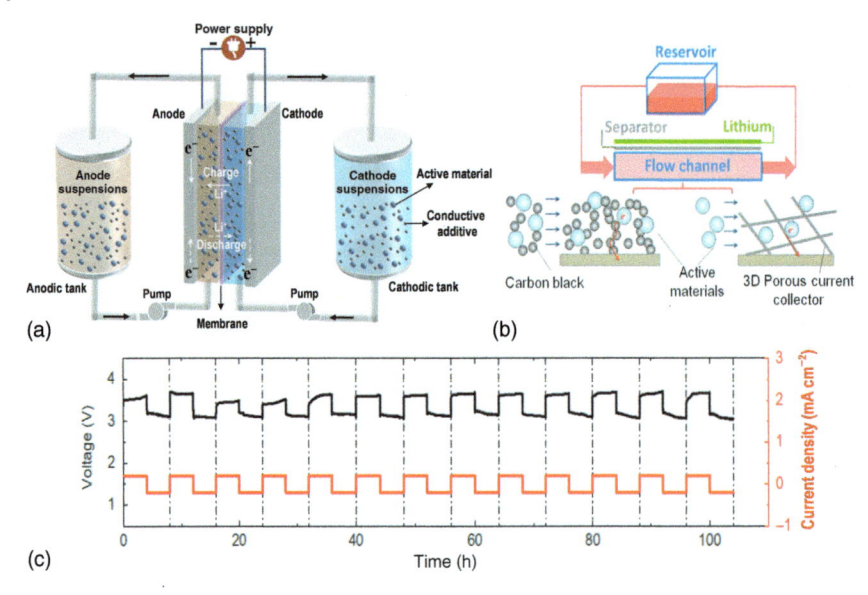

Figure 42.1 (a) Configuration of a SSFC system using flowing lithium-ion cathode and anode suspensions. (b) Schematic illustration of two-component/single-component slurry-based lithium-ion flow batteries without/with a 3D porous current collector. (c) Cycling characteristics of 20 vol% LFP slurry-based Li-LiFePO$_4$ flow cell at 0.2 mA cm^{-2} under continuous flow modes without an initial discharge. Source: (b, c) Reproduced from Chen et al. [3] with permission from Elsevier.

species were used in a flow cell configuration (Figure 42.1a) [1], with which the traditional charge collection on the electrode sheet is replaced with charge percolation through the conductive carbon nanoparticle networks within the flowing suspension. Thus the energy of SSFC is stored in fluids of suspensions containing solid electroactive materials which are stored in separate tanks and charges transport via dispersed while continuous percolating networks of conducting materials in the suspension. Compared with conventional FBs, as the electroactive materials in SSFCs are in solid form, a porous separator would generally suffice to prevent the mixing of active materials from the two electrode compartments while still allowing the transport of Li$^+$ in the electrolyte through the pores. Using a solid electroactive material suspension instead of solution addresses the energy density limitations of conventional FBs, whose capacity is limited by the solubility of redox species in electrolyte solutions (generally <2 M concentrations). In SSFC, assuming a solid content of 50% due to the addition of conducting additives, the volumetric capacity of the suspensions can achieve a concentration of 10–40 M, which is 5–20 times greater than that of the liquid redox solution counterparts [4]. When applied to non-aqueous LIB materials, the energy density is further increased by another factor of 1.5–3, in direct proportion to the cell voltage. A wide variety of Li$^+$ storage electroactive materials have been employed in SSFC, based on the different reaction mechanisms such as those of Li$^+$ insertion–extraction, precipitation-dissolution chemistries, and multiple redox semi-solid-liquid active materials, etc.

42.2.1 Electroactive Materials Based on Li+ Insertion-Extraction Chemistry

Analogue to that of the conventional solid electrode LIBs, the main reactions in the semi-solid suspensions in SSFCs resemble the electrochemical Li+ ion insertion and extraction reactions when the commonly used LIB electrode materials are employed. These electroactive lithium storage materials generally contain a high concentration of lithium. The theoretical molar concentrations of reversibly stored lithium in cathode materials such as the layered $LiCoO_2$ (in terms of 0.5 Li), olivine $LiFePO_4$, spinel $LiMnPO_4$, and $LiNi_{0.5}Mn_{1.5}O_4$ are about 26.6, 22.8, 21.7, and 24.1 mol l^{-1}, respectively, while in anode materials such as $Li_4Ti_5O_{12}$, graphite, and Si are 22.6, 21.4, and 87 mol l^{-1}, respectively [5]. The high Li+ ion concentration of solid electroactive materials in the suspension and the wide potential window of non-aqueous electrolytes enable SSFCs to deliver a high-energy density, while still keeping the operational flexibility and scalability of flow batteries.

For the cathode materials, the well-studied LIBs cathode materials, such as $LiFePO_4$ (~170 mAh g^{-1} and 3.45 V vs. Li/Li+), $LiCoO_2$ (~272 mAh g^{-1} and 4.00 V vs. Li/Li+), $LiNi_{1/3}Co_{1/3}Mn_{1/3}O_2$ (~278 mAh g^{-1} and 4.00 V vs. Li/Li+) and $LiNi_{0.5}Mn_{1.5}O_4$ (~147 mAh g^{-1} and 4.70 V vs. Li/Li+) [6, 7] were studied as cathode in SSFC. While for the anode, the most used anode material is $Li_4Ti_5O_{12}$ (~175 mAh g^{-1} and 1.55 V vs. Li/Li+), in addition to Si material (~4200 mAh g^{-1} and 0.30 V vs. Li/Li+) [8, 9]. Chiang's group demonstrated the first SSFC using $LiCoO_2$, $LiNi_{0.5}Mn_{1.5}O_4$, and $Li_4Ti_5O_{12}$ as the electroactive materials. The electrochemical charge and discharge were investigated using both continuous and intermittent flow modes. The $LiCoO_2$ slurry (22.4 vol%, 11.5 M, and 0.7% Ketjen black, KB) had a reversible capacity of 127 mAh g^{-1} under continuous flow mode. However, there was a huge energy loss from pumping due to the high viscosity of the dispersion. To optimize the operating efficiency, an intermittent flow mode was also demonstrated which significantly reduced the parasitic loss compared with continuous flow, with which a pumping-dissipation less than 1% of the total discharge energy has been achieved. A full cell under intermittent mode was demonstrated with $LiCoO_2$ (20 vol%, 10.2 M, and 1.5% KB) as the cathode, $Li_4Ti_5O_{12}$ (10 vol%, 2.3 M and 2% KB) as the anode, and both suspensions contained the same organic Li+ electrolyte. Although the Coulombic efficiencies of the first two iterations for the intermittent flow (73% and 80%) were lower than that of the static one (98%), there are many possible optimization opportunities, for example, a closer matching of the cathodic and anodic suspension capacities and tuning flow rates to match capacities. In addition, the transformation of $Li_4Ti_5O_{12}$ from insulator to conductor upon lithiation shows asymmetric overpotentials during charge/discharge as fluid electrodes [10], for which Ventosa et al., demonstrated that hydrogen annealing of $Li_4Ti_5O_{12}$ would significantly improve the electrochemical performance of SSFC.

For both electrically insulating and conducting electroactive materials, an addition of adequate amount of carbon is required for optimal electrical percolation through the suspensions. The main function of these conducting additives is to form a dynamic network for electron conduction ensuring the redox reaction of solid active particles and to prevent them from rapid sedimentation. Qi and Koenig investigated the rheological and electrochemical behaviors of a carbon-free suspension of $Li_4Ti_5O_{12}$ [11] or $LiCoO_2$ [12]. While a single-component suspension offers electron conduction by direct collision/contact with the current collector or indirectly by inter-particle interactions, the electrochemical performance of such a flow cell was poor. A common conducting additive used in the reported semi-solid flow batteries is Ketjenblack (KB), which is a highly conductive carbon black. The addition of conducting additives could significantly affect the overall energy density of a suspension electrode and thus has to be tailored to the specific active material and electrolyte system. Hamelet et al. assessed the performance of a series of $LiFePO_4$ and KB suspensions in 1 M $LiPF_6$ electrolyte, with which a wide range of energy density can be achieved by varying the loading at different compositions [13]. For instance, by increasing the active material volumetric percentage to 12.6%, the suspensions obtained an energy density of 50 Wh kg^{-1} (or 67 Wh l^{-1}), similar to that of vanadium flow battery (VFB) electrolyte. Biendicho et al. investigated the effect of KB content on $LiNi_{1/3}Co_{1/3}Mn_{1/3}O_2$-based suspensions by galvanostatic charge/discharge and electrochemical impedance spectroscopy [14]. The cell with low-volume percentage of KB in the suspension showed a poor electrical and electrochemical performance owing to huge internal resistance (>5000 Ω). For instance, a flow cell containing a suspension with 9.53 vol% of KB and 13.90 vol% of $LiNi_{1/3}Co_{1/3}Mn_{1/3}O_2$ exhibited overpotentials of 0.30 and 0.70 V at 0.33 and 1 mA cm^{-2}, respectively. Besides the addition of conducting additive, coating of the electroactive materials with conductive materials, for example, carbon, metal, or through doping, is another strategy to enhance charge transport. Chen et al. reported monodispersed Si–C nanocomposite with ultrathin graphitic carbon coating revealed high reversible capacity (>1200 mAh g^{-1}), high Coulombic efficiency (>90%), and long-cycle life (>100 cycles) when used as SSFC anolyte in a Li half-cell [8]. In conjunction with a highly concentrated catholyte (5.0 M LiI), a lithium metal-free 3 V full flow cell was demonstrated with high Coulombic efficiency (>90%) and good cycle life (>60 cycles). The Si–C nanocomposite effectively enhances the electrical conductivity and suppresses the volume expansion and labile surface of the Si-containing material, which are critical to the functioning of semisolid suspensions.

In order to enhance the volumetric energy density and power density, it is always desired to maximize the content of solid active materials and meanwhile the conductive agent for better conductivity in the suspension electrolyte. However, its viscosity will concomitantly increase resulting in more energy consumption of pumping and intricate fluid dynamics. Furthermore, the colloidal particles rapidly aggregate when suspended in polar solvents under high ionic strength conditions due to van der Waals interactions. Lewis and coworkers introduced a nonionic dispersant, polyvinylpyrrolidone (PVP), with an appropriate amount to selectively stabilize the $LiFePO_4$ particles [15]. The biphasic suspension showed attenuated predicted pressure drops and enhanced electronic conductivity. A biphasic $LiFePO_4$ suspension (20 vol% $LiFePO_4$, 1.25 vol% KB, 0.3 vol% PVP) was tested in an intermittent-flow cell and obtained an overall capacity of 131 mAh g^{-1} (93 Wh l^{-1}) at a current density of 1.67 mA cm^{-2}. Liao and coworkers also used PVP to stabilize Si anolyte [16]. They found that PVP weakens the van der Waals force between solid particles, thus decreases the viscosity and yield stress of the highly concentrated Si slurry. An optimal Si suspension anolyte (30 vol% Si, 10 vol% KB, 0.4 wt% PVP) with a low viscosity of 240 mPas at a shear rate of 40 s^{-1} exhibited a high capacity of 1300 mAh g^{-1}, stable cycling performance for >100 cycles and a high Coulombic efficiency of >98% in a static cell.

Table 42.1 summarizes the various LIBs active materials that have been reported in semi-solid battery systems. The composition of the slurry and the testing mode affect the electrochemical performance of the semi-solid LFBs. An alternative approach to increase the volumetric energy density and power density is to maintain a high content of solid active materials while minimizing the conducting additives with the help of 3D current collectors in flow channels. Chen et al. proposed such a concept in a single-component SSFC (Figure 42.1b) [3]. The elimination of conductive carbon particles significantly decreased the viscosity of the suspension without reducing the content of electroactive material, thus enhanced the volumetric capacity and energy density. Both intermittent-flow and continuous-flow modes were tested with carbon black-free LiFePO$_4$ slurry. With 20 vol% LiFePO$_4$ slurry as the catholyte, the continuous-flow cells can reversibly cycle for more than 100 hours at 0.2 mA cm^{-2} (Figure 42.1c).

42.2.2 Electroactive Materials Based on Precipitation-Dissolution Chemistry

Lithium–sulfur (Li–S) batteries, which use low-cost and non-toxic sulfur as the cathode material, have a high theoretical capacity (1675 mAh g^{-1}) and specific energy density (2600 Wh kg^{-1}), and have thus been propelled to a position of paramount interest in recent years [19–21]. Unlike other lithium battery chemistries based on Li$^+$ insertion/extraction into/from the host materials, Li–S batteries undergo precipitation-dissolution reactions. The configuration of a Li–S semi-solid flow battery is illustrated in Figure 42.2a. While materials based on the host-guest chemistries only involve the reaction of solid phases, sulfur compounds undergo a transformation from a solid to a liquid phase during operation [4]. As shown in Figure 42.2, in general, sulfur is slightly soluble in many polar electrolyte solvents, while the polysulfides intermediate with different chain lengths produced during charge and discharge exhibit varied solubilities in organic solvents [22]. The long-chain lithium polysulfides (Li$_2$S$_x$, $4 \leq x \leq 8$), exhibit relatively high solubility in organic solvents and have been widely used as the main redox-active species in Li–S LFBs. However, the insoluble and insulating short-chain lithium sulfides (Li$_2$S, Li$_2$S$_2$) usually precipitate on the cathode, leading to the loss of active materials and capacity decay during long-term cycling.

Cui and coworkers firstly proposed a membrane-free hybrid Li–S FBs with lithium polysulfide (Li$_2$S$_8$) in 1,3-dioxolane (DOL)/1,2-dimethoxyethane (DME) and passivated lithium metal as the catholyte and the anode, respectively [23].

Table 42.1 A summary of the semi-solid type LFBs.

System	Cathode	Anode	Battery type	Current density	Capacities (mAh g^{-1})	Average voltage (V)	References
LiCoO$_2$/Li$_4$Ti$_5$O$_{12}$	20 vol% LiCoO$_2$ suspension	10 vol% Li$_4$Ti$_5$O$_{12}$ suspension	Intermittent flow cell	C/8	121	2.35	[1]
LiFePO$_4$/LiTi$_2$(PO$_4$)$_3$	10 vol% LiFePO$_4$ suspension	18 vol% LiTi$_2$(PO$_4$)$_3$ suspension	Intermittent flow cell	7C	122	0.90	[17]
Li$_4$Ti$_5$O$_{12}$/Li		5.7 vol% Li$_4$Ti$_5$O$_{12}$ suspension	Static cell (half-cell, vs. Li/Li$^+$)	C/15	165	1.60	[18]
LiNi$_{1/3}$Co$_{1/3}$Mn$_{1/3}$O$_2$/Li	13.9 vol% LiNi$_{1/3}$Co$_{1/3}$Mn$_{1/3}$O$_2$		Static Cell (half-cell, vs. Li/Li$^+$)	5 mA cm^{-2}	130	3.60	[14]
Li$_4$Ti$_5$O$_{12}$/Li		25 wt% Li$_4$Ti$_5$O$_{12}$ suspension	Flow cell (half-cell, vs. Li/Li$^+$)	3 mA cm^{-2}	17	1.90	[10]
Li$_4$Ti$_5$O$_{12}$/Li		10 vol% Li$_4$Ti$_5$O$_{12}$ suspension (carbon free)	Dispersion flow cell (half-cell, vs. Li/Li$^+$)	0.0064 mA cm^{-2}	1.1 × 10^{-3}	1.55	[11]
LiCoO$_2$/Li$_4$Ti$_5$O$_{12}$	0.5 vol% LiCoO$_2$ suspension (carbon free)	1.0 vol% Li$_4$Ti$_5$O$_{12}$ suspension (carbon free)	Dispersion flow cell	0.001 mA	—	2.20	[12]
Si-C/LiI	5.0 M LiI	10 vol% Si-C suspension	Flow cell (half-cell, vs. Li/Li$^+$)	0.1 mA cm^{-2}	1100 (10th)	2.85	[8]
LiFePO$_4$/Li	20 vol% LiFePO$_4$ suspension		Flow cell with 3D current collectors (half-cell, vs. Li/Li$^+$)	0.5 mA cm^{-2}	50 mAh l^{-1}	3.30	[3]

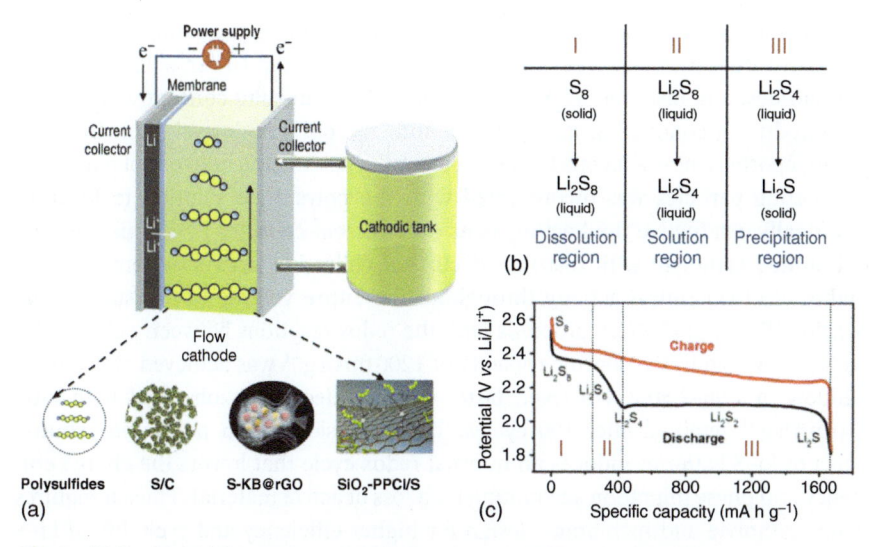

Figure 42.2 (a) A Li–S based semi-solid flow battery. Different types of flowable sulfur-based catholytes are indicated. (b) Summary of different polysulfide products that form during discharge and the respective active states (solid or liquid). (c) Breakdown of voltage curves by reaction products. Source: Figure (c) adapted from Yang et al. [21].

To avoid the formation of insoluble Li_2S_2 and Li_2S, the catholyte was only cycled between sulfur and Li_2S_4 by controlling the voltage. This battery demonstrated a high-energy density of 97 Wh kg^{-1} or 108 Wh l^{-1} due to the high solubility of polysulfide species. As the lithium anode was well passivated through the addition of lithium nitrate in ether solvent, the parasitic reactions between lithium and polysulfides were suppressed and thus excellent performance over 2000 cycles was achieved with a 2.5 M catholyte at a limited capacity of 200 mAh g^{-1}. However, due to the precipitation of the insoluble short-chain sulfides the battery capacity decayed quickly when further increasing the mass loading of catholyte. To address this challenge, the authors proposed a strategy to activate these sulfide species by reacting with sulfur powder under stirring and heating (70 °C) [24]. These Li–S LFBs demonstrated a high volumetric energy density of 135 Wh l^{-1}, good cycle life, and high single-cell capacity even at a high mass loading of 0.125 g cm^{-3}. Liu and coworkers proposed another approach to solve the solubility issue by using dimethyl sulfoxide-based electrolyte, which enables a high solubility of lithium polysulfides, especially for the short-chain species [19]. As the specific capacity of reactions between sulfur and Li_2S_4 is only 418 mAh g^{-1}, extending the reactions from sulfur to Li_2S remarkably boosts the capacity to 1675 mAh g^{-1}. A high reversible capacity of 1450 mAh g^{-1} at C/5 was delivered at the first cycle in a non-flowing static cell due to the largely reduced deposition of Li_2S/Li_2S_2 species. Unfortunately, the performance of Li–S flow cell is not on par with the static cell due to the poor kinetics and high viscosity of the catholyte when it operates in flow mode. In addition, the volumetric capacity is limited by the overall solubility of the polysulfide species.

To further boost the volumetric capacity by fully utilizing the storage capacity of the Li–S FBs, which involves multistep reactions between sulfur, soluble lithium polysulfides, and insoluble Li_2S/Li_2S_2 species, the semi-solid concept was devised by introducing conducting nanoparticles into the polysulfides solution. Chiang's group reported a non-aqueous Li–S FBs where the percolating network of nanoscale conducting particles was incorporated with the polysulfide solution to form an electrically conductive fluid acting as its own current collector [25]. This flowable polysulfide catholyte with distributed current collecting networks demonstrated higher electrochemical activity throughout the entire volume of polysulfide suspension. When the battery proceeds with the redox reactions between polysulfides and Li_2S, a high initial specific capacity of 1200 mAh g^{-1} was achieved at a voltage window of 1.90–2.50 V. However, faster capacity decay was observed with only 610 mAh g^{-1} retained after 100 cycles. The intrinsic lithium polysulfide shuttle effect of Li–S batteries induces an internal redox cycle that lowers the charge efficiency and phase migration accounting for a loss of active material. Thus, it requires better catholyte and membrane design for higher efficiency and cycle life of Li–S FBs. Lu and coworkers employed a high concentration sulfur-impregnated carbon (S/C) composite instead of polysulfides as the flow cathode [26]. The use of S/C increases the volumetric capacity of the flow cathode by 22.5% and simultaneously decreases the viscosity compared with polysulfides phase, further addressing the issues of sulfur and Li_2S. This battery achieved a catholyte volumetric capacity of 294 Ah l^{-1} with good cycle life, high Coulombic efficiency (>90%, 100 cycles), and energy efficiency (>80%, 100 cycles). It demonstrated a >5 times higher capacity than the state-of-the-art VFB electrolytes (60 Ah l^{-1}).

Zhang and coworkers investigated different sulfur composites, such as sulfur-KB@reduced graphene oxide (S-KB@rGO) composite [27], functionally designed SiO_2-tethered 1-methyl-1-propylpiperidinium chloride (SiO_2-PPCl) ionic liquid nanoparticle [28], and PVP functionalized sulfur-KB-graphene (S-KBG@P) composite [29] as sulfur catholytes. The well-dispersed nanocarbons, such as graphene and KB, construct a continuous conducting network, which facilitates the transport of both electron and Li$^+$ and then promotes the redox reactions of dispersed sulfur species. The functional groups of the composites exert effective affinity with polysulfides and thus suppress the shuttle effect of polysulfides during the charge and discharge processes. The S-KB-G@P catholyte demonstrated a high sulfur utilization of 89.5% and energy density of 718 Wh l^{-1}. Moreover, a high-energy density of 445 Wh l^{-1} and excellent cycling stability were achieved at −30 °C. Overall, the semi-solid system provides a promising and implementable solution to accomplishing high-energy density Li-S LFBs. For the semi-solid catholyte, the sulfur host and conducting additives are critical as they not only provide a continuous percolation network for electron and Li$^+$ transport between the active materials and current collector, but also suppress the shuttling effect by retaining the soluble polysulfide species within the host. The design of multi-architectural and multi-functional host materials as the conventional Li-S batteries would provide more choices to further improve the electrochemical performance of Li-S based SSFC.

42.2.3 Electroactive Materials Based on Multiple Redox Reactions

Semi-solid suspensions including those based on Li$^+$ insertion-extraction or precipitation-dissolution chemistry, are composed of just one active material. Lu and coworker [30] proposed a concept of multiple redox semi-solid-liquid (MRSSL) flow battery, which employed a biphasic MRSSL electrolyte consisting of a highly soluble active material in the liquid phase and a high-capacity active material in the solid phase. This battery takes advantage of active species in both liquid and solid phases. The insoluble active material undergoes redox reactions coupling with Li metal anode, which provides most capacity and energy of the suspension. The liquid phase not only functions as Li$^+$-containing electrolyte and makes the suspension a uniform mixture with reduced viscosity, but also contributes extra capacity beyond the insoluble active materials. The MRSSL was firstly demonstrated by using liquid-phase LiI electrolyte and semi-solid S/C composite. This battery achieved a high catholyte volumetric capacity of 550 Ah l^{-1} and superior energy density of 580 Wh l^{-1} with a high Coulombic efficiency of above 95%. Chen and coworkers [31] reported a high solubility 2,2,6,6-tetramethylpiperidine-1-oxyl (TEMPO, 3.35 V vs. Li/Li$^+$) in the liquid phase and high reversibility 10-methylphenothiazine (MPT, 3.45 V vs. Li/Li$^+$) composite in the solid phase as catholyte in an organic MRSSL-based LFB. As the voltage gap between the liquid and solid phases is smaller than 0.1 V, the system could operate at a stable cell voltage of around 3.40 V. Moreover, the cycling stability of TEMPO electrolyte was improved and the viscosity of the MPT suspension was reduced due to the synergistic interactions between them. The organic TEMPO-MPT based MRSSL suspension presented a stable cycling performance and a high volumetric capacity of 50 Ah l^{-1} under an intermittent-flow mode test.

Potentially, SSFC has great advantages over the prevailing energy storage technologies: Compared with conventional FBs, it extends the capacity well beyond the solubility limits of electrolyte solutions without the critical demand for an ion-exchange membrane (for systems without having dissolved redox-active species in the suspension). It could drastically enhance the effective concentration of electroactive species by 5–20 times. When combined with high capacity and high voltage lithium storage materials, the energy density could be 20–30 times as high as the conventional FBs. Compared with the solid electrode battery technologies, SSFC retains the salient features of FBs, that is, decoupled energy storage and power generation which endows the system with great operation flexibility and scalability. However, the semi-solid method inherently has several hurdles to overcome before it becomes a credible and viable approach, such as the high viscosity of electroactive material suspension leading to high pump loss, phase separation, and low electrical conductivity and the requirement for adequate conducting additives to facilitate charge transport and reduce ohmic loss, which severely compromise the energy efficiency and impose challenges in scaling up the system and maintenance. In addition, similar to the static devices, the growth of SEI at the surface of both the anodic and cathodic electroactive materials in the suspensions, as a result of electrolyte decomposition during the first charge process, would further hinder charge transport and cause capacity and power fading.

42.3 Redox Targeting-based LFBs

42.3.1 Principles of Redox Targeting-based LFBs

The redox-targeting concept was first proposed by Wang et al., based on a closed-loop electrochemical-chemical cycle, to eliminate the need of conducting additives of insulating lithium storage materials for conventional battery applications [32]. When implementing the redox-targeting concept into FBs, it breaks the boundary of liquid and solid energy storage through the redox-mediated chemical reactions, providing a new approach for large-scale high-density electrochemical energy storage. The configuration of a redox targeting-based LFBs largely follows the conventional FBs and has three major components (Figure 42.3a). The first component is the power-generation unit, an electrochemical cell consisting of two electrodes separated by a Li$^+$-conducting membrane. The second component is the energy storage unit, consisting of two energy tanks in which the solid active energy-storage materials are stored and imbued with suitable redox electrolytes in

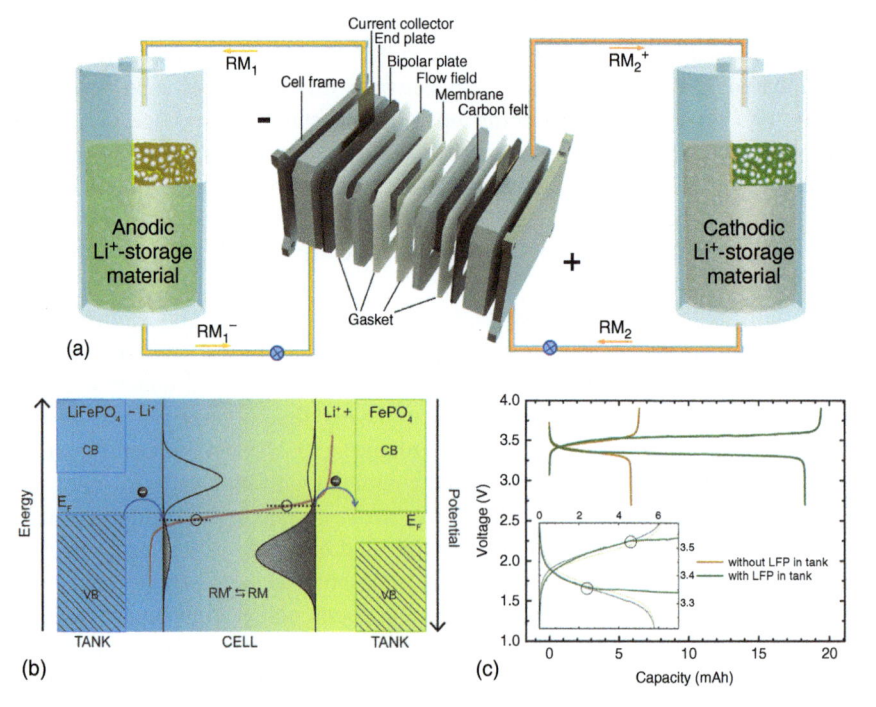

Figure 42.3 (a) Schematic illustration of a SMRT-based redox flow lithium-ion battery during discharge process. Source: Figure (a) reproduced from Zhou et al. [33] with permission from Wiley-VCH. (b) Energy diagram and charge transfer of the SMRT reactions of RM$^+$ with LiFePO$_4$ upon charging and RM with FePO$_4$ upon discharging. The thick dashed line marks the formal potential of RM/RM$^+$, and the thin dashed line indicates the Fermi level of solid material. (c) Voltage profiles of flow cells with 0.20 M FcIL in the catholyte and 0.37 M equivalent LiFePO$_4$ granules in the tank. The inset is the enlarged voltage profiles of the flow cells after IR correction. Source: Figure (b, c) reproduced from Zhou et al. [34] with permission from Elsevier.

their pores. Different from the SSFCs, the solid materials remain immobile in the tanks and are chemically charged and discharged via reversible redox-targeting reactions, which do not require direct electrical contact with the current collector or addition of conducting additives. The exchanges of electrons and Li$^+$ ions between the electrodes and active materials are mediated by circulating the redox mediators (RMs) dissolved in electrolytes. The third unit is the control system, which includes two pumps for circulating the redox electrolytes (not the solid materials) between the electrochemical cell and energy tanks. The control unit, therefore, feeds the recharged RMs from the energy tanks to the electrode compartments in the power unit, where electrochemical reactions occur to produce electricity.

As the solid battery materials are reversibly charged/discharged via chemical reactions, bulky conducting additives can be eliminated, and the tank capacity of the flow battery can be as high as the solid electroactive materials stored in the tanks. In addition, as the system only circulates electrolyte solutions with dissolved RMs in moderate concentrations, it could overcome the inherent issues faced by SSFC. In the meantime, the battery retains the configuration of FBs, the energy storage unit is decoupled from the power generation unit, thus offering greater operational flexibility and scalability with a modular design.

42.3.2 Development of Redox Targeting-based LFBs

At first, the reversible redox-mediated lithiation/delithiation of lithium storage materials are generally realized by two redox mediators, whose potentials straddle that of the solid lithium storage material. The potential difference between the redox molecule and lithium storage material provides the driving force for the lithiation or delithiation in the tank. During the charge process, the redox molecule RM$_1$ with a higher potential is firstly oxidized at the cathode (reaction (42.1)) and then flows into the solid material storage tank, in which it oxidizes the solid material (here used LiFePO$_4$ as an example) for effective chemical delithiation (reaction (42.2)). During the discharge process, the redox molecule RM$_2$ with a lower potential is firstly reduced at the cathode (reaction (42.3)) and then flows into the tank to reduce the solid material for chemical lithiation (reaction (42.4)).

Charging process:

Electrochemical reaction on cathode:

$$RM_1^{red} \rightarrow RM_1^{ox} + e^- \qquad (42.1)$$

Chemical reaction in the tank:

$$RM_1^{ox} + LiFePO_4 \rightarrow RM_1^{red} + FePO_4 + Li^+ \qquad (42.2)$$

Discharge process:

Electrochemical reaction on cathode:

$$RM_2^{ox} + e^- \rightarrow RM_2^{red} \qquad (42.3)$$

Chemical reaction in the tank:

$$RM_2^{red} + FePO_4 + Li^+ \rightarrow RM_2^{ox} + LiFePO_4 \qquad (42.4)$$

In 2013, Huang et al. reported the first redox targeting-based LFB half-cell [35]. Olivine LiFePO$_4$ was chosen as the Li$^+$-storage material in the cathodic tank and lithium metal as the anode separated by a LISICON Li$^+$-conducting membrane. Dibromoferrocene (FcBr$_2$) and ferrocene (Fc) were selected as the redox mediators to chemically delithiate and lithiate LiFePO$_4$, respectively. LiFePO$_4$ has a potential of 3.45 V vs. Li/Li$^+$, while Fc is at 3.25 V and FcBr$_2$ is at 3.55 V. Driven by the potential differences, more than 70% LiFePO$_4$ was reversibly charged and discharged by the two molecules via redox targeting reactions. In 2014, the first anodic half-cell was demonstrated using TiO$_2$ as Li$^+$-storage material in the anodic tank [36]. The lithiation/delithiation potential of TiO$_2$ is around 1.80 V vs. Li/Li$^+$. Two redox molecules, cobaltocene (CoCp$_2$) and bis(pentamethylcyclopentadienyl)cobalt (CoCp*$_2$), whose potentials are 1.90 and 1.36 V vs. Li/Li$^+$, respectively, were selected as the redox-targeting mediators. In 2015, a LFB full cell was developed by using LiFePO$_4$ and TiO$_2$ as the solid active materials by pairing with Fc/FcBr$_2$ and CoCp$_2$/CoCp*$_2$, respectively, which obtained a high-energy density up to 238 Wh l^{-1}, approximately five times higher than that of a VFB [37].

The voltage efficiency of FB is defined as the ratio of cell voltage between discharge and charge processes. In redox targeting-based LFB system, the voltage loss is mainly caused by the potential difference between two RMs. To enhance the voltage efficiency, it is desired to target the solid electroactive material with RMs of smaller potential difference. The redox couples, bis(pentamethylcyclopentadienyl)chromium (CrCp*$_2$, 1.77 V vs. Li/Li$^+$) and bis(cyclopentadienyl)cobalt (CoCp$_2$, 1.92 V vs. Li/Li$^+$), whose potential difference is only 0.15 V were employed as redox targeting mediators for the lithiation/delithiation of TiO$_2$ [38]. The voltage efficiency of this half-cell was increased to 84%. However, the use of two redox molecules and one solid material involving two stages of reactions on one side undoubtedly brings greater complexity to the operation and cycling stability of the FBs.

Later, a single bifunctional redox molecule with two-way redox reactions was selected as RM to reduce the complexity of LFB brought by two molecules. For instance, iodide, which has two redox reactions with potentials (I$^-$/I$_3^-$, 3.15 V and I$_3^-$/I$_2$, 3.70 V vs. Li/Li$^+$) just straddling that of LiFePO$_4$, was chosen as a single RM for both lithiation and delithiation of LiFePO$_4$ [39]. This system could in theory deliver a high-energy density of 670 Wh l^{-1} and has demonstrated a good capacity retention during cycling. Another bifunctional redox molecule, 2,3,5,6-tetramethyl-p-phenylenediamine (TMPD), presents two pairs of redox reactions at 3.20 and 3.60 V vs. Li/Li$^+$, was employed to target LiFePO$_4$ [40]. This cell can potentially achieve a tank energy density as high as 1023 Wh l^{-1} and power density of 61 mW cm^{-2}. Owing to the matched potentials of RM and solid material, a high voltage efficiency of 91% was delivered at a controlled utilization of LiFePO$_4$. Nevertheless, redox molecules with a smaller potential difference or single molecules with multiple redox reactions have been studied to improve the electrochemical performance of the FB; the complex electrolyte composition and redox reactions, as well as the resulting large voltage losses have not been completely solved.

Zhou et al. reported an intriguing approach to drastically boost the voltage efficiency and reduce the operational complexity of LFB by a single-molecule redox-targeting (SMRT) reaction [34]. Here, the redox molecule is required to have an identical redox potential to the solid electroactive material and high solubility to ensure fast redox reactions. The redox targeting reaction between the solid material and soluble mediator is driven by the Nernstian potential difference induced by the activity changes of the mediator during the charge and discharge cycles. Therefore, the redox materials and reactions in SMRT systems are reduced to "one molecule, one solid material and one reaction," as shown below:

Electrochemical reaction on cathode:

$$RM \leftrightarrow RM^+ + e^- \tag{42.5}$$

Chemical reaction in the tank:

$$RM^+ + LiFePO_4 \leftrightarrow RM + FePO_4 + Li^+ \tag{42.6}$$

As shown in Figure 42.3b, when RM^+ is predominant in the electrolyte during the charge process, electrons are transferred from the valence band of $LiFePO_4$ to the lowest unoccupied molecular orbital (LUMO) of RM^+, concomitantly with Li^+ extraction from $LiFePO_4$. Conversely, when RM is predominant in the electrolyte during the discharge process, electrons are transferred from the highest occupied molecule orbital (HOMO) of RM to the conduction band of $FePO_4$, accompanied by Li^+ insertion into $FePO_4$. The cell voltage is determined by the electrochemical reaction of the single RM on the electrode and the capacity is provided by the solid material via the reversible chemical reaction in the tank. 1-Ferrocenylmethyl-3-methylimidazolium bis(trifluoromethanesulfonyl)amide, a ferrocene-grafted ionic liquid (FcIL) with nearly identical redox potential to that of $LiFePO_4$ was designed and synthesized as the RM. Driven by the Nernstian potential difference, a near-unity utilization yield of $LiFePO_4$ is achieved, with which the volumetric tank energy density could potentially be optimized up to 942 Wh l^{-1}. Moreover, benefitting from the SMRT reactions, the flow cell exhibited only one voltage plateau with a voltage efficiency over 94% (Figure 42.3c). Table 42.2 summarizes the key performances of the various reported redox targeting-based LFBs. The SMRT-based LFBs show superior voltage efficiency and solid material utilization, making them more appealing for practical applications.

Redox-targeting reactions employed to boost the volumetric capacity of LFB are not limited to the previously studied materials and systems. This concept can be applied to a broad range of active lithium storage materials when paired with suitable redox molecules with matched potentials. With Li^+ as the charge balancing ion, a variety of cathodic materials, such as $LiFePO_4$, $LiCoO_2$, $LiMn_2O_4$, and $LiNi_{1-x-y}Mn_xCo_yO_2$, and different sorts of anodic materials, such as transition metal oxides, sulfides, selenides, nitrides, and phosphides, as well as alloying-type materials with low lithiation potentials, could be employed as the energy storage materials in the cathodic and anodic tanks, respectively. Yu et al. reported an aqueous lithium flow battery system based on the SMRT reactions of $LiFePO_4$ and $LiTi_2(PO_4)_3$ as the cathodic and anodic energy storage material, respectively [41].

Table 42.2 A summary of the redox targeting-based LFBs. All the batteries were tested in flow mode.

System	Catholyte (V vs. Li/Li$^+$)	Anolyte (V vs. Li/Li$^+$)	Current density (mA cm^{-2})	Voltage efficiency	Solid material utilization	References
LiFePO$_4$/Li	20 mM FcBr$_2$ (3.55) and 20 mM Fc (3.25)		0.31	81%	70%	[35]
TiO$_2$/Li		5 mM CoCp*$_2$ (1.36) and 5 mM CoCp$_2$ (1.90)	0.10	56%	69%	[36]
LiFePO$_4$/TiO$_2$	5 mM FcBr$_2$ (3.55) and 5 mM Fc (3.25)	5 mM CoCp*$_2$ (1.36) and 5 mM CoCp$_2$ (1.90)	0.075	45%	36%	[37]
LiFePO$_4$/Li	10 mM LiI (3.15/3.70)		0.025	75%	38%	[39]
TiO$_2$/Li		5 mM CrCp*2 (1.77) and 5 mM CoCp2 (1.92)	0.025	84%	48%	[38]
LiFePO$_4$/Li	20 mM TMPD (3.20/3.60)		0.125	91%	73%	[40]
LiFePO$_4$/Li	0.5 M FcIL (3.43)		0.025	95%	95%	[34]

With [Fe(CN)$_6$]$^{3-/4-}$ and S^{2-}/S$_2^{2-}$ as the redox mediators in the catholyte and anolyte, the cell revealed an anodic and cathodic volumetric capacity up to 305 and 207 Ah l^{-1} at 5 mA cm^{-2}, when LiFePO$_4$ and LiTi$_2$(PO$_4$)$_3$ were respectively loaded into the cathodic and anodic tank. These are four to six times as high as that of the VFB. In addition, with water-based electrolytes, the system presents notably enhanced Li$^+$ conductivity in the membrane and consequently much-improved power performance as compared to its nonaqueous counterpart.

Besides the lithium battery chemistry, those of Na$^+$, K$^+$, H$^+$, NH$_4^+$, OH$^-$, Cl$^-$, or other earth-abundant species could also be implemented in the redox targeting-based flow batteries [42–45]. In 2019, Self et al. demonstrated a redox targeting-based sodium battery half-cell [44]. Reversible sodium storage in red phosphorus loaded in an external packed bed reactor was achieved with mediated reactions by a pair of soluble anion radical species biphenyl and pyrene in the absence of binders and conducting additives. This effectively decouples the capacity of the battery from the solubility of redox species in the electrolyte and can conceivably achieve an energy density exceeding 200 Wh kg^{-1}. Recently, a sodium ion flow battery was demonstrated by Zhou et al. [45]. They employed the NASICON-type Na$_3$V$_2$(PO$_4$)$_3$ (NVP), which has two redox reactions at 3.40 and 1.63 V vs. Na due to the multivalent reactions of vanadium, as both the anodic and cathodic energy storage materials. 10-Methylphenothiazine (MPTZ), with a robust aromatic core that can readily be oxidized to its radical cation state, was selected as the RM to mediate the SMRT reaction with NVP ($V^{3+} \leftrightarrow V^{4+}$) in the cathodic side. 9-Fluorenone (FL), which has been investigated as an anolyte molecule for organic FB, was chosen to mediate the SMRT reaction of NVP ($V^{3+} \leftrightarrow V^{2+}$) in the anodic side. With the addition of NVP granules in the tanks, the cathodic and anodic capacities are 17 and 3 times higher than that of the catholyte and anolyte at their solubility limits. Such a full cell attained an energy density of 88 Wh l^{-1}. It operates upon lean electrolyte compositions beyond the lithium-based battery chemistry, paving an intriguing path for cost-effective large-scale energy storage.

42.3.3 Redox Targeting-based Lithium–Sulfur Flow Batteries

Lee and coworkers applied the redox-targeting concept to address the issues encountered in Li–S batteries and demonstrated the first redox-targeting-based Li–S flow battery. Due to the multistep reactions of polysulfides, a pair of RMs CrCp*$_2$ (2.01 V vs. Li/Li$^+$) and bis(pentamethyl-cyclopentadienyl) nickel (NiCp*$_2$, 2.43 V vs. Li/Li$^+$) were selected for the redox-targeting reactions [46]. When the sulfur-based copolymer-poly(S-r-DIB) was used in lieu of elemental sulfur in the cathodic tank, the discharge capacity was extended nearly by eight folds due to the chemical reaction between CrCp*$_2$ and poly(S-r-DIB). With the assistance of RMs, the poly(S-r-DIB) electroactive material could be progressively lithiated to discharged organosulfur DIB products and insoluble short-chain lithium sulfides during the discharge process and delithiated reversibly in the following charge cycle. This redox targeting-based flow cell demonstrated a high Coulombic efficiency above 99.5% and energy efficiency above 75% after the first five cycles. Unfortunately, it showed a capacity decay of 2.43% per cycle for the first 20 cycles due to the permeation of RMs and polysulfides through the membrane. In addition, the driving force between CrCp*$_2$ and S/LiS$_2$ is only 100 mV which may be too low to effectively reduce the polysulfide species. The cell components should be further optimized to suppress the parasitic processes, such as the crossover of mediators and polysulfides through the membrane upon prolonged operation.

More recently, Lu and coworkers have proposed a "self-mediated" FB concept employing the inherently present redox shuttles (soluble polychalcogenides) to access charges stored in solid forms (e.g. S and Se) [47]. The electrochemical reactions proceed via self-mediated reaction of the dissolved polychalcogenides generated during S and Se redox reactions, which can freely diffuse in the catholyte and transfer electrons via various comproportionation and disproportionation reactions. Meanwhile, the full self-mediated reactions are completed via a combination of electrochemical, comproportionation, and disproportionation reactions of the inherently dissolved polychalcogenides in the system at different stages of charge/discharge. Such a self-mediated flow battery demonstrated a high degree of material utilization ($\leq$99%) and ultra-high catholyte volumetric capacities of 1268 and 1096 Ah l^{-1} for Li–S and Li–Se, respectively. For practical application, a good understanding of the self-mediated process and controlled formation of the insoluble species during the charge process are critically desired in future studies.

42.3.4 Redox Targeting-based Lithium–Oxygen Flow Batteries

The Li–O_2 battery is a promising energy storage system because of its high-energy density of ~3505 Wh kg^{-1}, which is nearly 10 times the state-of-the-art LIBs [48–51]. It is composed of a metallic lithium anode, a cathode fed with O_2 as the active mass, and a Li$^+$-containing electrolyte solution. Upon discharging, oxygen is reduced to form Li_2O_2 on the cathode when aprotic electrolyte is employed, while it is electrochemically decomposed evolving oxygen upon charging during a rechargeable cycle [52]. However, Li_2O_2 is insoluble in the aprotic solvents and thus deposits as an insulating film on the surface of cathode during the discharge process, leading to passivation and pore-clogging of the air electrode [53], which results in large overpotential, low-rate performance and even premature termination of discharge [54]. Upon charging, the poor Li_2O_2 reaction kinetics imposes a large voltage hysteresis, resulting in a low round-trip energy efficiency [55]. In addition to the conventional electrocatalysts grafted on the electrode, some redox mediators have been introduced to the electrolyte to reduce the overpotentials for the reactions of O_2/Li_2O_2. However, since the reactions remain to take place in the vicinity of the electrode, the adverse effects of surface passivation and clogging problem are not entirely resolved, and the capacity remains dependent on the accessible volume of the cathode.

Redox flow Li–O_2 battery (RFLOB) was thus devised to tackle the above critical issues confronted by aprotic Li–O_2 batteries. With the assistance of RMs, the electrochemical reactions of O_2/Li_2O_2 in a conventional Li–O_2 battery could likewise proceed with redox-mediated chemical reactions off the electrode as those demonstrated in redox targeting-based LFB. During the discharge process, one redox mediator RM_1 with lower potential than the standard redox potential of O_2/Li_2O_2 couple is firstly reduced at the cathode, and subsequently flows into the external gas diffusion tank (GDT) where it is oxidized by O_2 gas in the presence of Li$^+$ ions:

Electrochemical reaction on cathode:

$$RM_1^{ox} + e^- \rightarrow RM_1^{red} \tag{42.7}$$

Chemical reaction in the GDT:

$$RM_1^{red} + O_2 + Li^+ \rightarrow RM_1^{ox} + Li_2O_2 \tag{42.8}$$

Then, the regenerated RM_1 flows back to the cell for a subsequent cycle of reactions. During this process, the formation of Li_2O_2 takes place in an external GDT spatially separated from the cathode. As a result, the electrode surface is protected and stays fresh. During the charge process, another redox mediator RM_2 with a higher potential than that of O_2/Li_2O_2 couple is oxidized at the cathode, and subsequently flows into the GDT tank where it further oxidizes Li_2O_2 and releases O_2.

Electrochemical reaction on cathode:

$$RM_2^{red} \rightarrow RM_2^{ox} + e^- \tag{42.9}$$

Chemical reaction in the GDT:

$$RM_1^{ox} + Li_2O_2 \rightarrow RM_2^{red} + O_2 + Li^+ \tag{42.10}$$

In such a battery system, the 3-phase oxygen evolution (OER) or oxygen reduction (ORR) reaction on the air cathode is reduced to two 2-phase reactions, involving one electrochemical reaction of RMs on the electrode and one chemical redox-targeting reaction of RMs with oxygen species in a GDT. As a result, the formation and decomposition of Li_2O_2 proceed chemically and are shifted to GDT, spatially separated from the cathode of the cell, which elegantly solves the surface passivation and pore-clogging problems, and consequently promotes the capacity and round-trip energy efficiency. Such a decoupled configuration for independent power generation (electrochemical reaction) and energy storage (chemical reactions) renders RFLOB most of the merits of FBs such as operation flexibility and scalability, in addition to the primary advantages of $Li-O_2$ battery.

The first RFLOB was reported by Wang and coworkers in 2015 [56]. As the redox potential of O_2/Li_2O_2 is ~2.96 V (vs. Li/Li^+), iodine species (3.10/3.70 V vs. Li/Li^+ for $I^-/I_3^-/I_2$) and ethyl viologen (EV, 2.65 V vs. Li/Li^+), which redox potentials straddle that of Li_2O_2, were used as RMs for OER and ORR, respectively. Later in 2016, another pair of soluble RMs were adopted for an optimized RFLOB to address the corrosive issue of iodine species. Tris{4-[2-(2-methoxyethoxy)ethoxy]phenyl}amine (TMPPA, 3.63 V vs. Li/Li^+) and 2,5-di-*tert*-butyl-1,4-benzoquinone (DTBBQ, 2.63 V vs. Li/Li^+) were used to catalyze the OER and ORR processes, respectively, and exhibited better cycling stability [57]. More recently, dual RMs have been employed to boost the power of RFLOB, in which duroquinone (DQ) was introduced as a RM into the electrolyte to enhance the ORR process while ethyl viologen (EV) as another RM to eliminate the formation of soluble superoxide [58]. As illustrated in Figure 42.4a, a spraying nozzle was integrated at the inlet to enhance the ORR reaction in the GDT. When the EV-DQ (0.2 M/0.2 M) electrolyte was sprayed with O_2 into the reactor tank, the cell could be discharged at a constant current density of 15 mA cm^{-2} with a >80% utilization of Li metal loaded on the anode. When the cell was fed with O_2 and dry air, an areal power of 60 and 34 mW cm^{-2} has been achieved, respectively (Figure 42.4b). To recharge the above RFLOB and examine the robustness of the ORR process, the authors proposed an approach by mechanically refueling Li metal to realize continuous operation. As described in Figure 42.4c, the fuel can just be fed into the cell by simply replenishing fresh Li metal into the anode compartment after it is consumed. With the help of redox-targeting reactions, the discharge products only accumulate in the GDT, the rest cell components, including the cathode and electrolyte, can be used uninterrupted until the GDT tank is full, at which it can simply be replaced with an empty one.

In summary, the redox-targeting reactions, as exhibited in the various battery chemistries, i.e. LIBs, Li–S, and Li–O_2 LFBs, enable swift charge exchanges between the RMs and solid lithium storage materials in the tanks. As a result, the capacity of the battery is no longer dependent on the concentration of redox species in the electrolyte but the quantity of solid materials loaded in the tanks. In addition, different from the SSFCs, the viscosity of electrolyte in redox targeting-based systems is remarkably reduced and remains unchanged during operation, and the presence of conducting additives becomes less critical. Thus, the energy density of redox targeting-based LFBs can be enormously increased with less complex electrolyte fluid dynamics and less pumping energy consumption compared with SSFC. Furthermore, the redox-targeting LFBs have greater tolerance to overcharging/ overdischarging and are hence safer, since the solid lithium storage materials are not directly charged or discharged. Despite the great promise that the redox targeting-based LFB systems have exhibited to address the grand challenges

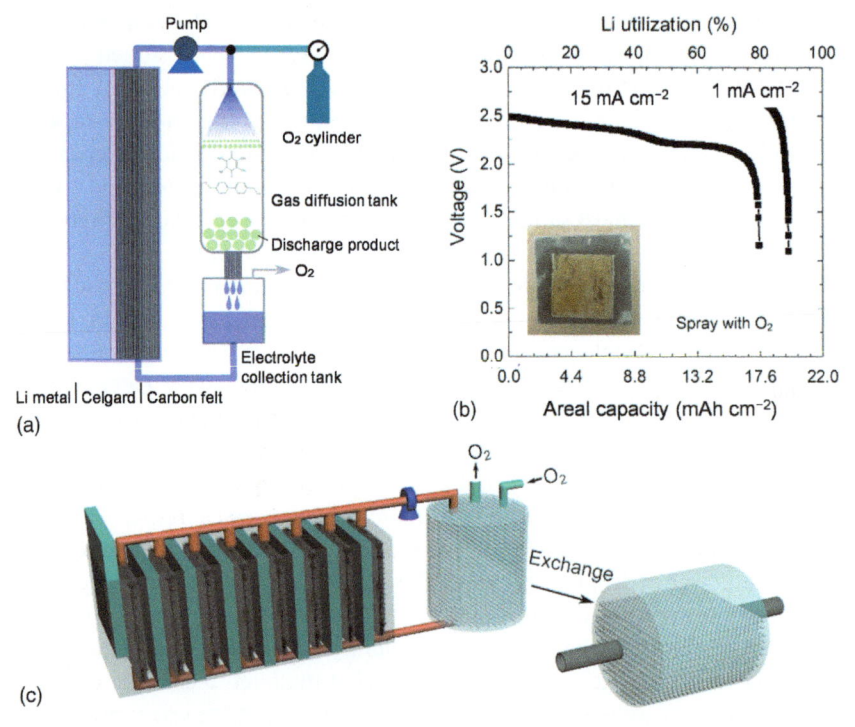

Figure 42.4 (a) Configuration of a redox flow Li-O_2 battery (RFLOB) (single cell) with integrated electrolyte spray. Source: Zhu et al. [58] with permission from Royal Society of Chemistry. (b) Discharge voltage profiles of RFLOB operated with electrolyte spray. The cells were fed with oxygen. The electrolyte was EV-DQ (0.2 M/0.2 M) in 1 M LiTFSI/DMSO. (c) Illustration of the operation of an RFLOB cell system (stack) with exchangeable gas diffusion tank (GDT). The discharge products collected from GDT are recycled by capturing CO_2 to form a value-added product Li_2CO_3. Source: Figure (b, c) reproduced from Zhu et al. [58] with permission from Royal Society of Chemistry.

confronted by other battery technologies, there have been new challenges, such as the choice of suitable and robust redox mediators, optimization of the loading of solid electroactive materials in the energy storage tanks and Li-ion conducting membranes, etc., for these novel systems to overcome before it becomes viable for practical applications.

42.4 Challenges and Outlook

This chapter presents an overview of the recent progress in two different types of non-aqueous LFBs: (i) SSFCs with flowable semi-solid electrolytes consisting of solid Li-storage electroactive materials and conducting additives dispersed in electrolyte solution, (ii) redox targeting-based LFBs with the solid electroactive materials kept statically in energy storage tanks while redox electrolyte circulating through the system. Both the LFBs retain the advantages of operation flexibility (decoupled energy storage and power generation) and scalability of FBs, while in the meantime have considerably enhanced energy density leveraging different lithium battery chemistries. In addition, due to the fact that the electroactive species in electrolyte are in solid phase, another advantage for SSFCs is that they could eliminate the use of ion-exchange membranes, which has been a critical issue for organic FBs. However, the main challenge for SSFCs is the use of viscous semi-solid slurry of active materials, which may undergo phase separation, sedimentation, and clog the electrolyte flow, for which large pumping energy would be consumed and sometimes requires judicious optimization of the slurry composition. In addition, a large quantity of conducting additive would be required for materials with mediocre conductivity so that a percolating network would be formed to facilitate charge transport, which inevitably compromises the volumetric capacity of the slurry. For redox targeting-based LFBs, given the many systems demonstrated with various lithium battery chemistries, the most critical challenges lie in the Li^+-conducting membranes for high power operation and the loading of solid electroactive materials in the tank. The former is also shared by the conventional non-aqueous FBs, for which the ceramic membranes commonly used in the reported work generally have inferior conductivity and mechanical stability to their polymeric counterparts, while most of the polymeric ion-exchange membranes fail to operate due to severe swelling in the organic electrolytes. Such an issue could be addressed by using aqueous electrolytes as those demonstrated for near-term applications, however, at a price of reduced energy density. The latter is more of an engineering issue. Knowledge of packed-bed reactors could be useful to the tank design and the loading of solid active materials so that the redox-targeting reactions between the redox mediators and solid material would be facilitated when the electrolyte solution flows through the tank.

Although the SSFC and redox targeting-based LFB systems are still in their infancy with many challenges to overcome, given the merits demonstrated and the great demand for large-scale energy storage in the near future, we believe LFB as an emerging technology is worth continued studies until the materials challenges are eventually adequately addressed.

References

1 Duduta, M., Ho, B., Wood, V.C. et al. (2011). Semi-solid lithium rechargeable flow battery. *Advanced Energy Materials* 1 (4): 511–516.
2 Zhao, Y., Ding, Y., Li, Y. et al. (2015). A chemistry and material perspective on lithium redox flow batteries towards high-density electrical energy storage. *Chemical Society Reviews* 44 (22): 7968–7996.
3 Chen, H., Liu, Y., Zhang, X. et al. (2021). Single-component slurry based lithium-ion flow battery with 3D current collectors. *Journal of Power Sources* 485: 229319.
4 Hatzell, K.B., Boota, M., and Gogotsi, Y. (2015). Materials for suspension (semi-solid) electrodes for energy and water technologies. *Chemical Society Reviews* 44 (23): 8664–8687.

5 Leung, P., Li, X., Ponce de León, C. et al. (2012). Progress in redox flow batteries, remaining challenges and their applications in energy storage. *RSC Advances* 2 (27): 10125.

6 Julien, C.M., Mauger, A., Zaghib, K., and Groult, H. (2014). Comparative issues of cathode materials for Li–Ion batteries. *Inorganics* 2 (1): 132–154.

7 Li, W., Song, B., and Manthiram, A. (2017). High-voltage positive electrode materials for lithium–ion batteries. *Chemical Society Reviews* 46 (10): 3006–3059.

8 Chen, H., Lai, N.-C., and Lu, Y.-C. (2017). Silicon–carbon nanocomposite semi-solid negolyte and its application in redox flow batteries. *Chemistry of Materials* 29 (17): 7533–7542.

9 Youssry, M., Madec, L., Soudan, P. et al. (2013). Non-aqueous carbon black suspensions for lithium-based redox flow batteries: rheology and simultaneous rheo-electrical behavior. *Physical Chemistry Chemical Physics* 15 (34): 14476–14486.

10 Ventosa, E., Skoumal, M., Vazquez, F.J. et al. (2015). Electron bottleneck in the charge/discharge mechanism of lithium titanates for batteries. *ChemSusChem* 8 (10): 1737–1744.

11 Qi, Z. and Koenig, G.M. (2016). A carbon-free lithium-ion solid dispersion redox couple with low viscosity for redox flow batteries. *Journal of Power Sources* 323: 97–106.

12 Qi, Z., Liu, A.L., and Koenig, G.M. (2017). Carbon-free solid dispersion $LiCoO_2$ redox couple characterization and electrochemical evaluation for all solid dispersion redox flow batteries. *Electrochimica Acta* 228: 91–99.

13 Hamelet, S., Tzedakis, T., Leriche, J.B. et al. (2012). Non-aqueous Li-based redox flow batteries. *Journal of the Electrochemical Society* 159 (8): A1360–A1367.

14 Biendicho, J.J., Flox, C., Sanz, L., and Morante, J.R. (2016). Static and dynamic studies on $LiNi_{1/3}Co_{1/3}Mn_{1/3}O_2$-based suspensions for semi-solid flow batteries. *ChemSusChem* 9 (15): 1938–1944.

15 Wei, T.-S., Fan, F.Y., Helal, A. et al. (2015). Biphasic electrode suspensions for Li–ion semi-solid flow cells with high energy density, fast charge transport, and low-dissipation flow. *Advanced Energy Materials* 5 (15): 1500535.

16 Wu, Y., Cao, D., Bai, X. et al. (2020). Effects of non-ionic surfactants on the rheological, electrical and electrochemical properties of highly loaded silicon suspension electrodes for semi-solid flow batteries. *ChemElectroChem* 7 (17): 3623–3631.

17 Li, Z., Smith, K.C., Dong, Y. et al. (2013). Aqueous semi-solid flow cell: demonstration and analysis. *Physical Chemistry Chemical Physics* 15 (38): 15833–15839.

18 Madec, L., Youssry, M., Cerbelaud, M. et al. (2014). Electronic vs ionic limitations to electrochemical performance in $Li_4Ti_5O_{12}$-based organic suspensions for lithium-redox flow batteries. *Journal of the Electrochemical Society* 161 (5): A693–A699.

19 Pan, H., Wei, X., Henderson, W.A. et al. (2015). On the way toward understanding solution chemistry of lithium polysulfides for high energy Li–S redox flow batteries. *Advanced Energy Materials* 5 (16): 1500113.

20 Wang, C., Li, K., Zhang, F. et al. (2018). Insight of enhanced redox chemistry for porous MoO_2 carbon-derived framework as polysulfide reservoir in lithium–sulfur batteries. *ACS Applied Materials & Interfaces* 10 (49): 42286–42293.

21 Yang, X., Li, X., Adair, K. et al. (2018). Structural design of lithium–sulfur batteries: from fundamental research to practical application. *Electrochemical Energy Reviews* 1 (3): 239–293.

22 Zhang, S., Guo, W., Yang, F. et al. (2019). Recent progress in polysulfide redox-flow batteries. *Batteries & Supercaps* 2 (7): 627–637.

23 Yang, Y., Zheng, G., and Cui, Y. (2013). A membrane-free lithium/polysulfide semi-liquid battery for large-scale energy storage. *Energy & Environmental Science* 6 (5): 1552.

24 Jin, Y., Zhou, G., Shi, F. et al. (2017). Reactivation of dead sulfide species in lithium polysulfide flow battery for grid scale energy storage. *Nature Communications* 8 (1): 462.

25 Fan, F.Y., Woodford, W.H., Li, Z. et al. (2014). Polysulfide flow batteries enabled by percolating nanoscale conductor networks. *Nano Letters* 14 (4): 2210–2218.

26 Chen, H., Zou, Q., Liang, Z. et al. (2015). Sulphur-impregnated flow cathode to enable high-energy-density lithium flow batteries. *Nature Communications* 6: 5877.

27 Xu, S., Zhang, L., Zhang, X. et al. (2017). A self-stabilized suspension catholyte to enable long-term stable Li–S flow batteries. *Journal of Materials Chemistry A* 5 (25): 12904–12913.

28 Xu, S., Cheng, Y., Zhang, L. et al. (2018). An effective polysulfides bridgebuilder to enable long-life lithium-sulfur flow batteries. *Nano Energy* 51: 113–121.

29 Xu, S., Zhang, L., Zhang, H. et al. (2020). A high-energy, low-temperature lithium-sulfur flow battery enabled by an amphiphilic-functionalized suspension catholyte. *Materials Today Energy* 18: 100495.

30 Chen, H. and Lu, Y.-C. (2016). A high-energy-density multiple redox semi-solid-liquid flow battery. *Advanced Energy Materials* 6 (8): 1502183.

31 Zhang, X., Zhang, P., and Chen, H. (2021). Organic multiple redox semi-solid-liquid suspension for Li-based hybrid flow battery. *ChemSusChem* 14 (8): 1913–1920.

32 Wang, Q., Zakeeruddin, S.M., Wang, D. et al. (2006). Redox targeting of insulating electrode materials: a new approach to high-energy-density batteries. *Angewandte Chemie* 118 (48): 8377–8380.

33 Zhou, M., Chen, Y., Salla, M. et al. (2020a). Single-molecule redox-targeting reactions for a pH-neutral aqueous organic redox flow battery. *Angewandte Chemie International Edition* 59 (34): 14286–14291.

34 Zhou, M., Huang, Q., Pham Truong, T.N. et al. (2017). Nernstian-potential-driven redox-targeting reactions of battery materials. *Chem* 3 (6): 1036–1049.

35 Huang, Q., Li, H., Gratzel, M., and Wang, Q. (2013). Reversible chemical delithiation/lithiation of $LiFePO_4$: towards a redox flow lithium-ion battery. *Physical Chemistry Chemical Physics* 15 (6): 1793–1797.

36 Pan, F., Yang, J., Huang, Q. et al. (2014). Redox targeting of anatase TiO_2 for redox flow lithium–ion batteries. *Advanced Energy Materials* 4 (15): 1400567.

37 Jia, C., Pan, F., Zhu, Y.G. et al. (2015). High–energy density nonaqueous all redox flow lithium battery enabled with a polymeric membrane. *Science Advances* 1 (10): e1500886.

38 Pan, F., Huang, Q., Huang, H., and Wang, Q. (2016). High-energy density redox flow lithium battery with unprecedented voltage efficiency. *Chemistry of Materials* 28 (7): 2052–2057.

39 Huang, Q., Yang, J., Ng, C.B. et al. (2016). A redox flow lithium battery based on the redox targeting reactions between $LiFePO_4$ and iodide. *Energy & Environmental Science* 9 (3): 917–921.

40 Zhu, Y.G., Du, Y., Jia, C. et al. (2017a). Unleashing the power and energy of $LiFePO_4$-based redox flow lithium battery with a bifunctional redox mediator. *Journal of the American Chemical Society* 139 (18): 6286–6289.

41 Yu, J., Fan, L., Yan, R. et al. (2018). Redox targeting-based aqueous redox flow lithium battery. *ACS Energy Letters* 3 (10): 2314–2320.

42 Chen, Y., Zhou, M., Xia, Y. et al. (2019). A stable and high-capacity redox targeting-based electrolyte for aqueous flow batteries. *Joule* 3 (9): 2255–2267.

43 Páez, T., Martínez-Cuezva, A., Palma, J., and Ventosa, E. (2019). Mediated alkaline flow batteries: from fundamentals to application. *ACS Applied Energy Materials* 2 (11): 8328–8336.

44 Self, E.C., Delnick, F.M., Ruther, R.E. et al. (2019). High-capacity organic radical mediated phosphorus anode for sodium-based redox flow batteries. *ACS Energy Letters* 4 (11): 2593–2600.

45 Zhou, M., Chen, Y., Zhang, Q. et al. (2019). $Na_3V_2(PO_4)_3$ as the sole solid energy storage material for redox flow sodium–ion battery. *Advanced Energy Materials* 9 (30): 1901188.

46 Li, J., Yang, L., Yang, S., and Lee, J.Y. (2015). The application of redox targeting principles to the design of rechargeable Li–S flow batteries. *Advanced Energy Materials* 5 (24): 1501808.

47 Zhou, Y., Cong, G., Chen, H. et al. (2020b). A self-mediating redox flow battery: high-capacity polychalcogenide-based redox flow battery mediated by inherently present redox shuttles. *ACS Energy Letters* 5 (6): 1732–1740.

48 Bruce, P.G., Freunberger, S.A., Hardwick, L.J., and Tarascon, J.-M. (2012). $Li–O_2$ and Li–S batteries with high energy storage. *Nature Materials* 11 (1): 19–29.

49 Lee, J.-S., Tai Kim, S., Cao, R. et al. (2011). Metal–air batteries with high energy density: Li–air versus Zn–air. *Advanced Energy Materials* 1 (1): 34–50.

50 Lu, Y.-C., Gallant, B.M., Kwabi, D.G. et al. (2013). Lithium–oxygen batteries: bridging mechanistic understanding and battery performance. *Energy & Environmental Science* 6 (3): 750–768.

51 Zhu, Y.G., Liu, Q., Rong, Y. et al. (2017b). Proton enhanced dynamic battery chemistry for aprotic lithium–oxygen batteries. *Nature Communications* 8: 14308.

52 Ogasawara, T., Débart, A., Holzapfel, M. et al. (2006). Rechargeable Li_2O_2 electrode for lithium batteries. *Journal of the American Chemical Society* 128 (4): 1390–1393.

53 Chen, Y., Freunberger, S.A., Peng, Z. et al. (2013). Charging a Li–O$_2$ battery using a redox mediator. *Nature Chemistry* 5 (6): 489–494.

54 Cheng, F. and Chen, J. (2012). Metal–air batteries: from oxygen reduction electrochemistry to cathode catalysts. *Chemical Society Reviews* 41 (6): 2172–2192.

55 Lim, H.-D., Song, H., Kim, J. et al. (2014). Superior rechargeability and efficiency of lithium–oxygen batteries: hierarchical air electrode architecture combined with a soluble catalyst. *Angewandte Chemie International Edition* 53 (15): 3926–3931.

56 Zhu, Y.G., Jia, C., Yang, J. et al. (2015). Dual redox catalysts for oxygen reduction and evolution reactions: towards a redox flow Li–O$_2$ battery. *Chemical Communications (Cambridge)* 51 (46): 9451–9454.

57 Zhu, Y.G., Wang, X., Jia, C. et al. (2016). Redox-mediated ORR and OER reactions: redox flow lithium oxygen batteries enabled with a pair of soluble redox catalysts. *ACS Catalysis* 6 (9): 6191–6197.

58 Zhu, Y.G., Goh, F.W.T., Yan, R. et al. (2018). Synergistic oxygen reduction of dual redox catalysts boosting the power of lithium–air battery. *Physical Chemistry Chemical Physics* 20 (44): 27930–27936.

43

Nonaqueous Metal-Free Flow Batteries

Kathryn Toghill and Craig Armstrong

Lancaster University, Department of Chemistry, Lancaster, LA1 4YB, UK

Dedicated to the memory of Dr Susan A. Odom (1980 – 2021).
An outstanding synthetic chemist and leader in this field.

Preamble

In this chapter we consider the somewhat nascent field of nonaqueous organic flow batteries (NAOFBs), evaluating highlights in the state of the art, assessing their limitations and design challenges, but also their promise as better structure-performance relationships develop. Within the field of nonaqueous flow batteries, metal coordination complexes dominated early research interest due to their simplicity and multiple oxidation states [1]. It is only recently that metal-free 'organic' redox species have started to emerge as viable candidates with the promise of molecular diversity, structure designability, and the potential for low-cost and sustainable redox-active materials [2]. Much of the redox-active organic molecule (ROM) literature to date has spilled over from lithium-ion batteries (LIBs) research, whereby such ROMs have been applied as overcharge-protection molecules [3, 4]. Trial and error with existing and established, commercially available nonaqueous compounds has provided a lot of baseline insight into the field, allowing performance benchmarking and indications of room to improve. More recent works of the late 2010s have shown more sophistication as function-oriented molecular design of organic redox-active molecules has developed. Furthermore, as a means to account for and mitigate premature discharge, one of the major challenges of the NAOFB, the redox molecule and solvent interaction is being established. Many excellent review articles exist in this space focussed on many of these aspects of NAOFBs, to which the reader is directed [5–13]. Here we will attempt to contextualise developments rather than report on all literature in this space.

43.1 Introduction

In all flow-battery applications, the objective is to find cost-effective redox materials that offer high-capacity solutions of high stability in the charged and discharged states, whilst keeping their molecular weights low [14]. In nonaqueous solutions, there are additional challenges of intrinsically higher solution resistances and the limited solubility of supporting electrolyte salts. In the following sections, the range of catholytes and anolytes that have been developed for FB applications are considered. In some cases, these are hybrid systems in which lithium metal is the anode. For the most part, these organic molecules form radical species in their charged state, some neutral, but mostly radical cations and anions giving rise to catholytes and anolytes respectively. These radical cationic/anionic states are typically formed by injection of an electron into an antibonding (or nonbonding) orbital or withdrawal of an electron from a bonding orbital, respectively [9, 15]. This results in the formation of a singly occupied molecular orbital (SOMO), which is not an energetically favoured state, and consequently, we associate radicals with fast chemical decay. Fast chemical decay is a major problem in flow batteries as it means rapid self-discharge or irreversible destruction of the ROMs. Nevertheless, there are rational design strategies for realising persistent ROM radical states. The reactivity of radicals can be diminished by delocalisation of the radical within a conjugated π system, the addition of electron-withdrawing or –donating groups, and steric protection by the proximity of bulky groups near the radical locus [15]. Hence researchers have developed numerous derivatives possessing promising ROM cores over the past decade as well as exploring their redox environment. Concurrently, ROMs have been pursued in aqueous FB research, however problematic radicals are typically not encountered because the protic electrolyte results in proton-coupled electron transfers.

Figure 43.1 and 43.2 aim to illustrate the range of ROMs capable of 100 cycles with respect to the formal redox potential. Notably, there is little consistency in the literature on how to report the potential of these redox couples. In nonaqueous solvents, the IUPAC recommendation is to use a ferrocene/ferrocenium (Fc/Fc$^+$) standard [17], but the ferrocene potential is susceptible to shift in different solvents [16]. So too is the lithium potential to an even greater extent, especially if the redox process is complicit in the overall redox reaction. Often authors report their potentials vs. the quasi-reference electrode, Ag/Ag$^+$, again not taking heed of the solvent differences and the significant variation in that redox potential with time, cycling, and electrolyte conditions. It is therefore a challenge to directly compare the potentials at which these catholytes and anolytes operate. An attempt to correct the potentials for the IUPAC standard of Fc/Fc$^+$ has been made in the figures and tables, though this is not possible for values reported against the Ag/Ag$^+$ quasi-reference electrode. Data reported against the Li/Li$^+$ redox potential have been converted by use of the 3.25 V vs. Fc/Fc$^+$ correction [16]. However, this should be common practice by the experimentalists and not undertaken by those evaluating the literature.

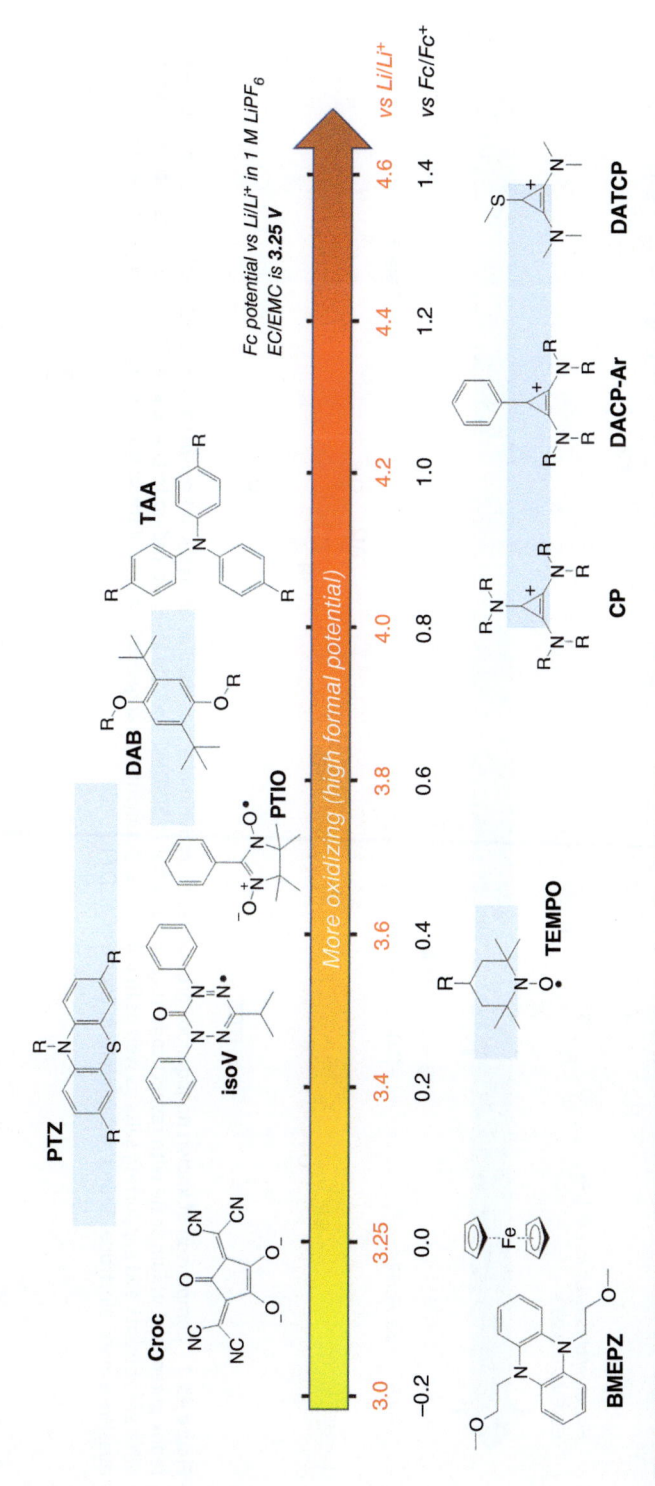

Figure 43.1 Comparison of known nonaqueous organic redox-active molecules used as catholytes in flow-battery research with respect to their **estimated** redox potential. The blue bars indicate the potential range for a particular ROM family. Lithium scale with respect to ferrocene potential in EC/DMC and 1 M LiPF$_6$ according to Laoire et al. [16]. Potentials subject to shift in different solvents and supporting salts as well as in different experimental arrangements. Source: Adapted from Laoire et al. [16].

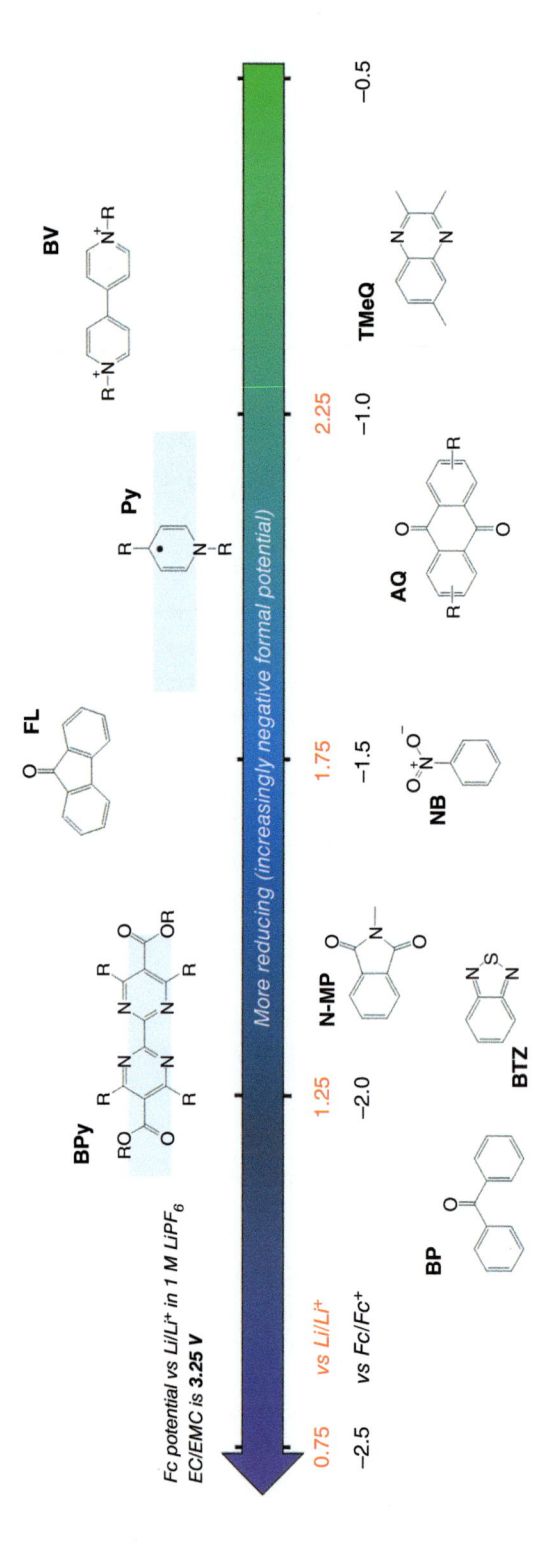

Figure 43.2 Comparison of known nonaqueous organic redox-active molecules used as anolytes in flow-battery research with respect to their **estimated** redox potential. Lithium scale with respect to ferrocene potential in EC/DMC and 1 M LiPF$_6$ according to Laoire et al [16]. Potentials subject to shift in different solvents and supporting salts as well as in different experimental arrangements. Many of these ROMs were recorded vs. the Ag/Ag$^+$ wire quasi-reference electrode. Source: Adapted from Laoire et al. [16].

43.2 Catholytes

Figure 43.1 compiles the key classes of catholyte compounds that have been used in NAOFBs to date. Some of these have been used extensively in multiple studies, others have had a single evaluation. The leading classes are considered below.

43.2.1 Nitroxyl Radical

Being perhaps the most familiar persistent radical in organic chemistry, it is unsurprising that 2,2,6,6-tetramethyl-1-piperidinyloxy (**TEMPO**) has featured in many FB designs. TEMPO possesses a nitroxyl group embedded within a six-membered ring that is surrounded by adjacent methyl substituents. Due to delocalisation of the unpaired electron in the N—O bond and steric protection by the methyl groups, the neutral radical is long-lasting and typically unreactive. At anodic potentials TEMPO can be reversibly oxidised into an oxoammonium cation, giving rise to applications in energy storage and conversion [18]. Indeed, TEMPO-based compounds have long been used in LIBs and organic radical batteries, wherein their polymeric structures give rise to tunable redox properties [19–22]. In a nonaqueous solvent, the underivatized TEMPO radical has an oxidation potential of c. 0.3 V vs. Ag/Ag$^+$ (c. 0.2 V vs. Fc/Fc$^+$, 3.4 V vs. Li/Li$^+$) making it applicable as a catholyte ROM. Featuring in the first nonaqueous all-organic flow battery, as demonstrated by Li et al. in 2011, TEMPO was paired with a methylphthalimide anolyte [23]. Only 20 cycles were evaluated owing to the low current density employed (0.35 mA cm^{-2}) but the preliminary results gave promising efficiencies; 90% CE, 82% VE, and 74% EE. Furthermore, the high solubility of the ROMs meant a tentative energy density of 12 W h l^{-1} was possible. Later works paired TEMPO with a lithium anode to give a hybrid-flow battery with considerably higher performance as expected of such an arrangement [18]. The work demonstrated that TEMPO could reach concentrations as high as 5.2 M (2.0 M with supporting LiPF$_6$ electrolyte) and attain high current densities up to 10 mA cm^{-2} with only a modest energy efficiency loss; 76 to 69% over 100 cycles for TEMPO concentrations of 0.8–2.0 M in EC:propylene carbonate (PC):dimethylcarbonate (DMC) solvent mixture. Further TEMPO derivatives have also been employed that feature derivatisation at the *para* position. Park et al. evaluated 4-oxo-TEMPO [24], and later Milshtein and coworkers evaluated acetamido TEMPO (**AcNH-TEMPO**) [25]. Both derivatives incorporated an electron-withdrawing group that prompted an advantageous positive shift in the reduction potential. More critically, the latter work demonstrated a single electrolyte flow-cell methodology to characterise the degradation of the electrolyte and quantify nonaqueous flow-cell performance. AcNH-TEMPO was found to be very stable over 20 cycles and consequently recommended as a benchmarking compound because both oxidation states are commercially available. The study is an exemplar investigation as to how thorough new flow-battery electrolytes should be evaluated, yet none of the studies have thus far demonstrated TEMPO to be of practical and likely use for NAOFBs going forward.

43.2.2 Dialkoxybenzene Derivatives

1,4-dialkoxybenzene derivatives are the most well studied category of ROM catholytes for NAOFBs. As with most of the early compounds considered in this field, alkoxybenzenes were being studied as overcharge protection molecules in Li-ion batteries. Most iconic of which, 2,5-di-*tert*-butyl-1,4-bis(2-methoxyethoxy)benzene (**DBBB**) developed by Argonne National Laboratory [26, 27], was proven to sustain over 200 cycles at c. 4.0 V vs. Li/Li$^+$, something equivalent to c. 10 000 charge–discharge cycles [28, 29]. In the context of flow batteries, DBBB was first studied in 2012 by Brushett et al. in coin cell experiments vs. a quinoxaline anolyte (though confusingly termed a Li-ion flow battery, it is not a Li/Li$^+$ anolyte couple) [28]. DBBB has a well-defined reversible redox potential of 3.9 V vs. Li/Li$^+$ in EC:DMC electrolyte, which in turn is seen to shift positively c. 90 mV with a solvent change to PC (see note on reference potentials in nonaqueous solvents). Further works considered dimethoxybenzene as a positive species in hybrid lithium flow-cell systems, evaluating and optimising the alkoxybenzene as a catholyte molecule [30].

With a vast number of derivatives possible, the 1,4-dimethoxybenzene redox core has been well studied in the context of rational design and in understanding radical cation persistence. In 2015, Huang et al. evaluated 11 derivatives aiming to balance steric protection from self-discharge to parasitic loss with the molecular weight of the ROM [31]. Despite modifications, DBBB is reported to decay to just 11% of the original radical cation concentration within five hours according to absorption spectroscopy, whereas the less engineered dimethoxybenzene is shown to retain 80% of the radical cation over a similar period [30]. More recent articles concerning the dialkoxybenzene ROM category have sought to identify properties that can stabilise the radical cation, and the mechanism by which parasitic side reactions or compound degradation take place [32]. Beyond this development, dialkoxybenzene molecules routinely feature as benchmark catholytes to cycle novel anolyte compounds against, or study new cell designs and approaches to flow batteries.

43.2.3 Phenothiazine and Phenazine

Introduced to FBs by the Odom group in 2015, phenothiazines (**PTZ**) comprise a central thiazine heterocycle unit with nitrogen (azo) and sulphur (thio) heteroatoms sandwiched between two phenyl moieties [33]. This was the first of several studies by the Odom group to rationally design and thoroughly evaluate the phenothiazine redox core as one suitable for nonaqueous FB electrolytes [34–38]. Initial work focussed on 3,7-bis(trifluoromethyl)-*N*-ethylphenothiazine (BCFEPT), which showed high solubility and stability in carbonate solvents as an overcharge protection molecule [39]. Relative to other catholytes, good stability, and cycling to 60 cycles vs. a trimethylquinoxaline anolyte was observed, sparking further development [33]. In a key paper of 2016, Milshtein assessed three *N*-derivatised phenothiazines in symmetric electrolyte experiments, yielding the superior N-(2-(2-methoxyethoxy)ethyl)phenothiazine (MEEPT) version. Indeed, MEEPT achieved over 100 cycles with little to no capacity fade [34]. Furthermore, as a noteworthy achievement of this work, the solubility of the MEEPT radical cation was measured giving a surprising hundredfold decrease in solubility compared to the neutral MEEPT species.

Analogously to PTZ, the heterocyclic cousin phenazine (**PZ**); possessing two nitrogen heteroatoms, has also been explored. Studied in recent years by Kwon et al. [40, 41], the researchers claim it possesses the highest energy density per mole as it can contribute two electrons per charge–discharge cycle. The initial molecule, dimethylhydrophenazine (DMPZ) was limited to a concentration of 60 mM and only stable over 30 cycles [40], but subsequent bio-inspired derivatisation led to 5,10-bis(2-methoxyethyl)-5,10-dihydrophenazine (**BMEPZ**), which has since shown capacity retention of >99% per cycle over 200 cycles for up to 0.4 M concentrations [41]. When combined with fluorenone as the anolyte in an asymmetric cell, this corresponds to the highest energy density of NAOFBs to date of 17 W h l^{-1}.

43.2.4 Cyclopropenium Derivatives

In 2017 a new class of catholyte ROM blossomed featuring a redox-active cyclopropenium (**CP**) core. Critical to these works by the Sanford group, structure–function design stimulated rapid development of these catholytes [38, 42–45]. The first works on this molecule evaluated five derivatives of tris (R-amino) cyclopropenium with the R groups (aryl or alkyl) modifying electronic and steric properties [42]. Introducing aryl groups positively shifted the redox potential of the compounds, but also destabilised their charged states resulting in rapid decay. However, molecules engineered to sterically block decomposition pathways using alkyl groups gave a species capable of over 200 cycles with <3% capacity fade in a low current density symmetric H-cell.

Computational density functional theory (DFT) studies informed the substitution of a nitrogen atom for a sulphur atom, giving a bis(dialkylamino)alkylthio CP (**DATCP**) derivative. This catholyte molecule is the most positive redox couple applied to date in NAOFBs with a redox potential of 1.4 V vs. Fc [44]. Coupled with phthalimide as the anolyte, an asymmetric flow cell achieved a 3.2 V potential, cycled at 10 mA cm^{-2}. It suffered rapid capacity fade, however, only achieving 17 cycles, though this was attributed to the anolyte decay, as at 30 cycles the catholyte retained over 90% of its material compared to just 42% of the anolyte [44]. Further DFT studies informed the design of bis(dialkylamino) cyclopropenium aryl (**DACP-Ar**) compounds, where the aryl group was varied extensively yielding a derivative with both high stability and potential [46]. Cycling in a symmetric half-cell, over 200 cycles were achieved at 35 mA cm^{-2} and 0.3 M concentration, retaining 100% CE and 92% capacity. Whereas demonstrated in an asymmetric flow battery using butyl viologen, 100 cycles were achieved however just 72% capacity was retained with concentrations limited to 50 mM due to the anolyte solubility. The prospects of CP derivatives in future work is evident, as high stability and potentials are achievable, yet solubility is presently low; at max 0.7 M.

43.2.5 Amines

A number of amine-based compounds have been used as catholyte molecules in NAOFBs. This is because tertiary amines can typically be reversibly oxidised and derivatisation of the amine group is easily accessible. Phenylenediamine, applied by Kim et al. in two studies, is a prime example that can be reversibly oxidised twice [47, 48]. Despite this, rather modest redox potentials were reported; with the first oxidation at a potential of c. −0.2 V vs. Fc, and cycling was performed in a coin cell at an excruciatingly low current density on 10 µA cm^{-2} (450 cycles). As such, the limited data on the material is presently insufficient to indicate its value, however further work may yield a more attractive derivative.

Triarylamines (**TAA**) on the other hand have been more explored with several derivatives documented [49–52]. Pasala combined the simple tris(4-bromophenyl) amine with a superoxide anolyte that used atmospheric O_2 in combination with the TBA$^+$ supporting electrolyte cation. The concept was very original however the design was only evaluated at very low concentrations in a static cell, giving mediocre performance [49]. Later, Romadina et al. evaluated several TAA derivatives reporting that very high solubilities of >2.2 M are possible, and some derivatives are miscible liquids at room temperature [52]. In addition, the TAA redox potential was tunable over a range of 600 mV, however cycling experiments showed limited capacity retention. Vastly improved stability was realised by Kwon et al. utilising tris(4-methoxyphenyl)amine (3MTPA). Cycling of 3MTPA in a symmetric electrolyte flow cell completed over 1400 cycles with 99.998% capacity retention per cycle, operating at 50 mM concentration [50].

43.3 Anolytes

The redox anolytes in NAOFBs to date are almost all dependent on the formation of a radical anion. Such a species has variable stability, so as with the radical cations of the catholytes evaluated, attempts to derivatize and stabilise the anions have been made. Unlike the radical cation, the radical anion is prone to undertaking nucleophilic attack of the solvent and supporting electrolyte ions, and as such careful selection of the battery conditions is necessary. Furthermore, the strong reduction potential of the molecules means they are highly susceptible to reaction with dissolved oxygen and trace water, as is the case with all FB anolytes. Figure 43.2 highlights anolytes that have been studied in full NAOFBs thus far with respect to their estimated reduction potential in aprotic media.

43.3.1 Phthalimide (N-MP)

In the first NAOFB reported in 2011, methylphthalimide (**N-MP**) was the anolyte used vs. a TEMPO catholyte [23]. It is a well-known ROM with a relatively persistent radical anion state and gives a redox potential in the region of −1.8 V vs. Fc [44, 48, 53]. It has been used in a number of flow-battery systems, simply to provide an anolyte in studies that are typically developing the catholyte. As such, the rapid capacity fade observed in these articles is typically attributed to the reactive and unoptimised phthalimide radical anion [44]. In one dedicated study by Kim et al. [48] the ROM performance was evaluated, whereby aliphatic chain lengths were modified to improve redox potentials and solubility. The performance of the N-MP is most affected by the supporting solvent and salt, however, In a study by Wei et al. [53] 2,5-di-*tert*-butyl-1-methoxy-4-[2'-methoxyethoxy]benzene (DBMMB) catholyte and N-MP anolyte were cycled for 50 cycles at relatively high current density (10 mA cm^{-2}) in dimethoxyethane (DME) solvent with LiTFSI supporting salt (Figure 43.3). The battery retained near 100% of its capacity with the solvent–salt combination, compared to systems containing

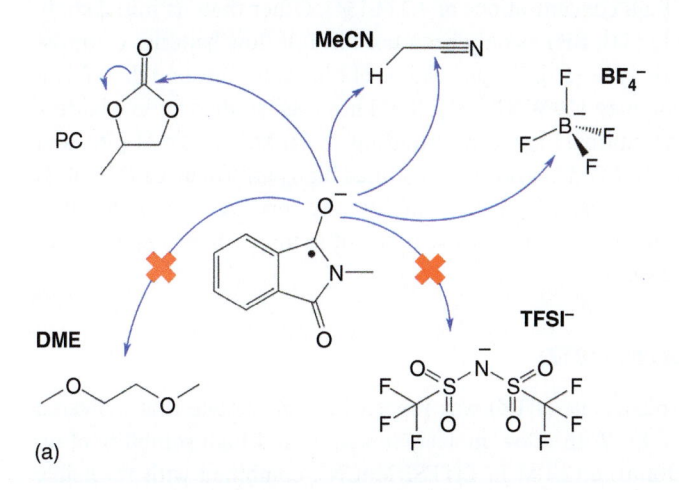

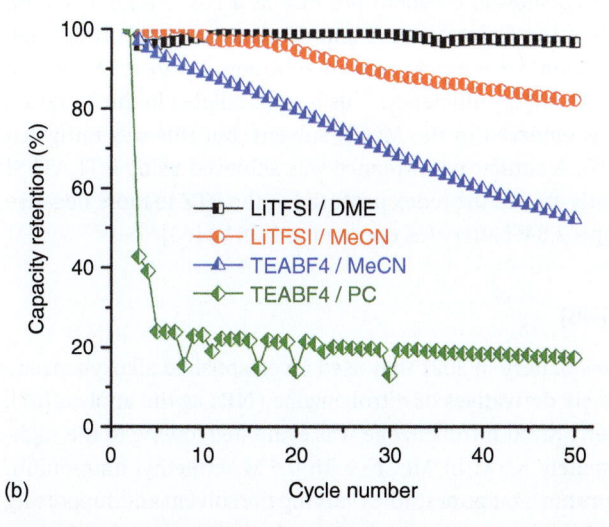

Figure 43.3 (a) Possible side reactions of the N-MP$^{•-}$ radical anion with PC, MeCN, and BF$_4^-$. (b) cycling capacity retention of the 0.1 M N-MP|DBMMB flow cells using different solvents and salts. Source: Reproduced with permission from Wei et al. [53].

electron-deficient substances. As such there were no losses to nucleophilic attack by the N-MP radical anion. It is therefore unclear why more recent works using the same anolyte have not adopted the DME/LiTFSI electrolyte given the apparent stabilisation.

43.3.2 Benzophenone (BP)

The ketone benzophenone (**BP**) delivers the most negative potentials of ROMs to date. It has over 300 naturally occurring derivatives, and a range has been studied in the context of flow batteries since 2017. In the right conditions and in the absence of a lithium salt, BP gives reversible voltammetry, shows high stability, and in pure acetonitrile can achieve high concentrations of 4.3 M [54]. Other than its initial study vs. a TEMPO catholyte [54], BP has only been used in full-flow batteries comprising alkoxybenzene catholytes [55]. Despite claims of high solubility and therefore a high energy capacity of up to 139 W h l^{-1}, the ROM has been studied in flow-battery arrangements at disappointingly low concentrations of 3 mM [54], 5 mM [56], and 0.1 M [57]. In general, this ROM has not demonstrated high performance in NAOFBs to date, with low energy efficiencies, low cyclability, and very low current densities encountered, despite the use of separators as opposed to ion exchange membranes (IEMs) in many of the studies.

43.3.3 Benzothiadiazole (BTZ)

Also in 2017, benzathiadiazole (**BTZ**) was reported as an anolyte that delivered improved cyclability [58]. With a low molecular weight and high solubility of up to 4.3 M in pure acetonitrile (2.1 M in LiTFSI/MeCN), combined with reversible and stable voltammetry, BTZ showed excellent promise as a NAOFB anolyte [59]. In mixed electrolyte cells using Celgard and Daramic separators, relatively high current densities of 60 mA cm^{-2} were achieved in solutions up to 0.5 M, whilst retaining approximately 70% energy efficiency. This is unparalleled in the literature to date. Capacity fade was observed in the MeCN solvent, but this was mitigated when substituted for DME. A similar performance was achieved using a TEATFSI salt, which also significantly shifted the redox potential of the BTZ to more negative potentials, thus delivering a 2.6 V battery (as opposed to 2.36 V) [60].

43.3.4 Nitrobenzene (NB)

Lui et al reported on a flow battery in 2021 that used an established alkoxybenzene catholyte and considered six derivatives of nitrobenzene (**NB**) as the anolyte [61]. In the flow cell, the underivatized nitrobenzene was evaluated, owing to the highest solubility of approximately 6.5 M in MeCN with 0.5 M tetraethyl ammonium (TEA) BF$_4$. The researchers ran a range of studies varying the solvent and supporting salt, identifying what combination gave the highest conductivities, but also the best cycling performance. Their work found that MeCN with TBA BF$_4$ gave the highest discharge capacity and slowest capacity decay, compared with lithium salts in

MeCN, and TBA BF$_4$ in PC and DME [61]. These results contrast other studies of radical anion anolytes in which nitrile/carbonate solvents and tetrafluoroborate salts were considered prone to nucleophilic attack [53]. While boasting a high concentration of 6.5 M in salt-modified MeCN, attempts to operate higher concentration flow batteries were challenged by poor material utilisation and high decay rates. Nuclear magnetic resonance experiments indicated that the NB was irreversibly dimerising when at higher concentrations. Nevertheless, this recent study holds considerable promise for NB as an anolyte in NAOFBs, as it attained high current densities of up to 50 mA cm^{-2} in a mixed electrolyte, has low molecular weight (123 g mol^{-1}), and is readily available at relatively low cost [53].

43.3.5 Fluorenone (FL)

The molecule 9-fluorenone (**FL**) is a known antimalarial drug, and is reduced to a radical anion at potentials below −1.39 V vs. Ag (there is ambiguity in the literature as to what this redox potential is, and only reported vs. the Ag/Ag$^+$). It was first reported in a NAOFB by Wei et al. in conjunction with DBMMB catholyte, showing a 0.9 M concentration in supported solvent [62]. In this initial study, the FL radical anion was found to be highly sensitive to the solvent and supporting salt, and that capacity losses could be mitigated by using a TEATFSI salt with DME solvent. Contrarily, later works combining with a phenazine catholyte have applied LiTFSI and MeCN finding this to be the better solvent–electrolyte combination [40, 41]. The earlier work using dimethyldihydroxyphenazine (**DMEPZ**) only achieved 30 cycles in a flow cell with **FL** [40], but the later phenazine derivative, 5,10-bis(2-methoxyethyl)-5,10-dihydrophenazine (**BMEPZ**) achieved over 200 cycles with a capacity retention of 99.6% [41], suggesting the FL radical anion was not decaying in the MeCN and lithium salt medium. Higher concentrations of 0.4 catholyte/0.8 M anolyte did see non-negligible capacity decay in the BEMPZ study, but this was one of the highest concentrations used in NAOFB to date. In all three studies, one notable feature is that the flow cells were cycled at relatively high current densities of 15–20 mA cm^{-2}.

43.3.6 Pyridine Derivatives (Py and BPy)

The Sanford group has systematically explored pyridine (**Py**) derivatives as viable anolytes for NAOFBs since their first article in 2015 [63–65]. Their inaugural publication iteratively studied 10 classes of pyridine eventually settling on *N*-methyl 4-acetylpyridinium tetrafluoroborate as the optimised derivative. Their focus was to operate reversibly in the presence of LiBF$_4$ supporting salt, as it gave promising electrochemical performance, and the reduced analogues showed stability for days [63]. A subsequent article developed a predictive tool to identify ROMs of particular properties, the prime candidate of which achieved 200 cycles with minimal capacity decay in a symmetric half-cell [65]. Neither work reported full flow-cell battery cycling, instead opting for diagnostic experiments only. Most recently the group published a molecular design paper wherein Py derivatives were cycled against CP derivatives and a design strategy to mitigate capacity fade was described. Specifically, the group advocated the use of oligomeric versions of the ROMs as a means of reducing crossover [45].

Chai et al. also systematically evaluated a range of bipyridine (**BPy**) derivatives as a family of NAOFB anolytes [66]. Here two candidates were evaluated in a full cell using a PEGylated phenazine compound. 0.2 M electrolytes were cycled up to 50 mA cm^{-2} whilst retaining c. 70% capacity retention, albeit at low energy efficiency of c. 55%. In another comprehensive study of iterative design, Griffin et al. synthesised and appraised 24 bipyridine derivatives in an attempt to produce a 2-electron transfer anolyte with long-term stability [67]. The study combined experimental and computational techniques to ascertain structure–function relationships that informed the synthesis of an anolyte of high stability and persistent charge. However, these were yet again not evaluated in a full-flow-battery system [67]. In general, pyridines offer a wide scope for extensive rational design and molecular tuning but are largely unexplored in real FB environments.

43.3.7 Viologen

Viologen is a long-established anolyte in aqueous organic FBs, as they can access 2 electrons in their reduction. Despite this, a poor solubility in nonaqueous solvents has meant they have barely been studied in NAOFBs. Their first application in non-aqueous flow batteries (NAFBs) was by Hu et al. in 2018 where the TFSI derivative of viologen was cycled vs. a ferrocene catholyte [68]. In the works by Yan [46] and Romadina [52] butylviologen (**BV**) was used as the anolyte that has limited solubility and could only be used at concentrations of c. 50 mM in MeCN, but it is the catholyte that is the focus of each paper and the anolyte is considered an aside. The redox potential is much more positive than other anolyte candidates at 0.81 V vs. Fc/Fc$^+$, but a second reduction is also accessible at c. −1.25 V vs. Fc, yet this has not been accessed in publications to date using free viologen [46]. This less negative potential is possibly preventing capacity fade by nucleophilic attack of the solvent and salts, however, allowing the use of MeCN and lithium salts.

Chai et al. used viologen in their flow-cell studies of bulky polyethyleneglycol (PEG) modified ROMs [69]. The typically aqueous soluble viologen was made soluble in aprotic solvent to 0.2 M by the process of appending PEG groups, but this also raised the viscosity considerably too, such that a detrimental polarisation was observed in the mixed electrolyte flow cell and so the VE and EE was 50% at best dropping to 40%. Furthermore, the catholyte and anolyte would interact and self-discharge in the charged states. Nevertheless, cycling over 300 times was achieved in 0.1 M concentrations with a 99.7% capacity retention per cycle (69% overall) [70]. Attanaykae followed the work of Hu et al. and modified the viologen to a TFSI salt form of the ROM (bis(2-(2-methoxyethoxy)ethyl)viologen bis(bis(trifluoromethanesulphonyl)imide) (**MEEV-(TFSI)$_2$**). In this study, 0.25 M solutions were cycled at 10 mA cm^{-2} in a mixed electrolyte solution of a phenazine-TFSI catholyte. At 100 cycles the system had CE >99.1% with EE 64.6–70.6%. Capacity fade was attributed to unequal species crossover and self-discharge as no other species were detected in post-cycling analysis, but electrolyte volumes in each half-cell did change significantly [71].

43.3.8 Quinones and Quinoxaline

Quinones are the leading ROM in aqueous organic FBs but do not feature heavily in NA research. Early works on NAOFBs evaluated quinoxaline (**TMeQ**) [28] and camphorequinone [24] as potential FB anolytes but did not study them in flow battery arrangements. Quinoxaline was later studied in an asymmetric flow cell vs. phenothiazine, in one of the earliest reports of a NAOFB cycling over 50 cycles [33]. The work focussed on the catholyte rather than the anolyte, however, with the latter being relatively unoptimized. Quinones were studied yet again in 2018 against a phenothiazine catholyte, with the availability of a 2-electron transfer from anthraquinone (**AQ**) being the research focus. While promising to offer a considerably higher energy density, the 2-electron system proved unstable and therefore unviable [72]. The authors did not evaluate a 1-electron charge–discharge of the same electrolytes however, therefore the long-term stability of the monoanionic quinone species is not known. Nevertheless, poor solubility also prevents anthraquinone from being a viable NAOFB anolyte.

43.4 Symmetric and Bipolar Redox Materials

Bipolar redox organic molecules are so-called due to their ability to function as both catholyte and anolyte. They are symmetric ROMs that can be used in a full symmetric flow battery (as opposed to the examples of symmetric half-cells so far discussed to evaluate the properties of a catholyte or anolyte independently). They are molecules that can both abstract and donate electrons on charge while retaining a persistent radical state, either by disproportionation from a neutral state or resonance of an ionic species.

The first demonstration of a symmetric ROM was the diaminoanthaquinone derivative 'Disperse blue 134' (**DB-134**), which is a commercial dye used in the fabrics industry [73]. Voltammetry of the compound gives an attractive four-electron redox series from dianion to dication in MeCN solvent giving a two-electron FB with maximal 2.72 V cell potential, making DB-134 a bipolar ROM. Unfortunately, the neutral DB-134 displays a low solubility of ~20 mM in MeCN, but addition of toluene increased the solubility fivefold. Galvanostatic cycling in an unoptimised glass cell gave limited performance, requiring 100 hours to perform six charge–discharge cycles with only the anionic and cationic states accessed. It is therefore difficult to evaluate the promise of DB-134 without flow-cell testing, however, the authors do explore the benefits of the symmetric FB and electrolyte reversal [73]. Another early example of a symmetric ROM was the neutral nitroxide radical 2-phenyl-4,4,5,5-tetramethylimidazoline-1-oxyl-3-oxide (**PTIO**) [74]. It has a high solubility of c. 2.6 M in its neutral state, and it electrochemically disproportionates to a negatively charged and positively charged state with the redox couples separated by 1.74 V in MeCN. The complex was cycled 100 times separated by Daramic allowing for a 20 mA cm^{-2} current density, though at 0.5 M due to detrimentally increasing viscosity at higher concentrations. By use of infrared spectroscopy, the authors also demonstrated easy state-of-charge determination during operation.

The macrocycle 5,10,15,20-tetraphenylporphyrin (**H$_2$TPP**) [75] possesses a four-electron series whereby H$_2$TPP can be oxidised and reduced twice as catholyte and anolyte, respectively, giving a two-electron NAOFB with 2.83 V cell potential. Unfortunately, H$_2$TPP is very insoluble and required dichloromethane as the battery solvent to achieve 10 mM concentration at highest saturation. Thus, symmetric batteries were conducted using a suspension of H$_2$TPP and conducting additive Ketjen Black. Experiments gave good capacity retention across 800 cycles and demonstrated the viability of employing the electrolyte for cold-climate energy storage down to −40 °C [75]. More recently TEMPO was studied in tailored ionic liquid such that the TEMPO was used as both catholyte and anolyte [76]. The ionic liquids served to stabilise the TEMPO cationic and anionic states such that 200 cycles were achieved, though at 100 times lower current density as the hybrid Li system due to high intrinsic resistances in the static cell used.

Over the past few decades, verdazyl radicals have developed into one of the largest families of neutral radical ROMs with high relative persistence. Most forms are considered indefinitely stable in both solid and solutions state, are air- and water-stable, avoid dimerisation, and are readily modified to adapt performance [77]. Despite these credentials, only a few papers have emerged proposing using verdazyl radicals in NAOFBs [78]. Despite the reported stability and promise of the oxo-verdazyl radical, the symmetric FB reported a poor voltaic efficiency over 150 cycles at a very low current density and a clear decay in the theoretical capacity such that at 40 cycles only 50% was accessed. The authors attribute the poor performance predominantly to the incompatible anion-exchange membrane and supporting electrolytes, rather than the instability of the charged ROM [78].

Similar to verdazyl the pseudo oxocarbon, croconate violet (**Croc**), was evaluated as a symmetrical bipolar ROM for NAOFBs [79]. Another promising compound theoretically, the molecule could attain multiple redox states that spanned 1.82 V and could achieve c. 1 M concentration in its dianionic form. However, full symmetric battery cycling demonstrated rapid decay. To identify the problematic electrolyte, each redox couple was cycled in a half flow cell, where the radical trianionic state of the anolyte was identified as the root cause of flow-cell decay due to radical disproportionation within 10 cycles. The catholyte forming a radical anion withstood 100 cycles at high Coulombic efficiency, but a significant loss in capacity retention indicating a sustained loss of active material. As a promising catholyte species the croconate couple was cycled asymmetrically vs. BTZ, but fast decay was observed yet again attributed to the membrane and rapid ROM crossover.

Most recently a helical carbenium species was evaluated as a symmetric redox species, achieving a high 2.12 V potential difference in charged states [80]. The anolyte, a neutral radical, showed robust performance for 800 voltammetric cycles, whereas the dication deteriorated and formed a film on the electrode. In an unusual approach, the researchers used an H-cell to twice polarise the electrolytes, such that each half-cell electrolyte oscillated from the fully reduced radical state to the fully oxidised dicationic state. In doing so 80 charge–discharge cycles of high CE were achieved before a sudden deterioration of the cell capacity, suggesting that by twice charging the ROM the gradual decay of one redox species could be avoided [80].

43.5 Limitations and Challenges

In retrospect of the development of NAOFBs, it is evident that a multitude of challenges remain as barriers to future commercialisation of the technology. Indeed, the overarching concept of utilising nonaqueous solvent imparts severe limitations that remain stubbornly unsurpassed.

43.5.1 Electrolyte Stability

Considering the success of the commercial vanadium FB, the exceptional lifetime of its electrolytes is key in offsetting the initially high cost of new battery installations. The high stability of vanadium is perhaps a consequence of its simplicity; a design so elegantly minimalist that it is difficult to envision many ways it could go awry. Unfortunately, the search for enhanced FB electrolytes has resulted in molecules being pursued with radically higher complexity; a property that is typically correlated with instability and low durability. It is therefore unsurprising that emerging ROMs both in aqueous and nonaqueous electrolytes have been plagued by high reactivity and short lifetimes. Therefore, improving the stability of these redox molecules is the most immediate challenge in the research field. This is more complicated for nonaqueous electrolytes, however, because the possibilities for degradation are more numerous. Understanding these mechanisms is therefore of critical importance but infrequently achieved in the literature. Despite this, research into ROMs has highlighted a trend whereby instability originates from charged oxidation states. Indeed, studies have repeatedly boasted compounds to have excellent solubility, stability, and redox behaviour in the discharged state (and over the timescale of a cyclic voltammogram) yet once charged they precipitate, self-discharge, or decompose within short timescales. In one particularly astute study by the Odom group, eight catholyte candidates including derivatives of phenothiazine, cyclopropenium and dialkoxybenzene were synthesised in their charged states and comprehensively studied with respect to their persistence [38]. Notably, the higher oxidation state species were more prone to self-discharge and molecular degradation by various mechanisms, and low concentration solutions were compromised by trace impurities in the electrolyte [38]. Similarly the radical cations of a phenothiazine and DBBB were studied by UV–vis spectroscopy to assess the stability of the charged state, demonstrating the rapid decay of the DBBB (Figure 43.4) [33]. Articles like this clearly show that the properties of redox materials must be evaluated in all oxidation states accessed during battery

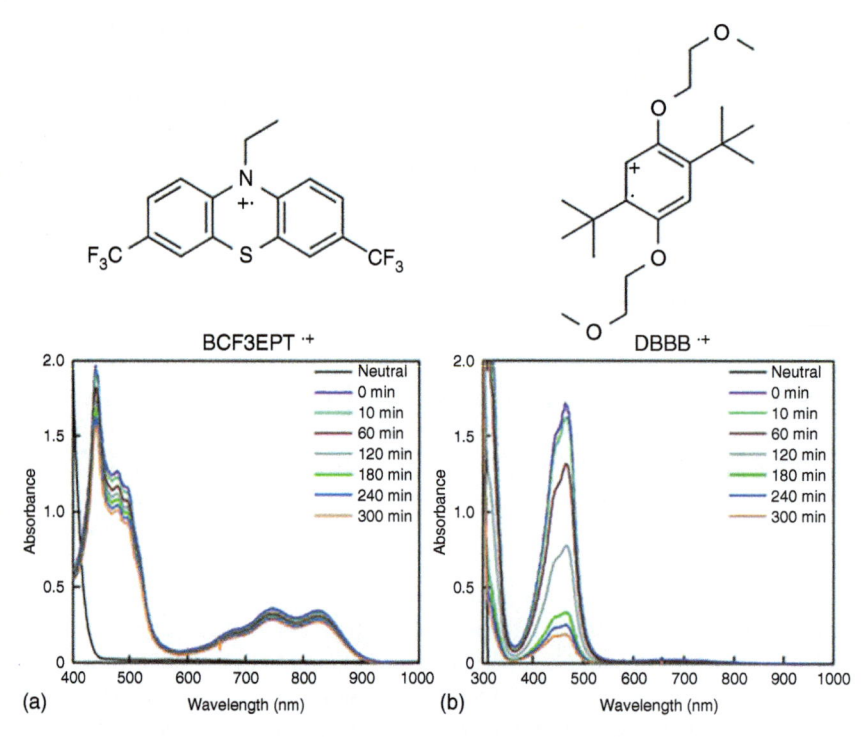

Figure 43.4 UV–Vis absorption spectra of radical cations BCF3EPT (a) and DBBB (b) recorded at various times from 0 to 5 hours after generation by bulk electrolysis at 5 mM in 0.2 M LiBF$_4$ in PC. Source: Reproduced with permission from Kaur et al. [33].

operation, and quantified independently of other reactions i.e. in a symmetric electrolyte configuration [25, 34, 64, 79]. Furthermore, it is of the utmost importance to characterise novel redox materials in realistic demonstrations such as in full cell cycling experiments featuring state-of-charge/capacity utilisation studies at high concentration. Evidently, it is impossible to predict the quality of a candidate molecule based on voltammetry alone [79]. Fortunately, few research papers in this field are claiming incredible performances on that basis alone.

Degradation mechanisms that are life-limiting for ROMs are dimerisation [33], disproportionation [79], self-discharge [38], and reactions with the solvent/supporting salt [62]. Given that molecular radicals are effectively covalent-deficient species, reactions that terminate or propagate the radical are energetically favoured. ROM dimerisation is in principle possible for all explored organic radicals, but it has been rarely encountered in practice. This is most likely because ROMs typically feature steric bulk, or because the resulting dimer would break planar conjugation. A noteworthy but unproven example of ROM dimerisation is however the pinacol coupling of ketyl radicals that have repeatedly featured as anolytes [54, 62, 79]. Similar to dimerisation, ROM radical disproportionation is also unlikely but does occur when electron pairing stabilisation is favoured, such as in the case of radical anions [9, 15, 79]. Few examples of these reactions in NAOFBs have been identified but it is important to note that most studies are conducted at low concentration, and dimerisation/disproportionation may become more problematic when operating at higher concentrations such as in high energy density electrolytes [81]. More frequently, unstable ROM states tend to react with other components of the electrolyte, in part because solvent and electrolyte molecules are in higher abundance and form solvation shells around ionic ROMs. More importantly, solvent/electrolyte molecules can act as nucleophiles and electrophiles or participate in radical propagation reactions. The outcome of these reactions is chemical degradation of the ROM that causes irreversible capacity loss during battery operation. In some instances, reactions of charged ROMs instead cause self-discharge whereby the ROM is not chemically degraded, but rather is oxidised or reduced back to its uncharged state [38]. This induces FB capacity loss by charge imbalance, but this may be recovered by use of chemical oxidants/reductants. The precise cause of self-discharge in many ROM systems remains unknown but some researchers have suggested that trace electrolyte impurities such as metal ions may be responsible. It is important to note here that self-discharge may become more problematic for low/high potential anolytes/catholytes as these molecules are intrinsically stronger reducing/oxidising agents, respectively.

It is challenging here to generalise such diverse chemistry, as electrolyte reactivity is case specific, so instead, we will highlight key studies that quantified lifetimes via battery experiments or spectroscopic methods and proposed degradation mechanisms. In 2015, Wei et al. studied a DBMMB and fluorenone NAOFB, initially in MeCN with TEATFSI, but this saw capacity loss in the first 10 cycles [62]. By use of electron paramagnetic resonance spectroscopy (EPR) the radical lifetimes were quantified as a function of time and electrolyte composition. This showed that reactivity with the solvent was responsible because radical retention was very dependent on the solvent and supporting electrolyte employed. Here MeCN with TEA BF_4 caused the fastest radical decay whereas, in DME with TEA TFSI, the decay was much slower [62]. A similar report found that DME solvent improved capacity retention in a BTZ flow battery compared to MeCN [58], but the effect was not studied. It is not clear why experiments with DME perform better, however, the answer may be due to the high polarity of MeCN. This may incur higher nucleophilic and electrophilic reactivity, or provide more easily abstracted methyl protons, or hydrogen atoms via bond homolysis. Beyond solvent reactivity, the choice of supporting electrolyte is also critical to the performance and longevity of the ROMs and flow batteries that comprise a number of studies have shown fluorinated anions (BF_4^- and PF_6^-) to be particularly reactive to charged ROMs. One theory is that these salts can hydrolyse in residual water solvent impurity, which creates a more acidic environment in which ROMs destabilise and decay. This was shown by Romadina et al. as they systematically changed the supporting electrolytes and gauged flow-cell performance over 50 cycles [52], as shown in Figure 43.5. The application of lithiated salts has also shown to be problematic as it can interact strongly with anionic ROMs. For example, BP and BTZ are both rendered irreversible and unstable in voltammograms when lithium salts are used, whereas equivalent voltammetry using tetraalkylammonium salts give highly reversible waves [54, 79]. Here it was hypothesised that this is due to the electrophilic lithium cation attacking the carbonyl of the ROM

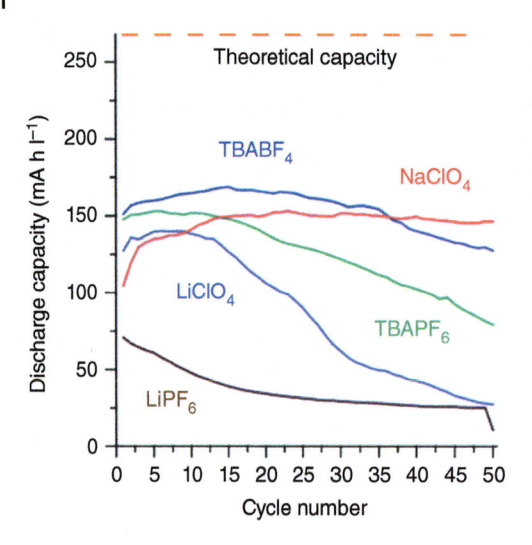

Figure 43.5 Dependence of the discharge capacity with the cycle number for 10 mM redox-flow cells containing triarylamine catholyte and viologen anolyte during 50 cycles for various supporting electrolytes. Source: Reproduced with permission from ref [52].

group and irreversibly forming side products [54]. This reactivity doesn't extend to sodium and potassium cations, however. Notably in the Romadina article and shown in Figure 43.5, it is the sodium perchlorate salt that gives the most sustained performance, yet it is a salt barely used in NAOFB research [52]. Furthermore, Hendricks et al. have identified Na and K salts give outstanding performance in NA solvents and can be used instead of expensive bulky tetraalkylammonium ions [64]. Without such studies being routinely undertaken in the literature, it is difficult to ascertain whether capacity fade in some instances could have been mitigated by a simple change of supporting salt.

43.5.2 Electrolyte Conductivity

A rather unappreciated advantage of the aqueous flow battery is the ability to drive high current densities by virtue of an intrinsically high conductivity electrolyte. Nonaqueous solvents immediately impose a barrier here, as even the most conductive exhibit 10s of ohms of resistance. In LIBs this is not a major problem, as the cells are inherently thin layered and only ion transfer is required. In contrast, FBs are solute in solvent systems, requiring supporting electrolytes to be added in preferably high concentration. This is problematic for NAFBs, because electrolytes such as tetraalkylammonium ions are large and expensive compared to aqueous salts, increasing system cost considerably. In addition, the increase in viscosity is more pronounced for nonaqueous solvents than water, and redox material solubility decreases as the electrolyte concentration is increased. This results in an undesirable compromise between electrolyte energy density and conductivity.

In principle, there are many solvents and supporting salts available for application in NAFBs. This affords numerous possible combinations to construct and tailor electrolyte properties toward system optimisation [82]. In practice, the choice of electrolyte composition is quite limited because of the requirement for high conductivity. This property is a consequence of a high dielectric constant and low viscosity. Unfortunately, nonaqueous solvents do not fare well compared to water, which is typically superior in both aspects. Conventionally, researchers have adopted MeCN, PC, DME, dimethyl sulfoxide (DMSO), DMF, and EC/DMC as they give best conductivity, however alternative low-conductivity solvents such as dichloromethane and *O*-dichlorobenzene, used to solubilise porphyrin [75] and fullerene [83] derivates, respectively, have also been demonstrated. Considering that solvent choice is also influenced by toxicity and cost, options like chlorinated solvents and DMF become especially unrealistic for practical application.

43.5.3 Membrane Incompatibility

One of the more derailing aspects of developing NAFBs is that purposely designed ionic membrane materials extensively do not exist for nonaqueous solvents. The application of fluoropolymer ion-exchange membranes, such as Nafion, in compartmentalised electrochemical cells have been instrumental to the success of aqueous FBs. While these membranes are not perfect, they show excellent proton conductivity whilst largely restricting metal ion crossover, and more importantly, without imparting crippling high cell resistances that limit current density. The same cannot be said for nonaqueous systems. Most researchers endeavour to still use the aqueous counterparts as a separator, sometimes with a degree of pre-treatment in common ions. Nevertheless, the fluoropolymer materials, and other emerging ionomers, are not really fit for purpose. Consequently, an undesirable combination of excessive membrane crossover of redox molecules, and high cell resistance is typically encountered. Compounding these issues are membrane durability and lifetime; neither of which is satisfactory for the aqueous membranes in organic solvents as excessive swelling and degradation can occur, which depends on the material and solvent combination.

To mitigate membrane incompatibility, many researchers have employed simplistic porous separators such as Celgard or Daramic [53, 58, 60, 64, 74, 79, 84] that feature in LiBs to prevent dendritic short-circuits [85]. These inexpensive polymer films boast exceptional chemical resilience and low cell resistance, allowing undoubtably higher currents. Unfortunately, crossover rates are disastrous at high electrolyte concentrations, which is particularly problematic for NAOFBs as most ROMs are not symmetric/bipolar redox molecules. Nevertheless, for early lab-scale experiments, which typically employ dilute electrolytes, crossover rates are manageable making these separators convenient for research. Recently, some researchers have combatted this crossover issue by designing purposefully large redox molecules that can be filtered by size exclusion. These bulky polymeric organic molecules (discussed in detail in Chapter 44) and oligomeric compounds offer a tantalising approach whereby crossover mitigation is engineered into the redox molecules rather than in the membrane chemistry. Noteworthy examples include oligomers of methoxybenzene [86], cyclopropenium [43], and phenothiazine [70].

43.5.4 Electrolyte Cost

While the organic molecules being developed in this field are arguably a low-cost and sustainable alternative to their inorganic metal counterparts, it remains that

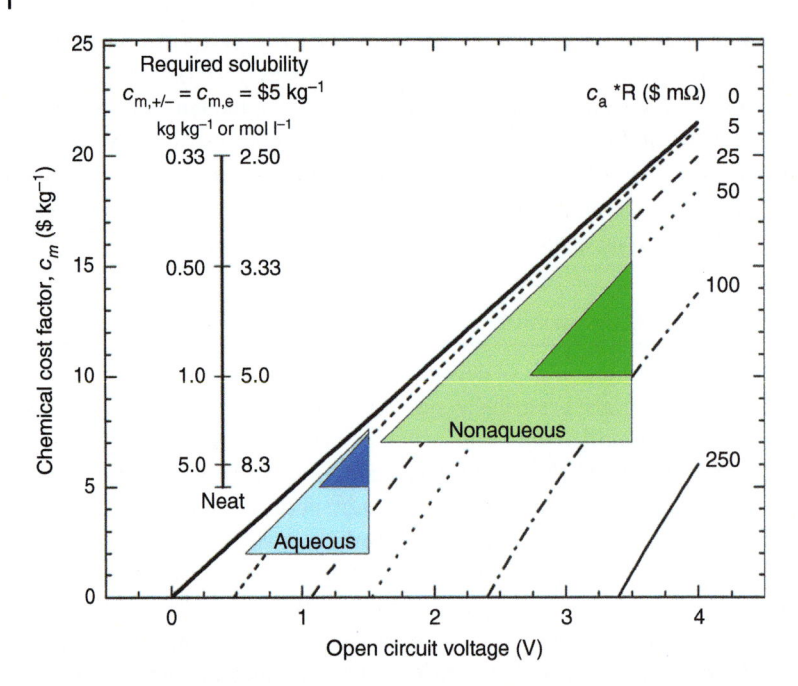

Figure 43.6 Allowable chemical cost factor on an active material basis (in $ kg^{-1}$) vs. open-circuit voltage for a range of reactor costs ($c_a R$ in $ m\Omega$). All points on a line give a system price of $120 per kWh with the parameters given in Table 2 of ref [87]. The region $U < 1.5$ V is considered to be available to aqueous systems. The dark shaded triangles are considered to have a higher likelihood of achievability compared to the larger lighter shaded triangles. The leftmost inset vertical scale shows the required solubility (in kg kg^{-1}) of a nonaqueous active species when solvent and solute cost $5 kg^{-1}. The rightmost inset vertical scale on the right shows the molar concentration, assuming specific volumes of 1 l kg^{-1}. Source: Reproduced with permission from Darling et al. [87]/Royal Society of Chemistry/Public Domain CC BY 3.0.

nonaqueous electrolytes impart higher costs and toxicity concerns than their aqueous counterparts. Balancing the advantages and disadvantages of aqueous and nonaqueous flow batteries was eloquently modelled by Darling et al., in which the authors conclude a challenging benchmark of properties, namely voltages over 3 V and solubilities of 5 M necessary to simply match the cost of a vanadium flow battery (VRB) [14, 87]. The key figure is shown in Figure 43.6. In NAOFBs the ROMs themselves may be comprised of sustainable and 'cheap' elements, but their synthetic route may be intensive, unsustainable, or not suited to the scale-up required to make thousands of litres of electrolyte. Extrapolating the costs of lab-scale fundamental synthesis to large-scale production is not advisable either, as the processes do not linearly correlate. It is a monumental challenge to balance size, stability, cost, and accessibility.

Finally, supporting salts are used in NAFBs with little regard for their cost and sustainability impact. In reflection of the literature, it is evident that tetraalkylammonium salts tend to impart higher stability to systems than their lithium salt analogues. But these bulky salts are considerably more expensive than the salts used in aqueous batteries, especially in the VRB where the acidic solvent serves as the electrolyte medium without additional supporting electrolyte. A move towards using low-cost sodium and potassium salts, with perchlorate anions may drop this cost significantly, however, without a considerable impact on the conductivity and performance of the most optimised ROMs to date [52, 64].

43.5.5 Solubility and Energy Density

One of the main selling points of NAOFBs is the promise of a high energy density fluid on account of a high voltage and, by means of derivatisation, high solubility. Redox materials operating in carbonates, glymes, and MeCN offer wide potential windows on a par with LIBs or higher, as well as higher solubilities than aqueous solvents for many classes of bulky, redox-active organic molecules. Yet the reality has been less rewarding due to disparity in stability and solubility when in charged states, the parasitic losses described above, and the difficulty in identifying ROMs that operate in the full window of the solvent. This is particularly the case for catholytes where in breaching the 1 V vs. Fc/Fc$^+$ mark has been very limited, despite some 1 V more of potential window to operate within. To date the highest cell voltages reported are of the 3 V mark, with Sanford et al. reporting in 2019 a 3.20 V asymmetric system of CP and N-MP redox species [44]. This unprecedented voltage was achieved by careful mechanism-based design informed by computational theory to optimise the CP structure. The curtailed performance of the full-flow battery was attributed to the unoptimised anolyte, which rapidly decayed after 15 cycles. Nevertheless, the work demonstrates a significant step towards realising the high voltage promise of NAOFBs.

It has been proposed in techno-economic assessments of flow batteries that for a nonaqueous candidate to make economic sense that it must maintain an intrinsic energy density above 178 mAh g^{-1} [87]. This is to achieve a specific energy density, but there are arguments that a FB does not need to strive to achieve a high energy density, as the stationary storage solution will scale as needed. Regardless, ambitious targets of 5 M solubility for NA ROMs have been touted to be competitive with their aqueous counterparts. But to date, NAOFB has been almost always operated at concentrations below 1 M, despite several ROMs boasting higher solubility including those which are miscible liquids [37, 55, 57, 61, 86]. A further issue is the reporting of ROM solubility. Researchers invariably report the solubility of the ROM with respect to pure solvent and not in a salt-modified form, but this can significantly impact solubility limits. This is worsened by the trend that only the solubility of the as-synthesised ROM is obtained rather than the associated charged states produced during battery cycling. In a study of phenothiazine derivatives, Milshtein et al. demonstrated the stark difference between solubility of the neutral and radical cation, where solubility dropped from >2.0 M to only 0.1 M upon oxidation [34]. Isolation of the radical ion ROM in the form of a salt is a challenge, but one that should be attempted when fully evaluating the performances of FBs, especially in studies that boast 'long-term' stability of the compounds.

43.5.6 Outlook – Transitional Developments

The rate at which NAOFBs have been developed over the past decade is astounding. The sophistication of recent works and some stand out earlier research really demonstrate that there is potential to achieve the targets laid out by the Department of Energy and associated research where NAFBs are concerned. In under 10 years, cycle lifetimes have been extended 10-fold from 30 to 300, molecules are demonstrating >99% capacity retention over hundreds of cycles, and cell voltages on par with LIBs have been achieved. With a wealth of ROMs now identified, the focus must shift towards making these work in scaled environments. With this comes a more nuanced approach to ROM design and study, looking to rationalise environmental relationships, increasing the concentrations to practical levels, and optimising the engineering of flow cells and stacks. The development of ion-selective and low-resistance membranes suited to nonaqueous solvents would provide a step change in this field as well.

Computational methods can be used highly effectively for rationalising the design of new ROMs by predicting various performance metrics. In recent works research groups have utilised DFT to propose a catalogue of molecules that are then whittled down to a few candidates, where in the redox potentials, solubility, and long-term persistence of the ROM is predicted to a high degree of accuracy [44, 46, 67]. This strategy has the advantage of evaluating numerous derivatives of core ROMs, reducing the time spent synthesising, and studying compounds that are ultimately not suited to FBs.

Where a range of ROMs are made, the challenge of ascertaining the optimum solvent and supporting salt can also impact the rate at which an electrolyte is developed. It can be a mammoth task to try to evaluate each iteration of redox molecule with numerous derivatives in a range of solvent and salt combinations. To try to circumvent this, some researchers have sought to use high-throughput, systematic bench testing methods [82, 88] to identify trends and optimise formulations. In the case of DBBB, an automated electrolyte synthesis and characterisation platform screened various alkali ion salts and carbonate [82] and ether [88] solvents to elucidate the optimum mixture with respect to solution conductivity and DBBB solubility. Automated platforms such as these are ideal for optimising electrolyte compositions before full-scale flow-cell testing is undertaken of the best performing candidates. Furthermore, computational methods are also beginning to elucidate the solvent-mediator relationship [55, 57, 67, 89, 90]. Some studies identify using lower dielectric solvents as key to enhance the redox properties of the ROMs [89], but this does not improve the performance of a FB as the area specific resistance (ASR) is then increased considerably to give a lower voltaic efficiency. Balancing redox performance and battery cycling performance will be a key challenge going forward.

Recent works have sought to address the low solubility of ROMs and the issue of membrane crossover by creating oligomeric forms of the redox species [43, 45, 51, 86]. The oligomer scaffolds are capable of holding high concentrations of redox species, and reportedly improve their long-term stability without compromising the redox kinetics. These bulky macromolecules also enable use of porous separators in the FB architecture rather than ion-exchange membranes. These works are very recent but highly promising, as the oligomeric structures are fully miscible in solvents achieving solubilities as high as 3.95 M with respect to the redox species, and greater than 300 fold reduction in membrane crossover [86, 91]. If NAOFB research can move to using Daramic or Celgard microporous separators effectively, then high current densities can more likely be achieved and additional VE losses from incompatible membranes reduced.

One of the leading challenges of all flow batteries is the intrinsic low energy density caused by having a solute in a solvent. Research reporting highly miscible and highly concentrated ROM solutions are increasing but there is a disparity in the subsequent research whereby flow-cell cycling is conducted at very low concentration. While a number of NAOFB researchers have now reported on >1 M concentrations from their ROMs, very few have evaluated their molecules at such high concentrations in a flow battery. For those who have attempted concentrated electrolytes in a FB, the results have been poor with enhanced capacity fade and irreversible parasitic side reactions taking place. The electrochemistry of a highly concentrated solution, aqueous and nonaqueous containing organic redox molecules, is considerably more complicated and different from what is understood at lower concentrations. Shkrob et al. [92] recently reported on the formation of self-assembled solute networks in concentrated electrolyte solutions that lead to pockets of neutral solute and ROMs and a heterogeneous ionic environment. The nanoconfinement of ROMs in these 'pockets' leads to a sharp increase in the decay rate. More research of this nature attempting to understand the physical–chemical properties of concentrated NAOFB electrolytes are critical to further advancing this field if it will ever be commercially viable.

Finally, studies of new and existing ROMs must be conducted using the right tools. The Brushett group developed a NA-FB cell design [93] that is open source and is now widely used by researcher working with NA chemistry, but not universally. It can vastly improve the performance of NAOFB as demonstrated with azole-bipyridine mixed ligand cobalt complexes, the redox molecules can deteriorate significantly in a simple aqueous cell but last 100 s of cycles in an optimised NA cell, and achieve 10 to 100 fold improvements in charge density [94, 95]. Many works still only evaluate their materials in H-cell stationary environments, and while there is some interest in this, it is not representative of the true performance of the ROMs being designed. The chemistry in a flow cell is markedly different from a bulk electrolysis cell, and it is too easy to attribute poor performance to unoptimised electrochemical cells.

43.5.7 Conclusions

NAOFBs are a challenging topic in which the advantages are not readily visible. While leaps and bounds have been made in the mere decade of their consideration, attaining cycling over 100 cycles is considered a significant milestone, but that has no practical relevance to the immediate future of flow batteries. The poor stability of these species in their charged states is compounded by high material and solvent costs, numerous sources of resistance, and crossover through incompatible separators. While they remain an academic interest and showcase extraordinary talent for rational design and targeted synthesis, it is difficult to imagine these systems being implemented on the scales required of large-scale stationary storage. It is perhaps better to consider such molecules and innovation on a smaller scale, to reimagine batteries, or return these species to their origins as overcharge protection molecules in LIBs.

The range of stable and high-performing catholytes is now extensive, and this bodes well for further advances in LIBs and overcharge protection. Stable anolytes are a greater challenge however, where lithium metal is shown to give good long-term performance, it defeats the purpose of a flow battery to retain independent capacity and power, and to diversify the field of electrochemical energy storage. To make any real-world progress in this subfield and permit practical scale-up, improvements are needed in all aspects of NAFB technology, including the ROMs, the component materials, and the flow-cell design. Automated high throughput screening is an effective way of rapidly evaluating ROMs in various electrolyte conditions. But it is the development of suitable separators that will truly be a breakthrough in this field, or a means of preventing crossover without compromising electrochemical kinetics and flow, at which point reducing the overall costs will become the focus. In this regards the recent use of oligomer scaffolds that can hold high concentrations of ROM show promise as they prevent crossover, but in turn, implicate the flow of the system.

References

1 Hogue, R.W. and Toghill, K.E. (2019). Metal coordination complexes in non-aqueous redox flow batteries. *Current Opinion in Electrochemistry* 18: 37–45. https://doi.org/10.1016/j.coelec.2019.08.006.

2 Winsberg, J., Hagemann, T., Janoschka, T. et al. (2017). Redox-flow batteries: from metals to organic redox-active materials. *Angewandte Chemie, International Edition* 56 (3): 686–711.

3 Odom, S.A. (2021). Overcharge protection of lithium-ion batteries with phenothiazine redox shuttles. *New Journal of Chemistry* 45 (8): 3750–3755.

4 Chen, Z., Qin, Y., and Amine, K. (2009 Oct). Redox shuttles for safer lithium-ion batteries. *Electrochimica Acta* 54 (24): 5605–5613. http://linkinghub.elsevier.com/retrieve/pii/S0013468609006392.

5 Khan, R. and Nishina, Y. (2021). Covalent functionalization of carbon materials with redox-active organic molecules for energy storage. *Nanoscale* 13 (1): 36–50.

6 Sikukuu, N.G. (2021). Organic molecules as bifunctional electroactive materials for symmetric redox flow batteries: a mini review. *Electrochemistry Communications* 127: 107052. https://doi.org/10.1016/j.elecom.2021.107052.

7 Ding, Y., Zhang, C., Zhang, L. et al. (2018). Molecular engineering of organic electroactive materials for redox flow batteries. *Chemical Society Reviews* 47 (1): 69–103.

8 Chen, H., Cong, G., and Lu, Y.C. (2018). Recent progress in organic redox flow batteries: active materials, electrolytes and membranes. *Journal of Energy Chemistry* 27 (5): 1304–1325. https://doi.org/10.1016/j.jechem.2018.02.009.

9 Armstrong, C.G. and Toghill, K.E. (2018). Stability of molecular radicals in organic non-aqueous redox flow batteries: a mini review. *Electrochemistry Communications* 91: 19–24.

10 Leung, P., Shah, A.A., Sanz, L. et al. (2017). Recent developments in organic redox flow batteries: a critical review. *Journal of Power Sources* 360: 243–283.

11 Kowalski, J.A., Su, L., Milshtein, J.D., and Brushett, F.R. (2016). Recent advances in molecular engineering of redox active organic molecules for nonaqueous flow batteries. *Current Opinion in Chemical Engineering* 13: 45–52. http://dx.doi.org/10.1016/j.coche.2016.08.002.

12 Huang, Y., Gu, S., Yan, Y., and Li, S.F.Y. (2015). Nonaqueous redox-flow batteries: features, challenges, and prospects. *Current Opinion in Chemical Engineering* 8: 105–113. http://linkinghub.elsevier.com/retrieve/pii/S2211339815000283.

13 Gong, K., Fang, Q., Gu, S. et al. (2015). Nonaqueous redox-flow batteries: organic solvents, supporting electrolytes, and redox pairs. *Energy & Environmental Science* 8 (12): 3515–3530. http://xlink.rsc.org/?DOI=C5EE02341F.

14 Dmello, R., Milshtein, J.D., Brushett, F.R., and Smith, K.C. (2016). Cost-driven materials selection criteria for redox flow battery electrolytes. *Journal of Power Sources* 330: 261–272. http://dx.doi.org/10.1016/j.jpowsour.2016.08.129.

15 Bauld, N.L. (1997). *Radicals, Ion Radicals, and Triplets*. Wiley-VCH, Inc.

16 Laoire, C.O., Plichta, E., Hendrickson, M. et al. (2009). Electrochemical studies of ferrocene in a lithium ion conducting organic carbonate electrolyte. *Electrochimica Acta* 54 (26): 6560–6564.

17 Izutsu, K. (2002). *Electrochemistry in Nonaqueous Solutions*. KGaA: Wiley-VCH Verlag GmbH & Co.

18 Wei, X., Xu, W., Vijayakumar, M. et al. (2014). TEMPO-based catholyte for high-energy density nonaqueous redox flow batteries. *Advanced Materials* 26 (45): 7649–7653. http://www.ncbi.nlm.nih.gov/pubmed/25327755.

19 Song, Z. and Zhou, H. (2013). Towards sustainable and versatile energy storage devices: an overview of organic electrode materials. *Energy & Environmental Science* 6 (8): 2280–2301.

20 Nishide, H., Iwasa, S., Pu, Y.J. et al. (2004). Organic radical battery: nitroxide polymers as a cathode-active material. *Electrochimica Acta* 50(2–3 SPEC. ISS.): 827–831.

21 Janoschka, T., Hager, M.D., and Schubert, U.S. (2012). Powering up the future: radical polymers for battery applications. *Advanced Materials* 24 (48): 6397–6409.

22 Liang, Y., Tao, Z., and Chen, J. (2012). Organic electrode materials for rechargeable lithium batteries. *Advanced Energy Materials* 2 (7): 742–769.

23 Li, Z., Li, S., Liu, S. et al. (2011). Electrochemical properties of an all-organic redox flow battery using 2,2,6,6-tetramethyl-1-piperidinyloxy and N-methylphthalimide. *Electrochemical and Solid-State Letters* 14 (12): A171. http://esl.ecsdl.org/cgi/doi/10.1149/2.012112esl.

24 Park, S.-K., Shim, J., Yang, J. et al. (2015). Electrochemical properties of a non-aqueous redox battery with all-organic redox couples. *Electrochemistry Communications* 59: 68–71. http://linkinghub.elsevier.com/retrieve/pii/S1388248115001939.

25 Milshtein, J.D., Barton, J.L., Darling, R.M., and Brushett, F.R. (2016). 4-Acetamido-2,2,6,6-tetramethylpiperidine-1-oxyl as a model organic redox active compound for nonaqueous flow batteries. *Journal of Power Sources* 327: 151–159. http://dx.doi.org/10.1016/j.jpowsour.2016.06.125.

26 Zhang, L., Zhang, Z., and Amine, K. (2013). Redox shuttles for overcharge protection of lithium-ion battery. *ECS Transactions* 45 (29): 57–66.

27 Zhang, L., Zhang, Z., Redfern, P.C. et al. (2012). Molecular engineering towards safer lithium-ion batteries: a highly stable and compatible redox shuttle for overcharge protection. *Energy & Environmental Science* 5 (8): 8204–8207.

28 Brushett, F.R., Vaughey, J.T., and Jansen, A.N. (2012). An all-organic non-aqueous lithium-ion redox flow battery. *Advanced Energy Materials* 2 (11): 1390–1396. http://doi.wiley.com/10.1002/aenm.201200322.

29 Leonet, O., Colmenares, L.C., Kvasha, A. et al. (2018). Improving the safety of lithium-ion battery via a redox shuttle additive 2,5-Di-tert-butyl-1,4-bis(2-methoxyethoxy)benzene (DBBB). *ACS Applied Materials & Interfaces* 10 (11): 9216–9219.

30 Huang, J., Pan, B., Duan, W. et al. (2016). The lightest organic radical cation for charge storage in redox flow batteries. *Scientific Reports* 6 (1): 32102. http://www.nature.com/articles/srep32102.

31 Huang, J., Su, L., Kowalski, J.A. et al. (2015). A subtractive approach to molecular engineering of dimethoxybenzene-based redox materials for non-aqueous flow batteries. *Journal of Materials Chemistry A* 3 (29): 14971–14976. http://xlink.rsc.org/?DOI=C5TA02380G.

32 Zhang, J., Shkrob, I.A., Assary, R.S. et al. (2017). Toward improved catholyte materials for redox flow batteries: what controls chemical stability of persistent radical cations? *Journal of Physical Chemistry C* 121 (42): 23347–23358. http://pubs.acs.org/doi/10.1021/acs.jpcc.7b08281.

33 Kaur, A.P., Holubowitch, N.E., Ergun, S. et al. (2015). A highly soluble organic catholyte for non-aqueous redox flow batteries. *Energy Technology* 3 (5): 476–480. http://doi.wiley.com/10.1002/ente.201500020.

34 Milshtein, J.D., Kaur, A.P., Casselman, M.D. et al. (2016). High current density, long duration cycling of soluble organic active species for non-aqueous redox flow batteries. *Energy & Environmental Science* 9 (11): 3531–3543. http://pubs.rsc.org/en/Content/ArticleLanding/2016/EE/C6EE02027E%5Cnhttp://xlink.rsc.org/?DOI=C6EE02027E.

35 Narayana, K.A., Casselman, M.D., Elliott, C.F. et al. (2015). N-substituted phenothiazine derivatives: how the stability of the neutral and radical cation forms affects overcharge performance in lithium-ion batteries. *ChemPhysChem* 16 (6): 1179–1189. http://doi.wiley.com/10.1002/cphc.201402674.

36 Kowalski, J.A., Casselman, M.D., Kaur, A.P. et al. (2017). A stable two-electron-donating phenothiazine for application in nonaqueous redox flow

batteries. *Journal of Materials Chemistry A* 5 (46): 24371–24379. http://pubs.rsc.org/en/Content/ArticleLanding/2017/TA/C7TA05883G.

37 Attanayake, N.H., Kowalski, J.A., Greco, K.V. et al. (2019). Tailoring two-electron-donating phenothiazines to enable high-concentration redox electrolytes for use in nonaqueous redox flow batteries. *Chemistry of Materials* 31 (12): 4353–4363.

38 Attanayake, N.H., Suduwella, T.M., Yan, Y. et al. (2021). Comparative study of organic radical cation stability and coulombic efficiency for nonaqueous redox flow battery applications. *Journal of Physical Chemistry C* 125 (26): 14170–14179.

39 Kaur, A.P., Ergun, S., Elliott, C.F., and Odom, S.A. (2014). 3,7-bis(trifluoromethyl)-N-ethylphenothiazine: a redox shuttle with extensive overcharge protection in lithium-ion batteries. *Journal of Materials Chemistry A* 2 (43): 18190–18193. http://xlink.rsc.org/?DOI=C4TA04463K.

40 Kwon, G., Lee, S., Hwang, J. et al. (2018). Multi-redox molecule for high-energy redox flow batteries. *Joule* 2 (9): 1771–1782. https://doi.org/10.1016/j.joule.2018.05.014.

41 Kwon, G., Lee, K., Lee, M.H. et al. (2019). Bio-inspired molecular redesign of a multi-redox catholyte for high-energy non-aqueous organic redox flow batteries. *Chem* 5 (10): 2642–2656. http://dx.doi.org/10.1016/j.chempr.2019.07.006.

42 Sevov, C.S., Samaroo, S.K., and Sanford, M.S. (2017). Cyclopropenium salts as cyclable, high-potential catholytes in nonaqueous media. *Advanced Energy Materials* 7 (5): 1602027. http://doi.wiley.com/10.1002/aenm.201602027.

43 Hendriks, K.H., Robinson, S.G., Braten, M.N. et al. (2018). High-performance oligomeric catholytes for effective macromolecular separation in nonaqueous redox flow batteries. *ACS Central Science* http://pubs.acs.org/doi/10.1021/acscentsci.7b00544.

44 Yan, Y., Robinson, S.G., Sigman, M.S., and Sanford, M.S. (2019). Mechanism-based design of a high-potential catholyte enables a 3.2 V all-organic non-aqueous redox flow battery. *Journal of the American Chemical Society* 141 (38): 15301–15306.

45 Shrestha, A., Hendriks, K.H., Sigman, M.S. et al. (2020). Realization of an asymmetric non-aqueous redox flow battery through molecular design to minimize active species crossover and decomposition. *Chemistry - A European Journal* 26 (24): 5369–5373.

46 Yan, Y., Vaid, T.P., and Sanford, M.S. (2020). Bis(diisopropylamino)cyclopropenium-arene cations as high oxidation potential and high stability catholytes for non-aqueous redox flow batteries. *Journal of the American Chemical Society* 142 (41): 17564–17571.

47 Kim, H., Lee, K.-J., Han, Y.-K. et al. (2017). A comparative study on the solubility and stability of p-phenylenediamine-based organic redox couples for non-aqueous flow batteries. *Journal of Power Sources* 348: 264–269. https://doi.org/10.1016/j.jpowsour.2017.03.019.

48 Kim, H., Hwang, S., Kim, Y. et al. (2018). Bi-functional effects of lengthening aliphatic chain of phthalimide-based negative redox couple and its non-aqueous

flow battery performance at stack cell. *APL Materials* 6 (4): 047901. http://aip.scitation.org/doi/10.1063/1.5010210.

49 Pasala, V., Ramachandra, C., Sethuraman, S., and Ramanujam, K. (2018). A high voltage organic redox flow battery with redox couples O_2/tetrabutylammonium complex and tris(4-bromophenyl)amine as redox active species. *Journal of the Electrochemical Society* 165 (11): A2696–A2702. http://jes.ecsdl.org/lookup/doi/10.1149/2.0661811jes.

50 Kwon, G., Lee, K., Yoo, J. et al. (2021). Highly persistent triphenylamine-based catholyte for durable organic redox flow batteries. *Energy Storage Materials* 42 (April): 185–192. https://doi.org/10.1016/j.ensm.2021.07.006.

51 Romadina, E.I., Komarov, D.S., Stevenson, K.J., and Troshin, P.A. (2021). New phenazine based anolyte material for high voltage organic redox flow batteries. *Chemical Communications* 57 (24): 2986–2989.

52 Romadina, E.I., Volodin, I.A., Stevenson, K.J., and Troshin, P.A. (2021). New highly soluble triarylamine-based materials as promising catholytes for redox flow batteries. *Journal of Materials Chemistry A* 9 (13): 8303–8307.

53 Wei, X., Duan, W., Huang, J. et al. (2016). A high-current, stable nonaqueous organic redox flow battery. *ACS Energy Letters* 1 (4): 705–711. http://pubs.acs.org/doi/10.1021/acsenergylett.6b00255.

54 Xing, X., Huo, Y., Wang, X. et al. (2017). A benzophenone-based anolyte for high energy density all-organic redox flow battery. *International Journal of Hydrogen Energy* 42 (27): 17488–17494. http://dx.doi.org/10.1016/j.ijhydene.2017.03.034.

55 Xing, X., Liu, Q., Xu, W. et al. (2019). All-liquid electroactive materials for high energy density organic flow battery. *ACS Applied Energy Materials* 2 (4): 2364–2369.

56 Huo, Y., Xing, X., Zhang, C. et al. (2019). An all organic redox flow battery with high cell voltage. *RSC Advances* 9 (23): 13128–13132.

57 Xing, X., Liu, Q., Wang, B. et al. (2020). A low potential solvent-miscible 3-methylbenzophenone anolyte material for high voltage and energy density all-organic flow battery. *Journal of Power Sources* 445 (June 2019): 227330. https://doi.org/10.1016/j.jpowsour.2019.227330.

58 Duan, W., Huang, J., Kowalski, J.A. et al. (2017). "Wine-Dark Sea" in an organic flow battery: storing negative charge in 2,1,3-benzothiadiazole radicals leads to improved cyclability. *ACS Energy Letters* 2 (5): 1156–1161. http://pubs.acs.org/doi/abs/10.1021/acsenergylett.7b00261.

59 Huang, J., Duan, W., Zhang, J. et al. (2018). Substituted thiadiazoles as energy-rich anolytes for nonaqueous redox flow cells. *Journal of Materials Chemistry A* 6 (15): 6251–6254. http://pubs.rsc.org/en/Content/ArticleLanding/2018/TA/C8TA01059E.

60 Yuan, J., Zhang, C., Zhen, Y. et al. (2019). Enhancing the performance of an all-organic non-aqueous redox flow battery. *Journal of Power Sources* 443 (August): 227283. https://doi.org/10.1016/j.jpowsour.2019.227283.

61 Liu, B., Tang, C.W., Zhang, C. et al. (2021). Cost-effective, high-energy-density, nonaqueous nitrobenzene organic redox flow battery. *Chemistry of Materials* 33 (3): 978–986.

62 Wei, X., Xu, W., Huang, J. et al. (2015). Radical compatibility with nonaqueous electrolytes and its impact on an all-organic redox flow battery. *Angewandte Chemie, International Edition* 54 (30): 8684–8687. http://www.ncbi.nlm.nih.gov/pubmed/25891480.

63 Sevov, C.S., Brooner, R.E.M., Chénard, E. et al. (2015). Evolutionary design of low molecular weight organic anolyte materials for applications in nonaqueous redox flow batteries. *Journal of the American Chemical Society* 137 (45): 14465–14472. http://pubs.acs.org/doi/10.1021/jacs.5b09572.

64 Hendriks, K.H., Sevov, C.S., Cook, M.E., and Sanford, M.S. (2017). Multielectron cycling of a low-potential anolyte in alkali metal electrolytes for nonaqueous redox flow batteries. *ACS Energy Letters* 2 (10): 2430–2435. http://pubs.acs.org/doi/10.1021/acsenergylett.7b00559.

65 Sevov, C.S., Hickey, D.P., Cook, M.E. et al. (2017). Physical organic approach to persistent, cyclable, low-potential electrolytes for flow battery applications. *Journal of the American Chemical Society* 139 (8): 2924–2927. http://pubs.acs.org/doi/10.1021/jacs.7b00147.

66 Chai, J., Lashgari, A., Wang, X., and Jiang, J. "Jimmy"(2020). Extending the redox potentials of metal-free anolytes: towards high energy density redox flow batteries. *Journal of the Electrochemical Society* 167 (10): 100556.

67 Griffin, J.D., Pancoast, A.R., and Sigman, M.S. (2021). Interrogation of 2,2′-bipyrimidines as low-potential two-electron electrolytes. *Journal of the American Chemical Society* 143 (2): 992–1004.

68 Hu, B., Tang, Y., Luo, J. et al. (2018). Improved radical stability of viologen anolytes in aqueous organic redox flow batteries. *Chemical Communications* 54 (50): 6871–6874.

69 Chai, J., Lashgari, A., Cao, Z. et al. (2020). PEGylation-enabled extended cyclability of a non-aqueous redox flow battery. *ACS Applied Materials & Interfaces* 12 (13): 15262–15270.

70 Chai, J., Lashgari, A., Wang, X. et al. "Jimmy"(2020). All-PEGylated redox-active metal-free organic molecules in non-aqueous redox flow battery. *Journal of Materials Chemistry A* 8 (31): 15715–15724.

71 Attanayake, N.H., Liang, Z., Wang, Y. et al. (2021). Dual function organic active materials for nonaqueous redox flow batteries. *Materials Advances* 2 (4): 1390–1401.

72 Huang, J., Yang, Z., Vijayakumar, M. et al. (2018). A two-electron storage non-aqueous organic redox flow battery. *Advanced Sustainable Systems* 2 (3): 1700131. http://doi.wiley.com/10.1002/adsu.201700131.

73 Potash, R.A., McKone, J.R., Conte, S., and Abruña, H.D. (2016). On the benefits of a symmetric redox flow battery. *Journal of the Electrochemical Society* 163 (3): A338–A344. https://doi.org/10.1149/2.0971602jes.

74 Duan, W., Vemuri, R.S., Milshtein, J.D. et al. (2016). A symmetric organic-based nonaqueous redox flow battery and its state of charge diagnostics by FTIR. *Journal of Materials Chemistry A* 4 (15): 5448–5456. https://doi.org/10.1039/c6ta01177b.

75 Ma, T., Pan, Z., Miao, L. et al. (2018). Porphyrin-based symmetric redox-flow batteries towards cold-climate energy storage. *Angewandte Chemie, International Edition* 57 (12): 3158–3162.

76 Wylie, L., Blesch, T., Freeman, R. et al. (2020). Reversible reduction of the TEMPO radical: one step closer to an all-organic redox flow battery. *ACS Sustainable Chemistry & Engineering* 8 (49): 17988–17996.

77 Hicks, R.G. (2020). *Stable Radicals: Fundamentals and Applied Aspects of Odd-Electron Compounds*. Wiley.

78 Korshunov, A., Milner, M.J., Grünebaum, M. et al. (2020). An oxo-verdazyl radical for a symmetrical non-aqueous redox flow battery. *Journal of Materials Chemistry A* 8 (42): 22280–22291.

79 Armstrong, C.G., Hogue, R.W., and Toghill, K.E. (2019). Application of the dianion croconate violet for symmetric organic non-aqueous redox flow battery electrolytes. *Journal of Power Sources* 440: 227037. https://doi.org/10.1016/j.jpowsour.2019.227037.

80 Moutet, J., Veleta, J.M., and Gianetti, T.L. (2021). Symmetric, robust, and high-voltage organic redox flow battery model based on a helical carbenium ion electrolyte. *ACS Applied Energy Materials* 4 (1): 9–14.

81 Carney, T.J., Collins, S.J., Moore, J.S., and Brushett, F.R. (2017). Concentration-dependent dimerization of anthraquinone disulfonic acid and its impact on charge storage. *Chemistry of Materials* 29 (11): 4801–4810.

82 Su, L., Ferrandon, M., Kowalski, J.A. et al. (2014). Electrolyte development for non-aqueous redox flow batteries using a high-throughput screening platform. *Journal of the Electrochemical Society* 161 (12): A1905–A1914. http://jes.ecsdl.org/cgi/doi/10.1149/2.0811412jes.

83 Friedl, J., Lebedeva, M.A., Porfyrakis, K. et al. (2018). All-fullerene-based cells for nonaqueous redox flow batteries. *Journal of the American Chemical Society* 140 (1): 401–405.

84 Montoto, E.C., Nagarjuna, G., Moore, J.S., and Rodríguez-López, J. (2017). Redox active polymers for non-aqueous redox flow batteries: validation of the size-exclusion approach. *Journal of the Electrochemical Society* 164 (7): A1688–A1694. http://jes.ecsdl.org/lookup/doi/10.1149/2.1511707jes.

85 Lee, H., Yanilmaz, M., Toprakci, O. et al. (2014). A review of recent developments in membrane separators for rechargeable lithium-ion batteries. *Energy & Environmental Science* 7 (12): 3857–3886.

86 Baran, M.J., Braten, M.N., Montoto, E.C. et al. (2018). Designing redox-active oligomers for crossover-free, nonaqueous redox-flow batteries with high volumetric energy density. *Chemistry of Materials* 30: 3861–3866. https://pubs.acs.org/sharingguidelines.

87 Darling, R.M., Gallagher, K.G., Kowalski, J.A. et al. (2014). Pathways to low-cost electrochemical energy storage: a comparison of aqueous and nonaqueous flow batteries. *Energy & Environmental Science* 7 (11): 3459–3477. http://pubs.rsc.org/en/content/articlelanding/2014/ee/c4ee02158d%5Cnhttp://pubs.rsc.org/en/Content/ArticleLanding/2014/EE/c4ee02158d#!divAbstract%5Cnhttp://pubs.rsc.org/en/content/articlepdf/2014/ee/c4ee02158d.

88 Su, L., Ferrandon, M., Barton, J.L. et al. (2017). An investigation of 2,5-di-tert-butyl-1,4-bis(methoxyethoxy)benzene in ether-based electrolytes. *Electrochimica Acta* 246: 251–258. https://doi.org/10.1016/j.electacta.2017.05.167.

89 Kim, H., Ryu, J.H., and Oh, S.M. (2020). Effect of radical–solvent interaction on battery performance in benzophenone-based charge storage systems. *Journal of the Electrochemical Society* 167 (16): 160526.

90 Xing, X., Liu, Q., Li, J. et al. (2019). A nonaqueous all organic semisolid flow battery. *Chemical Communications* 55 (94): 14214–14217.

91 Doris, S.E., Ward, A.L., Baskin, A. et al. (2017). Macromolecular design strategies for preventing active-material crossover in non-aqueous all-organic redox-flow batteries. *Angewandte Chemie, International Edition* 56 (6): 1595–1599. http://doi.wiley.com/10.1002/anie.201610582.

92 Shkrob, I.A., Li, T., Sarnello, E. et al. (2020). Self-assembled solute networks in crowded electrolyte solutions and nanoconfinement of charged redoxmer molecules. *The Journal of Physical Chemistry. B* 124 (45): 10226–10236.

93 Milshtein, J.D., Tenny, K.M., Barton, J.L. et al. (2017). Quantifying mass transfer rates in redox flow batteries. *Journal of the Electrochemical Society* 164 (11): E3265–E3275. http://jes.ecsdl.org/lookup/doi/10.1149/2.0201711jes.

94 Armstrong, C.G. and Toghill, K.E. (2017). Cobalt(II) complexes with azole-pyridine type ligands for non-aqueous redox-flow batteries: tunable electrochemistry via structural modification. *Journal of Power Sources* 349: 121–129. http://dx.doi.org/10.1016/j.jpowsour.2017.03.034.

95 Armstrong, C. G. (2020). Novel non-aqueous symmetric redox materials for redox flow battery energy storage. Lancaster University. http://www.research.lancs.ac.uk/portal/en/publications/novel-nonaqueous-symmetric-redox-materials-for-redox-flow-battery-energy-storage(1bf728ac-2cb9-44d9-b897-d6d220203c1f).html.

44

Polymeric Flow Batteries

Oliver Nolte[1,2,3], Martin D. Hager[1,2,3], and Ulrich S. Schubert[1,2,3]

[1]*Friedrich Schiller University Jena, Laboratory of Organic and Macromolecular Chemistry (IOMC), Humboldtstr. 10, 07743 Jena, Germany*
[2]*Friedrich Schiller University Jena, Center for Energy and Environmental Chemistry Jena (CEEC Jena), Philosophenweg 7a, 07743 Jena, Germany*
[3]*Friedrich Schiller University Jena, Jena Center for Soft Matter (JCSM), Philosophenweg 7, 07743 Jena, Germany*

44.1 Introduction

Of the several FB systems available, the vanadium flow battery (VFB) can be considered as the most studied system, which is also already commercially available [1]. This cell type comprises vanadium species as active materials, which are utilized in the electrolyte for both negolyte and posolyte (V(II)/V(III) and V(IV) and V(V), respectively; see Figure 44.1) [2, 3]. To keep the electrolytes separated, proton-conducting membranes are utilized (see Figure 44.2).

Besides metal-based FBs, organic flow batteries (OFBs) have been investigated more and more over the last years [4–7]. Organic materials are considered as promising alternatives, substituting critical metals and their environmentally and socially questionable footprints by organic compounds. These redox-active materials are based on organic elements, i.e. C, H, N, O, S, etc. Even though the main source of these compounds is mineral oil, recent research indicates the use of renewable resources in the future [8, 9]. Nevertheless, extraction, synthesis, and processing feature a much lower energy demand when compared to mineral mining. The structural diversity of organic molecules is very large; consequently, their resulting properties can be tuned by adjusting their molecular structure [10]. Hence, it is not surprising that a lot of organic molecules have been tested in FB electrolytes. Similar to metal-based FBs, organic FBs feature an ion-selective membrane (e.g. anion-selective membrane) to separate both electrolytes (see Figure 44.2) [11]. However, when significantly increasing the size of the redox-active molecules, both half-cells can be separated by a simple size-exclusion membrane. This increase in size can most easily be achieved by attaching the redox-active moieties to a polymer chain [12], e.g. a polymer-based on TEMPO as well as a polyviologen (see Figure 44.2) [13], leading to a polymeric flow battery (PFB).

Flow Batteries: From Fundamentals to Applications, First Edition.
Edited by Christina Roth, Jens Noack, and Maria Skyllas-Kazacos.
© 2023 WILEY-VCH GmbH. Published 2023 by WILEY-VCH GmbH.

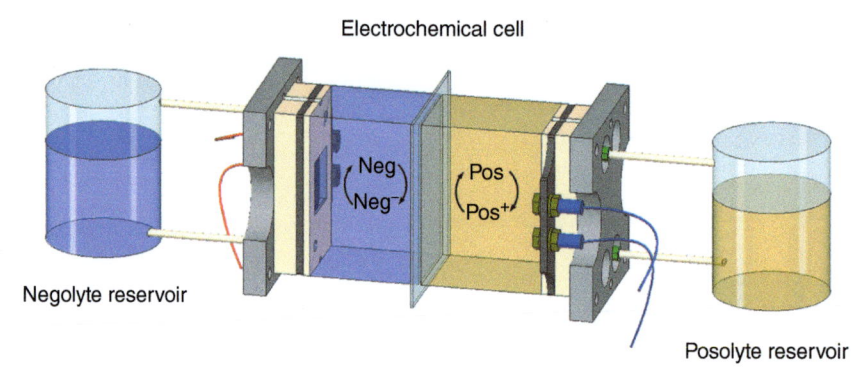

Figure 44.1 Schematic representation of a flow battery (FB). Neg = Negolyte; Pos = Posolyte.

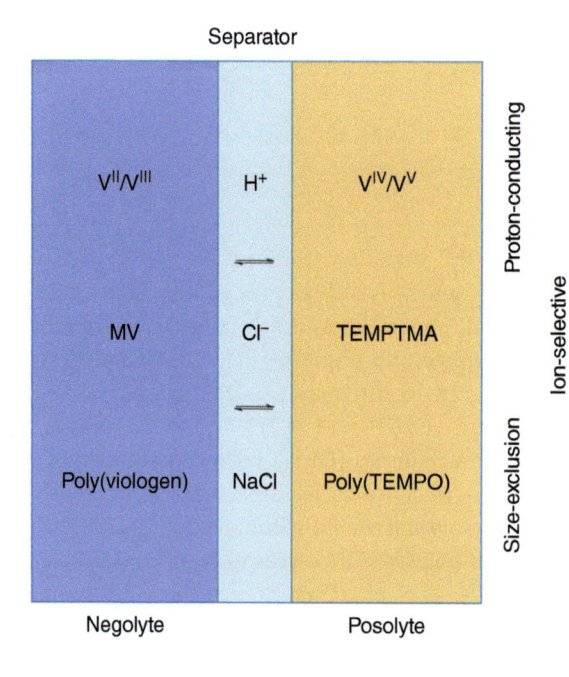

Figure 44.2 Comparison of different electrolytes and their corresponding membranes.

44.2 Basic Organic Redox Moieties

As detailed above, PFBs feature redox polymers, which consist of multiple organic redox moieties. Consequently, this paragraph will shortly describe the most important redox moieties utilized nowadays. Over the last two decades, a number of suitable organic redox moieties have been proposed, of which the most prominent are represented by quinones, viologens, aminoxyl radicals (e.g. TEMPO), and ferrocenes, among others [7].

Quinones are amongst the most intensively researched active materials for FBs, due to their synthetic availability, stability, and tunable redox potentials [14].

Nevertheless, there are only very few examples of polymeric quinones used in FBs. This fact might be attributed to the nature of quinone-based monomers, which are known to be badly polymerizable (quinones represent potent inhibitors for radical polymerizations) and are also known to feature a very low solubility. Thus, polymerization strategies that circumvent these pitfalls, such as polycondensation reactions, are often required [15]. To introduce solubility in aqueous media, the introduction of carboxylates and phenolic hydroxyl groups or sulfonic acid groups is often used (depending on the pH regime the electrolyte is used in) [16, 17]; for nonaqueous media, ammonium groups, and oligomeric ethylene oxides have been presented, depending on the polarity of the proposed solvent [18–20]. Furthermore, in aqueous media, proton-coupled redox equilibria are typically observed, placing higher demands on the structural stability of the polymer itself. Interestingly, however, a quinonoid polymer was one of the first organic materials to be ever proposed for a battery application [21].

An often-used negolyte material for organic FBs is represented by the viologen system (see Figure 44.3). Consequently, also polymers featuring viologen units in the polymer side chains have been prepared for the usage in FB electrolytes [13]. Due to their moderate redox potentials of approximately -0.45 V vs. NHE (first reduction step) [22], viologens still enable the utilization of aqueous electrolytes, which is beneficial for a potential large-scale application. Furthermore, the single-electron reduction leads to a stable radical species, which is not affecting the pH value of the solutions in aqueous media.

Further negolyte materials proposed for the use in OFBs include flavine and alloxazine for aqueous battery systems [9, 23], as well as systems based on camphorquinone and fluorenone for nonaqueous batteries [24, 25]. Quinoxalines have been investigated for both aqueous and nonaqueous systems [26, 27].

Ferrocenes are typically used as posolytes in the form of small molecules (see Figure 44.3) and have been studied intensively [28]. They have redox potentials

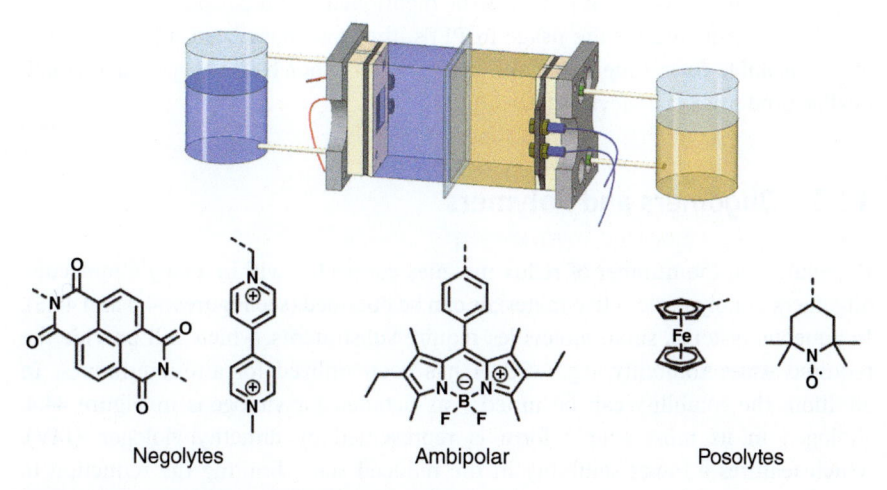

Figure 44.3 Schematic representation of redox-active moieties (row below) – negolytes (left), ambipolar (center), and posolytes (right).

of approximately 0.4 V against NHE [29]. Interestingly, ferrocenes are characterized by their remarkable chemical stability and are, thus, able to be used in aqueous media. Ferrocenes have also been incorporated into polymers for potential FB studies [30, 31].

Aminoxyl radicals are probably the most widely studied organic redox moiety for use as a posolyte material in OFB applications. The prominently used 2,2,6,6-tetramethylpiperidin-1-oxyl (TEMPO) derivatives have been frequently applied for FB studies due to their synthetic availability, low cost, good solubility, and stability [7, 28]. Furthermore, their redox potential is considerably positive (approximately 0.85 V vs. NHE); although it still lies within the aqueous stability window, allowing for a comparably high cell voltage [32]. Consequently, TEMPO-based polymers were tested in PFBs (see Figure 44.3) [13, 33–35]. Apart from the TEMPO-based materials, the structurally related PROXYL radical has been proposed for FBs as well [36, 37].

Other posolyte materials that have been presented in the literature thus far include methylene blue and various coordination complexes (e.g. ferrocyanide) for aqueous media as well as, e.g. diethoxybenzene-based molecules, phenothiazines, and phenazines for nonaqueous media [25, 26, 38, 39].

Additionally, a few symmetric organic FB concepts have been presented as well. These are mainly represented by substances that are able to be converted into at least three different redox states [18, 19, 39–45]. A further interesting approach is represented by the linkage of molecules of both negolyte and posolyte through a covalent bond, as presented by Janoschka and Winsberg [46, 47]. A symmetric polymer electrolyte based on boron–dipyrromethene (BODIPY)-based polymers has been utilized for a symmetric PFB with a potential of >2 V (see Figure 44.3) [48].

Basically, all of the aforementioned small molecular redox moieties represent interesting candidates for the incorporation into polymeric materials. One prerequisite is that either a direct polymerization or the functionalization of a respective polymer chain is possible [49]. It has to be mentioned that redox polymers have not only been considered for the usage in PFBs, they also have been utilized – often in an insoluble form (one major difference) – for other battery types, like small, flexible (and printable) devices [50–52].

44.3 Oligomers and Polymers

Depending on the number of redox moieties connected within a single molecule, oligomers or polymeric active materials can be obtained (see Figures 44.4 and 44.5). In aqueous systems, small molecules require substituents, which will provide the required water solubility, e.g., $-SO_3H$ has been utilized for anthraquinones. In addition, the solubility can be tuned – as detailed for viologens in Figure 44.4. Viologen in its most simple form is represented by dimethyl-viologen (MV), which features a lower solubility in the reduced state, limiting the reduction to the first single-electron step. The utilization of additional trimethyl-ammonium groups resulted in a viologen with enhanced solubility. Consequently, the second

Figure 44.4 Schematic representation of redox-active materials used in organic FBs. (a) Small molecules. (b) Well-defined oligomer based on viologen.

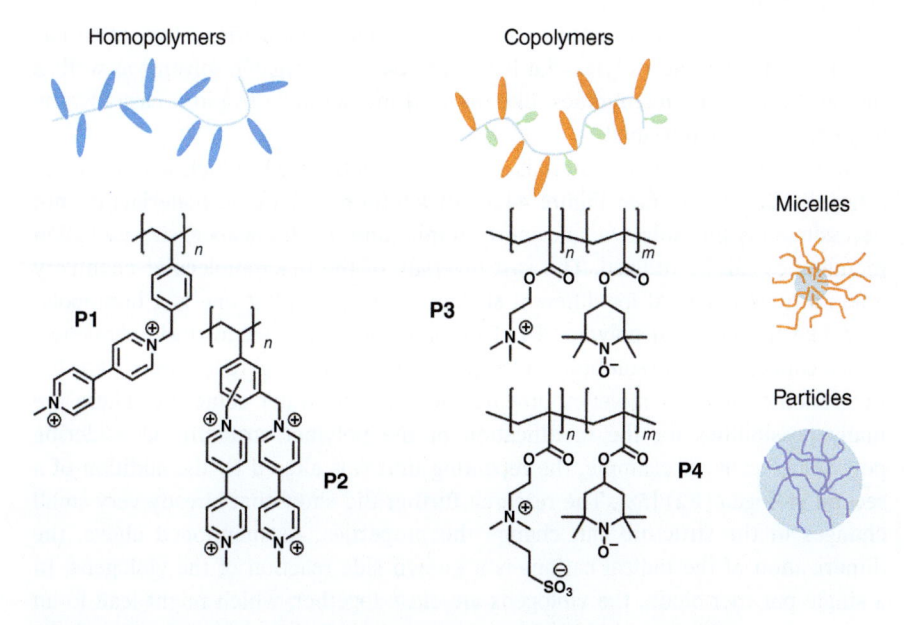

Figure 44.5 Schematic representation of polymeric materials used in FBs.

reduction became accessible. Furthermore, the addition of the pendant ammonium groups has been associated with a higher stability due to limiting the undesired dimerization of the viologen radical cations [53]. As seen in this example, the substituents will influence the overall charge of the applied redox moieties (for all redox states). The choice of the respective membrane, which has to be compatible with the utilized molecules, will be crucial. Both molecules within the posolyte, as well as the negolyte, should feature the same ionic charge (for all redox states).

For instance, a negolyte based on positively charged viologens will require an anion-exchange membrane. The suitable posolyte would be, e.g. TEMPTMA, which is also positively charged in all redox states [53]. A *bis*-sulfonate-substituted viologen, on the other hand, is neutral or even negatively charged in its reduced form and therefore requires a cation-exchange membrane [54].

To study the influence of the size of the molecules, well-defined oligomers have been prepared by Doris et al. (see Figure 44.4 for a viologen trimer) [55]. The series of monomer, dimer, and trimer was studied utilizing different membranes in nonaqueous electrolytes. For microporous membranes, a similar diffusion was observed. Only when the size of the pores in the membrane was limited by the usage of polymer membranes with intrinsic microporosity (PIM), an effect of the size could be observed, i.e. the larger trimer could not pass the membrane anymore. This effect was also observed for other oligomeric species, like cyclo-propenium-based tetramers [56]. Moreover, a three-arm oligo(ethylene glycol) featuring redox moieties (e.g. TEMPO) as end groups was utilized [57]. Interestingly, these oligomers are still liquid, allowing for the utilization of a redox-active liquid without any solvent and, consequently, providing a high theoretical volumetric energy density.

The aforementioned examples of (well-defined) oligomeric active materials commonly feature electrolytes that have been based on organic solvents as well as the need for special membranes, like the PIM membranes. To put it casually, the oligomers are still too small.

Thus, polymeric active materials have been synthesized, which feature a significantly larger size (see Figure 44.5). In addition, polymeric materials do not necessitate organic solvents, and more simple (and much cheaper) size-exclusion membranes can be utilized. The vast diversity of the macromolecular chemistry offers a huge potential for different structures. The simplest one is a homopolymer, like the viologen polymer **P1**. Homopolymers have an intrinsic challenge, which concerns a good solubility of the polymer chains (in all redox states). For **P1**, the cationic viologen moieties provide the required water solubility. There are many possibilities for the modification of the polymer structure. Considering polyviologens as an example, the repeating unit was altered by the addition of a second viologen (**P2**) [58]. The research further did show that already very small changes in the structure can change the properties. As mentioned above, the dimerization of the radical cations is a known side reaction of the viologens. In a single polymer chain, the viologens are close together, which might lead to an increased dimerization. Within this context, two different monomers have been compared – featuring either an *ortho* substitution or *meta* substitution. Whereas the latter materials result in stable polymers, the *ortho* substitution is promoting the undesired dimerization, resulting in less stable materials. Another important aspect of polymeric active materials is their molar mass (that will strongly influence the hydrodynamic radius of the macromolecule) as well as their molar mass distribution. The influence of the molar mass was demonstrated with polyviologens (**P1**) featuring different lengths [59]. As expected, a higher selectivity could be observed for the polymers with the highest molar masses.

Copolymers – polymers based on two (or more) monomers – are the material of choice if the solubility of the homopolymer is not sufficiently high. As a good example, the TEMPO-based homopolymer poly(2,2,6,6-tetramethylpiperidin-1-oxyl methacrylate) can be seen, which is not water-soluble. There are different comonomers available that can provide water solubility. These monomers can be selected from uncharged nonionic, negatively charged anionic, positively charged cationic, and double-charged zwitterionic comonomers, respectively. The choice will not be random, hence, the overall charge of the macromolecule (in both redox states) will be an important parameter – as already described above for their low-molecular pendants. To stay with the example of the TEMPO polymer, the active moiety is oxidized during charging. A cationic comonomer – like the trimethylammonium ethyl methacrylate for copolymer **P3** [13] – would be the monomer of choice to provide solubility. If a negatively charged comonomer would be selected, the unoxidized (and therefore negatively charged) polymer would feature a good solubility. However, during electrochemical charging, a formally neutral – less soluble – copolymer would be obtained. Comonomers can also be introduced to typical homopolymers to enhance specific properties of the resulting copolymer. An example can be seen in polyviologen **P1**, for which ((vinylbenzyl)trimethylammonium chloride) was utilized as a cationic comonomer resulting in a higher solubility of the charged state, which features less cationic charges on the redox moiety [13]. The comonomer can be used to optimize the properties of the resulting electrolytes. Within this context, the [((2-(methacryloxy)ethyl)dimethyl-(3-sulfopropyl)]ammonium hydroxide) was utilized as a zwitterionic comonomer for TEMPO polymers replacing the standard cationic comonomer resulting in electrolytes with a much lower viscosity [35].

Polymeric assemblies, based on several polymer chains, have also been applied in FB electrolytes (see Figure 44.5). Covalently linking two polymer blocks with strongly differing solution behavior enables the formation of different supramolecular structures. As an example, the block copolymer (PTMA-*b*-PS) was utilized in organic solvents. It was observed that spherical micelles (featuring a size of approximately 40 nm) were obtained. These micelles consist of a PS core and a surrounding redox-active PTMA corona [33]. The micellar electrolyte has been studied in combination with a zinc anode in a hybrid-flow battery.

The aforementioned examples are all based on soluble active materials/active polymers and the synthetic possibilities to improve their solubility have been detailed. In contrast, polymer dispersions will rely on completely unsoluble polymeric materials. The small polymeric particles are dispersed in aqueous solutions to ensure pumpability of the mixture. Polythiophene dispersions have been amongst the first examples described using this approach [60]. The polymer microparticles were utilized in propylene carbonate for both negolyte as well as posolyte, resulting in a high cell voltage of 2.5 V. Furthermore, a slurry-based FB was described, which is based on polyhydroquinone for the posolyte side and an NDI-based polymer for the negolyte side (see Figure 44.6a) [61]. Nevertheless, the low solubility of the negolyte material was limiting the performance of this system. This challenge was overcome by the usage of redox-active colloids, which were based on crosslinked polyviologens [62].

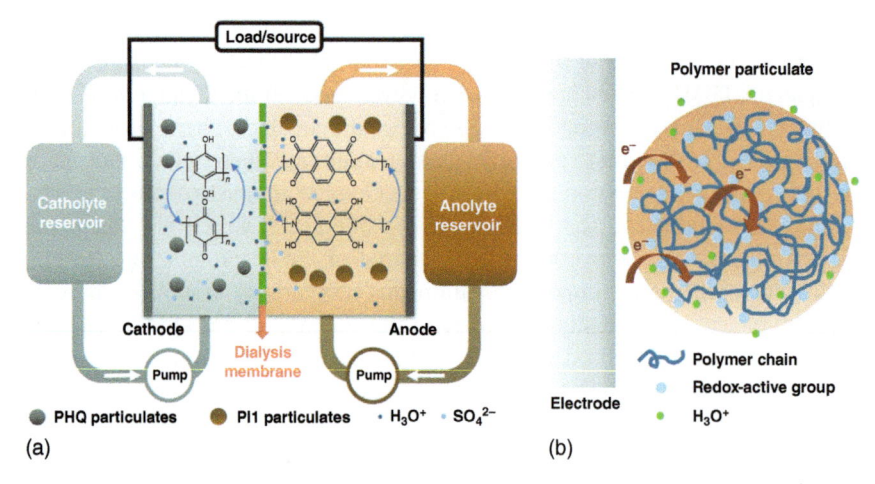

Figure 44.6 (a) Schematic representation of a slurry-based FB; (b) schematic representation of the electron transport within the particles. Source: Yan et al. [61]/ Springer Nature/Public Domain CC BY 4.0.

44.4 Examples of Polymeric FBs

Several of the active materials described above have been utilized for the fabrication of full cells based on polymeric electrolytes. The general aspects will be discussed using examples of different polymeric FBs (the interested reader is referred to a detailed overview in a recent review by Lai et al. [12]). Table 44.1 is summarizing the

Table 44.1 Overview on selected FBs.

Type	Negolyte	Posolyte	Electrolyte	Single cell voltage (V)	Current density (mA cm^{-2})	Capacity retention Number of cylces	References
VFB	V^{2+}/V^{3+}	VO^{2+}/VO_2^+	Sulfuric acid	1.26	80–100	80% after 20 000	
OFB	MV	TEMPTMA	Sodium chloride in water	1.3	100	100% 100	[13]
Soluble polymer	Viologen copolymer	TEMPO copolymer	Sodium chloride in water	1.1	40	86% 100	[13]
Soluble polymer	Phthalimide copolymer	TEMPO copolymer	DMF	2.15	0.5	60% 10 (static cell)	[63]
Slurry-based	NDI copolymer	Polyhydro-quinone	Sulfuric acid	0.53	20	74% 5000	[61]

important parameters of PFBs and their values are compared to other established FB systems.

Exchanging low-molecular mass organic redox-active materials with their soluble polymeric counterparts brings some interesting advantages. As mentioned before, the ion-selective membrane usually used to separate both half-cells is exchanged with a non-selective size-exclusion membrane. This has the advantage that the costs of size-exclusion membranes are much lower (approximately one order of magnitude) [12]. For instance, a Nafion 115 membrane used in a VFB has a share of roughly 60% of the area-normalized stack costs, being a significant cost-driving factor [64]. Moreover, to ensure good selectivity of the ion-selective membranes, an increase in thickness is usually preferred, which in turn increases the area resistance of the separator, resulting in energy losses. Due to the unique polymeric approach, however, the ability of the membrane to separate the electrolytes of the two half-cells can easily be tuned by the molar mass of the polymer. Furthermore, as the size-exclusion membrane is not selective for a single ionic species, all charge carriers of the supporting electrolyte are able to contribute to both conductivity and equalization of charges across the separator. Thus, the molecular design of the polymers is theoretically not bound to feature equally-charged polymers (i.e. cationic or anionic net charges of the polymers in both half-cells), overcoming a key limitation of their low-molecular counterparts. This, however, has so far not been shown experimentally.

As a typical PFB example, Janoschka et al. utilized a copolymer featuring TEMPO (**P3**) and a copolymer containing viologen in a PFB (see Figure 44.7) [13]. Both polymers were utilized in aqueous electrolytes with sodium chloride as supporting electrolyte resulting in capacities of around 10 Ah L^{-1}. The half cells were separated by a size-exclusion membrane, which had a molecular weight cut-off (MWCO) of around 6000 g mol^{-1}. The cycling tests of a pumped system revealed a capacity retention of 86% over 100 charging/discharging cycles (see Figure 44.7). Polymeric solutions feature higher viscosities compared to solutions of small molecules. Therefore, the concentration of the polymeric materials in the electrolyte is often more limited due to the viscosity rather than solubility. The higher viscosity is also the reason that lower current densities (in that case 40 mA cm^{-2}) have been obtained (small molecule-based FBs are easily reaching 100 mA cm^{-2} [11], see Table 44.1).

As the aforementioned example further highlights, water is the usual choice for the solvents used in PFB electrolytes, as it presents an economically viable, environmentally benign as well as nonflammable solvent. Furthermore, its ability to dissolve strongly polar solutes is paramount for the production of highly-conductive electrolytes, effectively lowering energy losses due to resistance and providing higher current densities.

As a relatively novel concept, the use of redox-active particles in dispersions has been focus of recent studies. NDI-based polymeric dispersions have been utilized and showed promising results [61]. Acceptable current densities (20 mA cm^{-2}) were obtained and the energy density was close to 2 Wh L^{-1}. Even though the cell voltage was comparably low (approximately 0.5 V), this is not a principle

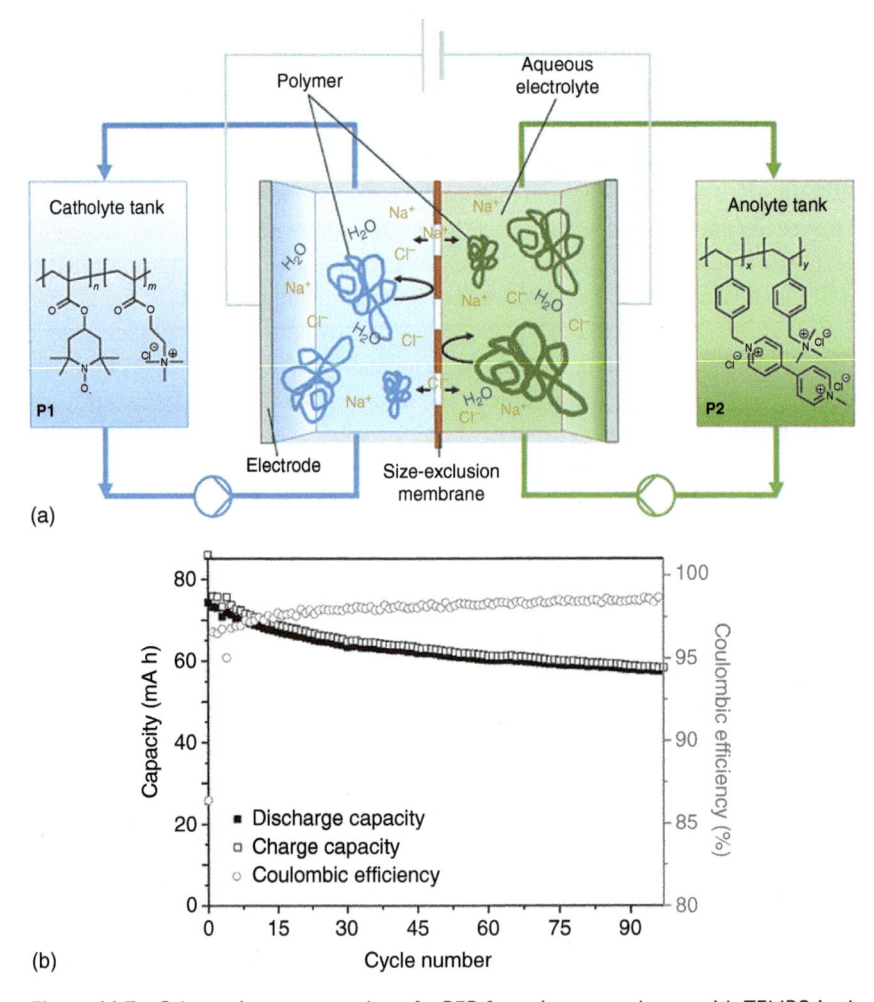

Figure 44.7 Schematic representation of a PFB featuring a copolymer with TEMPO in the posolyte and copolymer with viologen in the negolyte (a) and cycling performance over 100 cycles (b). Source: Reproduced with permission from Janoschka et al. [13].

limitation due to the dispersion approach, as it depends on the utilized active materials.

A similar approach that may feature polymers in FBs is conceptualized with the application of so-called redox targeting. The concept relies on an increase of the capacity of the whole battery system by the application of insoluble redox-active substances. Their redox state can be varied by the redox potential of a redox mediator that is electrochemically charged and discharged in the battery. Here, redox-active polymers may represent a smart and tunable possibility to produce insoluble materials. As no particulate matter has to be pumped through the electrochemical cell, a potential clogging of the electrodes is avoided [65]. In a recent example, a system comprising an insoluble redox-active polymer as well as a low-molecular redox-active material based on the same redox moiety (TEMPO).

44.5 Countering the Challenges

Even though the utilization of a polymeric design has several advantages, the approach still remains at a conceptual stage due to several drawbacks. Firstly, the incorporation of the redox-active moiety into a polymeric framework is in some cases not straightforward. As the nature of the propagating species in the frequently-used free radical polymerizations does often not tolerate redox-active moieties, these have to be incorporated/activated by a subsequent synthetic step. This, however, increases the complexity of the approach and limits the specific capacity of the materials to sub-theoretical levels due to nonquantitative yields of this step. Nevertheless, with a careful choice of the reaction parameters, high yields are achievable, as shown for instance by comparing the degree of oxidation of different TEMPO-containing polymers [35].

Another drawback is represented by the high molar masses of the macromolecules. As a result of the molecular friction, an increased solution viscosity is observable, which eventually leads to a high energy demand for the pumping of the electrolyte. This is particularly important at high molar masses and becomes paramount at high concentrations (= concentration being relevant for a practical application). Regardless, the molar mass of the polymers still has to be sufficiently high to allow a good retention by the size-exclusion membrane. Thus, efforts to limit the effective molar mass of the redox polymers and narrow their molar mass distribution have to be made, e.g. by the utilization of a chain-transfer agent, as shown by Janoschka et al. [13]. Additionally, the increased viscosity of the polymer solutions leads to a decrease in charge-carrier diffusion, thus supporting electrolytes are frequently used. The combination of both effects results in a limitation of the maximum current densities of PFBs, which are much lower than for organic FBs or VFBs (see Table 44.1).

One possible strategy for a partial mitigation of the performance losses caused by the higher viscosities is to produce polymers that are able to form secondary (supramolecular) structures, e.g. micelles or nanoparticles. The more spherical a particle is in shape, the less viscous a dispersion of a given concentration will be. A similar effect might be obtained by the introduction of crosslinks between polymer chains. Although these approaches have been partly utilized in dispersion-based FBs, a study comparing the viscosity of dispersions containing redox polymer-based particles of different shapes and sizes has not been conducted so far to the best of our knowledge.

Another approach was presented by the exchange of the purely cationic comonomer in a TEMPO-bearing polymer by a zwitterionic analog [35]. The study revealed that an effective decrease in viscosity can be achieved. Presumably, the polymer coil formed in solution is able to collapse to a denser structure due to the absence of the electrostatic repulsion in the formally neutral polymer (in its electrochemically uncharged state).

Hyperbranched polymers represent another promising strategy since these structures feature lower solution viscosities compared to their linear counterparts.

Nevertheless, for any electrolyte involving micro-/nanoparticles, these have to be circulated without causing a precipitation or a clogging of the porous electrodes. This is especially important for the frequently used porous electrodes, such as carbon felts, which provide a significantly higher reaction surface. To achieve a complete charging/discharging of all redox moieties within the particle, the particles need to be conductive, e.g. allowing for a hopping of electrons (see Figure 44.6b).

Lastly, in light of a stronger focus on the stability of proposed organic electrolytes, the integrity of the polymer backbone itself as well as any linker attaching the redox moiety to the polymer backbone has to be ensured. This may be particularly important in aqueous electrolytes due to the high nucleophilicity of water itself. Therefore, design strategies avoiding labile linkers and backbones have to be found. A strategy that was recently shown to be effective in preventing possible side reactions in small molecules is the reduction of the number of polar covalent bonds within the molecule. This can be seen by the comparison of an anthraquinone core linked to its pendant solubility-promoting group either *via* carbon–carbon bonds or an ether bond [66, 67]. Here, the nonpolar linker was able to significantly reduce the decomposition rate of the material.

44.6 Conclusion and Outlook

Despite being a rather new concept in FBs, PFBs have gained much interest over the recent years. Comparable to the very large variety of polymeric structures thinkable and accessible, there are also different concepts for the usage of polymeric active materials (from solution to dispersions). Recent years have also shown a tremendous increase in the research for organic FBs (based on small molecules). Up to now, only a smaller number of these compounds found their way into polymeric materials, while several known redox moieties have hitherto not been used within polymeric materials. The examples reported so far for different polymer-based electrolytes have clearly demonstrated that the polymer structure has a significant influence on the resulting electrochemical properties, showing that the polymer backbone represents more than an "innocent" carrier of the redox moieties. Thus, several effects have been identified, but a deeper knowledge of the influence of the polymer structure in solution (also in different charged states) on the properties is still almost completely missing. This challenge can only be solved through the use of different characterization techniques, which can access different length scales – from the molecular level to the device level – ideally also *in operando*.

Some challenges still remain, like the low obtainable volumetric capacities, which are mainly limited by solubility and viscosity of the respective electrolytes. 30–50 Ah L^{-1} can be considered as a benchmark for small molecules (for selected compounds even 100 Ah L^{-1}). The examples described have shown that PFBs often have low capacities (<10 Ah L^{-1} up to max. 20 Ah L^{-1}), which are still below their low-molecular counterparts. However, tuning the polymer architecture as well as the introduction of dispersions seem to be encouraging approaches, which will lead to higher capacities in the near future. As it is the case for the majority of the so far presented non-polymeric FB concepts, aqueous electrolytes seem to be the preferred choice for PFBs as well.

One additional aspect of polymeric materials, which has, thus far not been touched in the early investigations of PFBs, is the possibility to incorporate additional properties into the materials as it has been explored for many other applications of polymeric materials. Within this context, only a few properties should be mentioned, like switchability by external triggers or the stabilization of the redox moieties, which may enable further tuneability of polymeric redox-active materials in FB applications.

References

1 Shigematsu, T. (2019). The development and demonstration status of practical flow battery systems. *Current Opinion in Electrochemistry* 18: 55–60.
2 Lourenssen, K., Williams, J., Ahmadpour, F. et al. (2019). Vanadium redox flow batteries: a comprehensive review. *Journal of Energy Storage* 25: 100844.
3 Skyllas-Kazacos, M., Kazacos, G., Poon, G., and Verseema, H. (2010). Recent advances with UNSW vanadium-based redox flow batteries. *International Journal of Energy Research* 34 (2): 182–189.
4 Arenas, L.F., Ponce de León, C., and Walsh, F.C. (2019). Redox flow batteries for energy storage: their promise, achievements and challenges. *Current Opinion in Electrochemistry* 16: 117–126.
5 Wei, X., Pan, W., Duan, W. et al. (2017). Materials and systems for organic redox flow batteries: status and challenges. *ACS Energy Letters* 2 (9): 2187–2204.
6 Singh, V., Kim, S., Kang, J., and Byon, H.R. (2019). Aqueous organic redox flow batteries. *Nano Research* 12 (9): 1988–2001.
7 Winsberg, J., Hagemann, T., Janoschka, T. et al. (2017). Redox-flow batteries: from metals to organic redox-active materials. *Angewandte Chemie, International Edition* 56 (3): 686–711.
8 Schlemmer, W., Nothdurft, P., Petzold, A. et al. (2020). 2-Methoxyhydroquinone from vanillin for aqueous redox-flow batteries. *Angewandte Chemie* 132 (51): 23143–23146.
9 Orita, A., Verde, M.G., Sakai, M., and Meng, Y.S. (2016). A biomimetic redox flow battery based on flavin mononucleotide. *Nature Communications* 7: 13230.
10 Lee, S., Hong, J., and Kang, K. (2020). Redox-active organic compounds for future sustainable energy storage system. *Advanced Energy Materials* 10 (30): 2001445.
11 Janoschka, T., Martin, N., Hager, M.D., and Schubert, U.S. (2016). An aqueous redox-flow battery with high capacity and power: the TEMPTMA/MV system. *Angewandte Chemie, International Edition* 55 (46): 14427–14430.
12 Lai, Y.Y., Li, X., and Zhu, Y. (2020). Polymeric active materials for redox flow battery application. *ACS Applied Polymer Materials* 2 (2): 113–128.

13 Janoschka, T., Martin, N., Martin, U. et al. (2015). An aqueous, polymer-based redox-flow battery using non-corrosive, safe, and low-cost materials. *Nature* 527 (7576): 78–81.

14 Han, C., Li, H., Shi, R. et al. (2019). Organic quinones towards advanced electrochemical energy storage: recent advances and challenges. *Journal of Materials Chemistry A* 7 (41): 23378–23415.

15 Boone, H.W. and Hall, H.K. (1996). Novel polyaromatic quinone imines. 2. Synthesis of model compounds and stereoregular poly(quinone imines) from disubstituted anthraquinones. *Macromolecules* 29 (18): 5835–5842.

16 Gerken, J.B., Anson, C.W., Preger, Y. et al. (2020). Comparison of quinone-based catholytes for aqueous redox flow batteries and demonstration of long-term stability with tetrasubstituted quinones. *Advanced Energy Materials* 10 (20): 2000340.

17 Wedege, K., Drazevic, E., Konya, D., and Bentien, A. (2016). Organic redox species in aqueous flow batteries: redox potentials, chemical stability and solubility. *Scientific Reports* 6: 39101.

18 Geysens, P., Li, Y., Vankelecom, I. et al. (2020). Highly soluble 1,4-diaminoanthraquinone derivative for nonaqueous symmetric redox flow batteries. *ACS Sustainable Chemistry & Engineering* 8 (9): 3832–3843.

19 Pahlevaninezhad, M., Leung, P., Velasco, P.Q. et al. (2021). A nonaqueous organic redox flow battery using multi-electron quinone molecules. *Journal of Power Sources* 500: 229942.

20 Wang, W., Xu, W., Cosimbescu, L. et al. (2012). Anthraquinone with tailored structure for a nonaqueous metal-organic redox flow battery. *Chemical Communications (Cambridge, England)* 48 (53): 6669–6671.

21 Harding, M.S. (1958). Inventor Battery. US Patent US2831045A.

22 Bird, C.L. and Kuhn, A.T. (1981). Electrochemistry of the viologens. *Chemical Society Reviews* 10 (1): 49–82.

23 Lin, K., Gómez-Bombarelli, R., Beh, E.S. et al. (2016). A redox-flow battery with an alloxazine-based organic electrolyte. *Nature Energy* 1 (9): 16102.

24 Park, S.-K., Shim, J., Yang, J. et al. (2015). Electrochemical properties of a non-aqueous redox battery with all-organic redox couples. *Electrochemistry Communications* 59: 68–71.

25 Kwon, G., Lee, S., Hwang, J. et al. (2018). Multi-redox molecule for high-energy redox flow batteries. *Joule* 2 (9): 1771–1782.

26 Brushett, F.R., Vaughey, J.T., and Jansen, A.N. (2012). An all-organic non-aqueous lithium-ion redox flow battery. *Advanced Energy Materials* 2 (11): 1390–1396.

27 Milshtein, J.D., Su, L., Liou, C. et al. (2015). Voltammetry study of quinoxaline in aqueous electrolytes. *Electrochimica Acta* 180: 695–704.

28 Zhong, F., Yang, M., Ding, M., and Jia, C. (2020). Organic electroactive molecule-based electrolytes for redox flow batteries: status and challenges of molecular design. *Frontiers in Chemistry* 8: 451.

29 Gagne, R.R., Koval, C.A., and Lisensky, G.C. (2002). Ferrocene as an internal standard for electrochemical measurements. *Inorganic Chemistry* 19 (9): 2854–2855.

30 Burgess, M., Hernández-Burgos, K., Simpson, B.H. et al. (2015). Scanning electrochemical microscopy and hydrodynamic voltammetry investigation of charge transfer mechanisms on redox active polymers. *Journal of the Electrochemical Society* 163 (4): H3006–H3013.

31 Borchers, P.S., Strumpf, M., Friebe, C. et al. (2020). Aqueous redox flow battery suitable for high temperature applications based on a tailor-made ferrocene copolymer. *Advanced Energy Materials* 2001825.

32 Zhou, W., Liu, W., Qin, M. et al. (2020). Fundamental properties of TEMPO-based catholytes for aqueous redox flow batteries: effects of substituent groups and electrolytes on electrochemical properties, solubilities and battery performance. *RSC Advances* 10 (37): 21839–21844.

33 Winsberg, J., Muench, S., Hagemann, T. et al. (2016). Polymer/zinc hybrid-flow battery using block copolymer micelles featuring a TEMPO corona as catholyte. *Polymer Chemistry* 7 (9): 1711–1718.

34 Hagemann, T., Winsberg, J., Grube, M. et al. (2018). An aqueous all-organic redox-flow battery employing a (2,2,6,6-tetramethylpiperidin-1-yl)oxyl-containing polymer as catholyte and dimethyl viologen dichloride as anolyte. *Journal of Power Sources* 378: 546–554.

35 Hagemann, T., Strumpf, M., Schröter, E. et al. (2019). (2,2,6,6-tetramethylpiperidin-1-yl)oxyl-containing zwitterionic polymer as catholyte species for high-capacity aqueous polymer redox flow batteries. *Chemistry of Materials* 31 (19): 7987–7999.

36 Hu, B., Fan, H., Li, H. et al. (2021). Five-membered ring nitroxide radical: a new class of high-potential, stable catholyte for neutral aqueous organic redox flow batteries. *Advanced Functional Materials* 2102734.

37 Song, J., Hu, B., Fan, H., and Li, H. (2020). Inventors Aqueous organic flow battery based on pyrroline/alkanoxy radical compound. CN Patent CN 111628185 A.

38 Huang, J., Cheng, L., Assary, R.S. et al. (2015). Liquid catholyte molecules for nonaqueous redox flow batteries. *Advanced Energy Materials* 5 (6): 1401782–1401787.

39 Attanayake, N.H., Kowalski, J.A., Greco, K.V. et al. (2019). Tailoring two-electron-donating phenothiazines to enable high-concentration redox electrolytes for use in nonaqueous redox flow batteries. *Chemistry of Materials* 31 (12): 4353–4363.

40 Duan, W., Vemuri, R.S., Milshtein, J.D. et al. (2016). A symmetric organic-based nonaqueous redox flow battery and its state of charge diagnostics by FTIR. *Journal of Materials Chemistry A* 4 (15): 5448–5456.

41 Moutet, J., Veleta, J.M., and Gianetti, T.L. (2020). Symmetric, robust, and high-voltage organic redox flow battery model based on a helical carbenium ion electrolyte. *ACS Applied Energy Materials* 4 (1): 9–14.

42 Fornari, R.P., Mesta, M., Hjelm, J. et al. (2020). Molecular engineering strategies for symmetric aqueous organic redox flow batteries. *ACS Materials Letters* 2 (3): 239–246.

43 Liu, S., Zhou, M., Ma, T. et al. (2019). A symmetric aqueous redox flow battery based on viologen derivative. *Chinese Chemical Letters* 31 (6): 1690–1693.

44 Antoni, P.W., Bruckhoff, T., and Hansmann, M.M. (2019). Organic redox systems based on pyridinium-carbene hybrids. *Journal of the American Chemical Society* 141 (24): 9701–9711.

45 Wylie, L., Blesch, T., Freeman, R. et al. (2020). Reversible reduction of the TEMPO radical: one step closer to an all-organic redox flow battery. *ACS Sustainable Chemistry & Engineering* 8 (49): 17988–17996.

46 Janoschka, T., Friebe, C., Hager, M.D. et al. (2017). An approach toward replacing vanadium: a single organic molecule for the anode and cathode of an aqueous redox-flow battery. *ChemistryOpen* 6 (2): 216–220.

47 Winsberg, J., Stolze, C., Muench, S. et al. (2016). TEMPO/phenazine combi-molecule: a redox-active material for symmetric aqueous redox-flow batteries. *ACS Energy Letters* 1 (5): 976–980.

48 Winsberg, J., Hagemann, T., Muench, S. et al. (2016). Poly(boron-dipyrromethene)—a redox-active polymer class for polymer redox-flow batteries. *Chemistry of Materials* 28 (10): 3401–3405.

49 Rohland, P., Schröter, E., Nolte, O. et al. (2021). Redox-active polymers: the magic key towards energy storage – a polymer design guideline progress in polymer science. *Progress in Polymer Science* 101474.

50 Hager, M.D., Esser, B., Feng, X. et al. (2020). Polymer-based batteries-flexible and thin energy storage systems. *Advanced Materials* 32 (39): e2000587.

51 Friebe, C., Lex-Balducci, A., and Schubert, U.S. (2019). Sustainable energy storage: recent trends and developments toward fully organic batteries. *ChemSusChem* 12 (18): 4093–4115.

52 Muench, S., Wild, A., Friebe, C. et al. (2016). Polymer-based organic batteries. *Chemical Reviews* 116 (16): 9438–9484.

53 Hu, B., Tang, Y., Luo, J. et al. (2018). Improved radical stability of viologen anolytes in aqueous organic redox flow batteries. *Chemical Communications (Cambridge, England)* 54 (50): 6871–6874.

54 DeBruler, C., Hu, B., Moss, J. et al. (2018). A sulfonate-functionalized viologen enabling neutral cation exchange, aqueous organic redox flow batteries toward renewable energy storage. *ACS Energy Letters* 3 (3): 663–668.

55 Doris, S.E., Ward, A.L., Baskin, A. et al. (2017). Macromolecular design strategies for preventing active-material crossover in non-aqueous all-organic redox-flow batteries. *Angewandte Chemie, International Edition* 56 (6): 1595–1599.

56 Hendriks, K.H., Robinson, S.G., Braten, M.N. et al. (2018). High-performance oligomeric catholytes for effective macromolecular separation in nonaqueous redox flow batteries. *ACS Central Science* 4 (2): 189–196.

57 Baran, M.J., Braten, M.N., Montoto, E.C. et al. (2018). Designing redox-active oligomers for crossover-free, nonaqueous redox-flow batteries with high volumetric energy density. *Chemistry of Materials* 30 (11): 3861–3866.

58 Burgess, M., Chénard, E., Hernández-Burgos, K. et al. (2016). Impact of backbone tether length and structure on the electrochemical performance of viologen redox active polymers. *Chemistry of Materials* 28 (20): 7362–7374.

59 Nagarjuna, G., Hui, J., Cheng, K.J. et al. (2014). Impact of redox-active polymer molecular weight on the electrochemical properties and transport across porous separators in nonaqueous solvents. *Journal of the American Chemical Society* 136 (46): 16309–16316.

60 Oh, S.H., Lee, C.W., Chun, D.H. et al. (2014). A metal-free and all-organic redox flow battery with polythiophene as the electroactive species. *Journal of Materials Chemistry A* 2 (47): 19994–19998.

61 Yan, W., Wang, C., Tian, J. et al. (2019). All-polymer particulate slurry batteries. *Nature Communications* 10 (1): 2513–2524.

62 Montoto, E.C., Nagarjuna, G., Hui, J. et al. (2016). Redox active colloids as discrete energy storage carriers. *Journal of the American Chemical Society* 138 (40): 13230–13237.

63 Winsberg, J., Benndorf, S., Wild, A. et al. (2018). Synthesis and Characterization of a Phthalimide-Containing Redox-Active Polymer for High-Voltage Polymer-Based Redox-Flow Batteries. *Macromol. Chem. Phys.* 219: 1700267.

64 Cecchetti, M., Ebaugh, T.A., Yu, H. et al. (2020). Design and development of an innovative barrier layer to mitigate crossover in vanadium redox flow batteries. *Journal of the Electrochemical Society* 167 (13).

65 Schröter, E., Stolze, C., Saal, A., et al. (2022). All-organic redox targeting with a single redox moiety: Combining organic radical batteries and organic redox flow batteries. *ACS Applied Materials & Interfaces* 14 (5): 6638–6648.

66 Wu, M., Jing, Y., Wong, A.A. et al. (2020). Extremely stable anthraquinone negolytes synthesized from common precursors. *Chem* 6 (6): 1432–1442.

67 Kwabi, D.G., Lin, K., Ji, Y. et al. (2018). Alkaline quinone flow battery with long lifetime at pH 12. *Joule.* 2 (9): 1894–1906.

Part VII

Industrial and Commercialization Aspects of Flow Batteries

45

Inverter Interfacing and Grid Behaviour

John Fletcher and Jiacheng Li

University of New South Wales, Digital Grid Futures Institute, School of Electrical Engineering and Telecommunications, Kensington, NSW 2052, Australia

45.1 Introduction

DC sources of electrical energy, such as a flow battery, require conversion from DC to AC using an inverter to process and interface the electrical energy correctly to the grid. The inverter controller has to manage in real time many aspects of this vital interface between the grid and the source (or sink) of energy. Both the energy system on the DC-side of the inverter and the grid on the AC side of the inverter have specific needs and may have to satisfy certain standards and performance expectations. These needs will be discussed in Section 45.3 in relation to flow batteries.

In this chapter, the power electronic circuits and their control are discussed. These technologies are at the heart of the energy storage system and are really driving the adoption of storage technologies as well as underpinning new generation sources such as photovoltaics (PV), wind, and application areas like electric vehicles.

45.2 The Six-Switch, Three-Phase Inverter Circuit

At the core of a grid-connected inverter is typically a six-switch DC–AC inverter topology, Figure 45.1. The inverter can generate an AC voltage with variable magnitude and variable frequency by adjusting the modulation indices in a pulse-width modulation (PWM) scheme that controls the state of each of the three bridge legs that form the inverter.

Using such techniques, it is possible to control the output voltage magnitude, phase, and frequency of the inverter's three-phase voltages. Figure 45.2a illustrates the generation of pulse-width modulation waveforms where a high-frequency carrier waveform is compared with the modulating ('reference sine wave' in this example) waveform to generate a PWM gating signal, Figure 45.2b.

The gating signals are then fed to the gate drives to turn on or off individual IGBTs (T_a, T_b etc.) in the inverter circuit, Figure 45.3. When an upper switch is on, T_a is

Flow Batteries: From Fundamentals to Applications, First Edition.
Edited by Christina Roth, Jens Noack, and Maria Skyllas-Kazacos.
© 2023 WILEY-VCH GmbH. Published 2023 by WILEY-VCH GmbH.

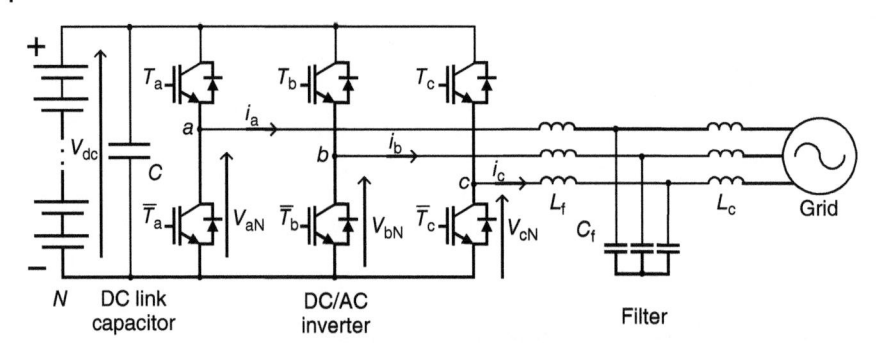

Figure 45.1 A typical six-switch inverter used to convert a DC voltage source, shown as V_{dc}, into three-line voltages that satisfy the requirements for a three-phase system. The line voltages, V_{ab}, V_{bc}, and V_{ca} are given by Kirchhoff's voltage law, $V_{ab} = V_{aN} - V_{bN}$.

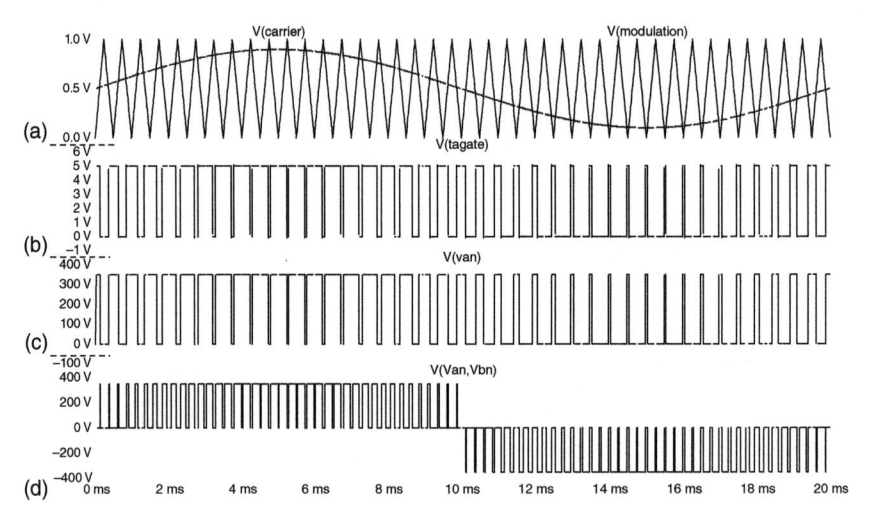

Figure 45.2 Pulse-width modulation signals for a two-level converter with $V_{dc} = 350$ V, $f_{sw} = 2$ kHz and a modulation reference $m = 0.8$ and $f_m = 50$ Hz. (a) shows the carrier and reference comparison for phase a, (b) shows the resulting gate drive signal for the upper switch of phase a, (c) shows the output voltage at the output of phase leg a referenced the 0 V (N) node of the DC link, and (d) shows the resultant line voltage V_{ab}.

'high' and V_{dc} is applied to the associated output (for example, $V_{aN} = V_{dc}$), whereas when T_a is 'low' switch $\overline{T}_a$ is on and $V_{aN} = 0$. The output line voltages, (V_{ab}, V_{bc}, V_{ca}) are the differences between the terminal voltages, for example, $V_{ab} = V_{aN} - V_{bN}$. The output line voltages are therefore switched waveforms, Figure 45.3d, and their average value varies according to the magnitude of the modulating waveform that is converted to a pulse width by the PWM generator. Note that in Figure 45.2a, a sine waveform is being used as the modulating waveform, however, the modulating waveform can be arbitrary in shape.

The combination of an output filter and six-switch inverter, which is controlled with the PWM modulation technique, underpins the integration of generation

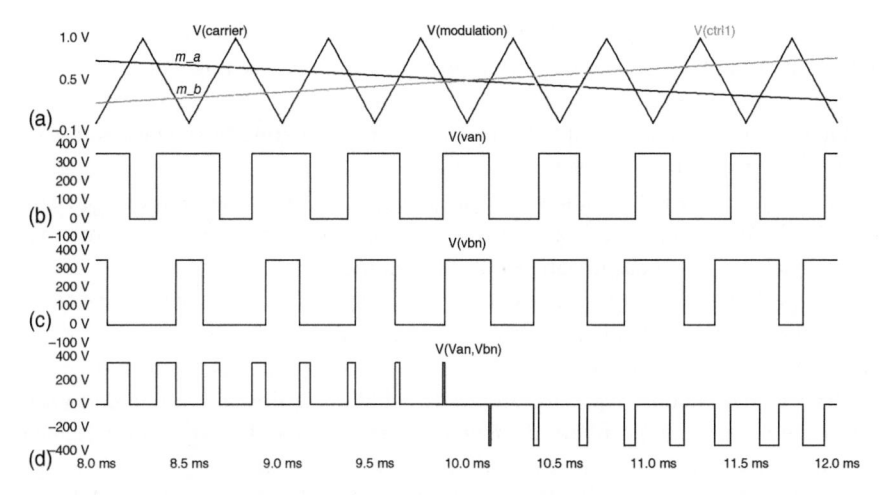

Figure 45.3 Pulse-width modulation signals for a two-level H-bridge converter with $V_{DC} = 350$ V, $f_{sw} = 2$ kHz and a modulation reference $m = 0.8$ and $f_m = 50$ Hz. (a) shows the carrier and reference comparison for phase leg a and b (m_a and m_b respectively), (b) and (c) shows the output voltage at the output of phase leg a and b referenced to the 0 V (N) node of the DC link (V_{an} and V_{bn}), and (d) shows the resultant line voltage V_{ab}. Notice how the polarity of the output voltage pulses changes at the zero crossing of m_a and m_b.

and storage to the grid. This is due to the inherent flexibility, simplicity, and the low-cost nature of this inverter topology, which can control the output voltage to allow for integration into the grid by satisfying the relevant local standards (e.g. in Australia AS4777). The ability to control voltage generated by the inverter near-instantaneously allows for high-bandwidth control of the line current (magnitude, frequency, and phase) relative to the grid voltage. Control of the current can then be used to adjust the real and reactive power which is generated or absorbed by the inverter, Figure 45.4. As the VA rating of the inverter increases (hence the power rating), the inverter switching frequency generally decreases, and consequently the LCL filter's low-pass cutoff frequency must decrease, reducing the bandwidth of the current control loop and the voltage control loop.

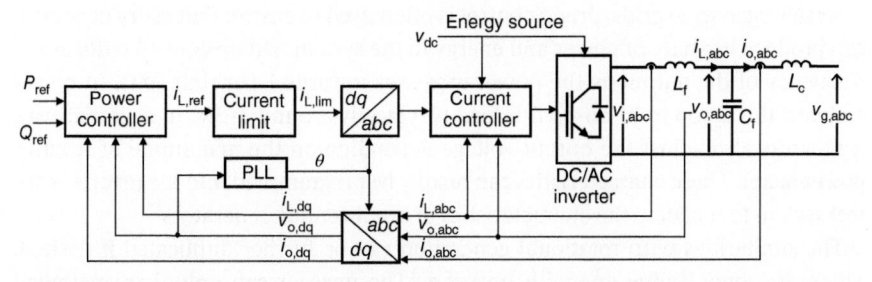

Figure 45.4 Grid-tie inverter control system for a three-phase system showing main control blocks and one of three phases of the inverter. The control system utilises a current controller that regulates the current injected into the grid, $i_{o,abc}$. This is used to regulate the real and reactive power, P and Q, respectively according to the references, P_{ref} and Q_{ref} and the instantaneous grid voltage. Source: Plet et al. [1]/IEEE.

In Figure 45.4, there are shown various blocks that perform the following functions. The phase-locked loop (PLL) determines the phase angle of the positive sequence fundamental component of the grid voltage, $v_{g,abc}$. The interfacing inductance, L_c, is used to enable control of $i_{o,abc}$ by regulating the voltage across L_c indirectly by regulating the voltage across C_f. L_f, C_f form a second-order low-pass filter that generates a near sine wave voltage $v_{o,abc}$ from switched output $v_{i,abc}$. The abc/dq blocks are reference frame transformations from stationary abc to rotating dq and vice versa that use the angle output from the PLL.

Such grid interfacing techniques are now widespread with the basic technology used for PV, wind, and battery storage systems. Included between the inverter and the grid is a filter of some description whose structure depends on the role of the inverter in the system. Typical examples in addition to the LCL filters discussed above are L filters and LC filters all of which offer low-pass filtering that is designed to remove the switching harmonics from the PWM-generated output. The voltage and current waveforms supplied by the inverter system then become sinusoidal with only modest amounts of distortion that will satisfy the relevant standards and codes. Figure 45.1 shows a three-phase inverter system that uses an LCL output filter.

To control the inverter typically a microprocessor or other programmable device is used. The microprocessor monitors a selection of voltages and currents to determine the grid voltage vector (magnitude and phase). This enables the microprocessor to correctly orientate the current vector (magnitude and phase) to inject the desired real power and reactive power. The control system must also respond to changes in the magnitude, frequency, and phase of voltages and current in the grid and must react to sags, swells, and outages in the grid.

When an inverter is grid feeding, its purpose is to inject real power into the AC grid. The magnitude of the injected power is determined by a high-level controller that requests potentially both real and reactive power to achieve an optimal operating point which may include maximising a certain revenue stream. This revenue stream may require absorbing real power at some times and days as well as releasing energy at times when generation is marginal compared to the load. There are some instances where the real power magnitude must be reduced or curtailed to match the power generation to the load (so called 'grid following').

As in larger-scale grids, droop control is often used to ensure that every generator can produce its share of power and energy to the system. Droop control reduces the frequency of the output as the power increases towards 1 per unit (pu). In power systems, the droop from no- to full-load is typically around 3–5%. In a similar way, generators also adjust the output voltage depending on the magnitude of reactive power factor. These characteristics can readily be programmed into the inverter control system to replicate the characteristics of synchronous generators.

The similarities with rotational generation can be further duplicated if desired. There are some limitations with inverters. The inverter can only deliver limited current, typically around 1.5–2 pu peak for a short time. This means that during network faults the inverter does not generate fault current to the same degree as synchronous generation as the inverter control system limits current to protect the power electronics. The inverter also has limited voltage generating capability which is fixed by the DC-link voltage at the input.

45.2.1 The Inverter DC-Link Model

The inverter typically has a DC-link capacitor on the DC input to the inverter. The following describes the DC-link capacitor voltage variation as a function of the input power to the DC link. The energy stored in the DC capacitor is

$$W_{dc} = \int P_{dc} dt = \frac{1}{2} C V_{dc}^2 \tag{45.1}$$

where C is the DC-link capacitance, V_{dc} is the voltage, W_{dc} is the stored energy, and P_{dc} is the input power to the DC link. The voltage and energy derivatives related to the DC-link capacitance are

$$\frac{dV_{dc}}{dt} = \frac{P_{dc}}{CV_{dc}}, \quad \frac{dW_{dc}}{dt} = P_{dc} \tag{45.2}$$

The P_{dc} is calculated as $P_{dc} = P_{in} - P_c$. Where P_{in} is the input power from the battery and P_c is the AC-side inverter output power. The DC-link voltage varies according to P_{dc} and is a constant when $P_{dc} = 0$, when the power $P_{dc} = P_i$.

45.2.2 Three-Phase Inverter and Pulse-Width Modulation

Figure 45.1 shows a six-switch inverter topology. It comprises three bridge legs in parallel. Figure 45.5 shows one bridge leg. When switch T_a is ON, the output voltage, V_{aN}, is V_{dc}. When switch $\overline{T}_a$ is ON, the output voltage is zero. (Note that both switches are not turned on at the same time). If the output is periodically switched between these two states, the output voltage, V_{aN}, averaged over each switching period, can be controlled between zero volts and V_{dc}. The switching period is usually fixed, and the width of the pulse of V_{dc} adjusted to change the output voltage.

Figure 45.6 shows an example of a pulse-width modulated signal and indicates how the width of the pulse is generated by comparing the modulating waveform with the carrier waveform – this is now mainly performed digitally but is also easy

Figure 45.5 One bridge leg of a voltage source inverter circuit.

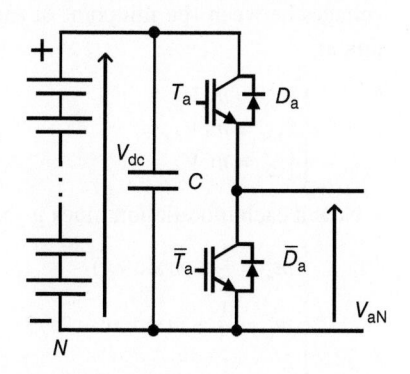

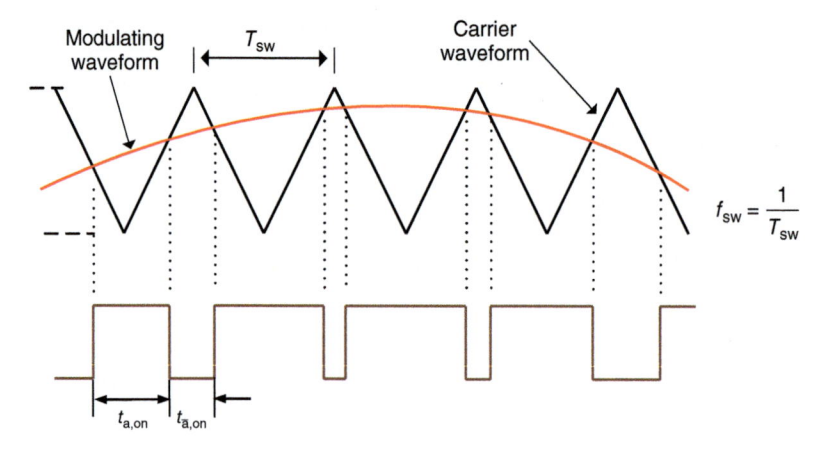

Figure 45.6 Example of carrier-based pulse-width modulated signal generation with a constant switching period T_{sw}.

to implement in analogue electronics. The average output voltage, at the terminals of the bridge leg, V_{aN}, is given by

$$V_{aN} = V_{dc} \frac{t_{a,on}}{T_{sw}} \tag{45.3}$$

where T_{sw} is the switching period, and $t_{a,on}$ is the on time of the switch T_a. The duty cycle, or modulation index, m is defined as

$$m = \frac{t_{a,on}}{T_{sw}} \tag{45.4}$$

Hence

$$V_{aN} = mV_{dc} \tag{45.5}$$

where m must be between 0 ($\overline{T}_a$ on continuously) and 1 (T_a on continuously). The modulation index, m, can be varied in time, therefore any desired voltage and frequency can be generated at the output terminals (within bounds determined by the switching frequency and V_{dc}). In the three-phase inverter shown in Figure 45.1, there are three-phase legs, hence three modulation indices, m_a, m_b, and m_c. The voltages between the midpoint of each phase leg and the 0 V (N) node of the DC link are

$$\begin{cases} V_{aN} = m_a V_{dc} \\ V_{bN} = m_b V_{dc} \\ V_{cN} = m_c V_{dc} \end{cases} \tag{45.6}$$

Now if each modulation index is made to vary according to

$$\begin{cases} m_a = \frac{1}{2} + m \sin(\omega t) \\ m_b = \frac{1}{2} + m \sin(\omega t - 2\pi/3) \\ m_c = \frac{1}{2} + m \sin(\omega t + 2\pi/3) \end{cases} \tag{45.7}$$

Then the resultant output line voltages will take the from

$$\begin{cases} V_{ab} = V_{aN} - V_{bN} = \sqrt{3}mV_{dc} \sin\left(\omega t - \dfrac{\pi}{6}\right) \\ V_{bc} = V_{bN} - V_{cN} = \sqrt{3}mV_{dc} \sin\left(\omega t - \dfrac{5\pi}{6}\right) \\ V_{ca} = V_{cN} - V_{aN} = \sqrt{3}mV_{dc} \sin\left(\omega t + \dfrac{\pi}{2}\right) \end{cases} \quad (45.8)$$

These are three-phase, balanced output line voltages, whose magnitude is controlled by m and whose output frequency and phase can be regulated by the frequency and phase of the modulating expression (45.8). The modulating waveforms can be manipulated digitally using high-performance microcontrollers or digital signal processors.

The inverter switches the input voltage supply so is commonly referred to as a voltage source inverter, VSI. Therefore, for low-frequency modelling, the VSI can be viewed and modelled as an ideal controllable voltage source whose bandwidth is usually much higher than the required excitation frequency required by the system. For example, Figure 45.7 shows a phasor diagram of a grid-connected inverter. In this case, the inverter is simply modelled as an ideal voltage source that is generating a balanced set of three-phase voltages whose magnitude and phase can be controlled relative to the grid voltage. This provides the capability to control the flow of real and reactive power to the grid as will be discussed later. Note that synchronisation to the grid frequency and phase is assumed.

45.2.3 Inverter Control Schemes – Grid Feeding

The inverter can be controlled by a variety of closed-loop current control schemes. These include representing and controlling the system using various possible reference frames that are stationary or are rotating and can be described, for example, as stationary abc phase quantities or synchronously-rotating $dq0$ quantities. Each has their relative merits. For example, using the $dq0$ frame makes the tuning of current controller gains relatively easy. This is because the dq quantities are DC quantities when operating in steady-state conditions and hence the proportional and integral (PI) compensator can reduce the steady-state error to zero. In the abc reference frame, PI phase current controllers must generate a tracking error as they follow a reference demand that is a fixed frequency. This tracking error can be eliminated with the use of a proportional and resonant (P + R) controller that effectively

Figure 45.7 Phasor diagram of inverter control in the dq reference frame to establish exported real power via $i_{o,d}$ and control of reactive power via $i_{o,q}$. Note that the grid voltage, $v_{g,dq}$, is aligned with the d axis of the transformation and the voltage across inductor L_c is leading the current flowing in L_c, $i_{o,dq}$, by 90°.

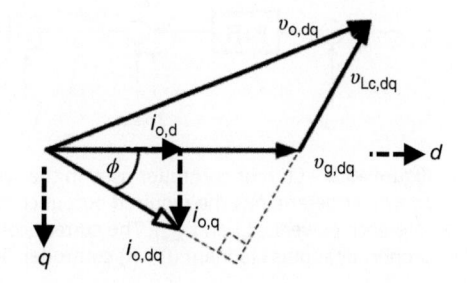

yields high gain at the resonant frequency, which is chosen as the grid frequency. The resonant gain helps minimise the error at the grid frequency but avoids excess gain at other frequencies. This reduces the effects of noise and the possibility of stability problems which are associated with large gains in negative feedback loops.

As an example of one realisation, the control system in Figure 45.4 utilises a combination of a stationary abc reference frame for the current control loop and a rotating dq reference frame to represent the grid voltage and for the determination of the current commands based on the real and reactive power references, P_{ref} and Q_{ref} respectively. Reference frame transformations are required between stationary and grid-synchronous reference frames. In this example, a phase-locked loop is used to track the synchronous reference frame angle, θ, using measured grid voltages, where knowledge θ allows the controller to correctly align the inverter voltages and currents to the grid (synchronisation).

The current controller associated with Figure 45.4 is shown in more detail in Figure 45.8. The inputs to the P + R compensator are the current errors in each of the three phases and the output from the P + R compensator is a set of phase voltage commands which reflect the voltage required across L_f to force $i_{L,abc}$ towards its reference value, $i_{L,abc}{}^*$. The voltage across the filter capacitor, $v_{o,abc}$, is a feedforward quantity and is added to the voltage command from the P + R compensator. The addition of these two terms equals the generated voltage required at the output of the inverter, $v_{i,abc}{}^*$, to force, $i_{L,abc}$, towards its reference value, $i_{L,abc}{}^*$. $v_{i,abc}{}^*$ is the required voltage to be applied by the inverter to the input of the LCL filter. The required modulation indices for each of the three-phases is calculated by dividing $v_{i,abc}{}^*$ by v_{dc} (the present DC-link voltage). The resultant modulation indices will then develop the necessary inverter output voltage.

$i_{L,abc}{}^*$ is the output from the current reference calculator. To decide these references, the first step is to convert the reference real and reactive powers, P_{ref} and Q_{ref}, into an output current reference, $i_{o,ref}$ by dividing by the grid voltage, which is expressed in the rotating dq reference frame.

$$\begin{bmatrix} i_{od,ref} \\ i_{oq,ref} \end{bmatrix} = \frac{1}{V_{od}^2 + V_{oq}^2} \begin{bmatrix} V_{od} & V_{oq} \\ V_{oq} & -V_{od} \end{bmatrix} \cdot \begin{bmatrix} P_{ref} \\ Q_{ref} \end{bmatrix} \tag{45.9}$$

Note that as the grid voltage magnitude varies so will the $i_{o,ref}$ value. As the grid voltage drops, the $i_{o,ref}$ will increase to deliver P_{ref} and Q_{ref}.

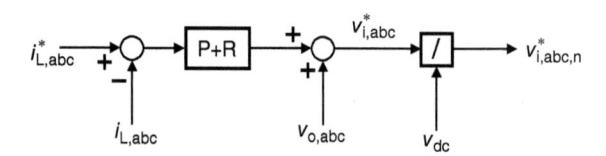

Figure 45.8 Current controller used in the system shown in Figure 45.4. The power controller determines the required output current from the inverter, $i_{L,abc}$, to inject the reference powers, P_{ref} and Q_{ref}. The current controller utilises an error amplifier and proportional plus resonant (P + R) controller. Source: Plet et al. [1]/IEEE.

Figure 45.9 Calculation of required reference currents using P_{ref} and Q_{ref} references and the measured grid voltage magnitude, $V_{o,dq}$. Source: Plet et al. [1]/IEEE.

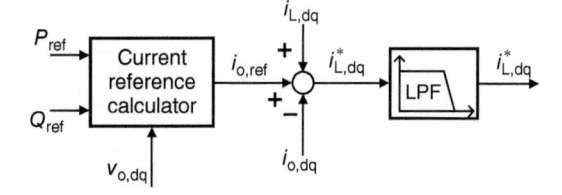

By examining the output filter, $i_C = i_L - i_o$ and $i_{L,ref} = i_{o,ref} + i_C$. These two expressions can be combined to yield $i_{L,ref} = i_{o,ref} + i_L - i_o$ which Figure 45.9 illustrates. The low pass filter has a cutoff frequency well below the power system frequency to avoid rapid sub-cycle variations in the current demand that would produce sub-cycle distortion in the injected currents and the supply voltage waveform depending on the impedance of the grid at the point of connection. The outputs from the low-pass filter (LPF) are the inductor current references in the dq frame. These are then converted back from the dq to abc reference frame.

With this control scheme, the inverter is capable of independently controlling real and reactive power (P_{ref} and Q_{ref}) and can make the real power both positive and negative, and the reactive power absorbing or generating. As the grid voltage varies, the control circuit will compensate by adjusting the injected current to track the reference commands. As the voltage drops at the inverter terminals, to develop the same power, the real and reactive components of the injected current must increase. Note that the equivalent impedance of the inverter at its terminals under these circumstances has a 'negative impedance' characteristic as the current rises as the applied voltage falls.

45.3 Inverter Control Modes for Energy Storage Applications

The primary aims of the inverter control are to regulate P_{ref} and Q_{ref}, according to the requirements of a high-level control. This high-level control must also coordinate with the battery to reflect on the P_{ref} and Q_{ref} demands any constraints such as a low state of charge (SOC) during power injection to the grid, or high SoC during battery charging from the grid.

If the energy storage system has a large change of voltage with output current (like a PEM fuel cell) or a large, rapid change in voltage from charge to discharge (like Li-ion close to complete discharge) then some form of DC–DC converter may be required to provide adequate voltage to the DC link as the output voltage from the energy storage system drops. Coordination between the DC–DC converter and the DC–AC inverter is necessary and typically uses the voltage across the DC-link capacitor to coordinate the injection of real power into the AC grid with the power delivered to or from the DC–DC converter on the energy storage side of the DC–DC converter. The function is realised, Figure 45.10, with two control loops as well: an outer regulation loop consisting of a DC voltage controller for v_{dc}. The output of the DC voltage controller is the reference current $i_{d,ref}$ for the current regulator. The inner current

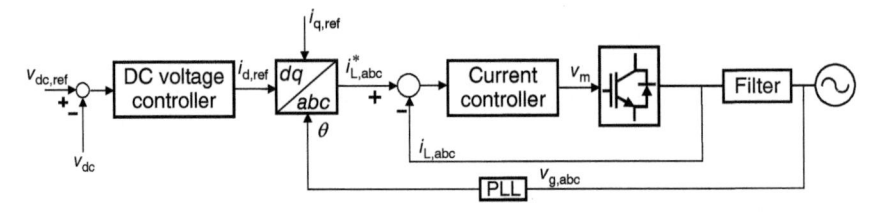

Figure 45.10 Block diagram of the inverter DC-link voltage, v_{dc}, outer control loop with an inner current control loop. In this example, a simple, single filter inductor L_f is assumed rather than an LCL filter. The latter would provide improved dynamic performance with lower harmonic current distortion.

regulation loop consists of a current controller with a feedback loop, that uses the measured $i_{L,abc}$ currents compared to the $i_{d,ref}$, to regulate the magnitude and phase of the voltage generated by the inverter to develop the necessary d-axis current ($i_{d,ref}$) required to maintain v_{dc} at its setpoint value, and a specified q-axis $i_{q,ref}$ reference (not shown for simplicity) that delivers a desired reactive power into the AC grid. The dynamics of the current controller depend somewhat on the reference frame adopted. Figure 45.11 shows two examples (a) using a rotating dq reference frame with feedforward voltage terms that decouple the filter inductor voltage drop, and (b) a stationary $\alpha\beta$ reference frame using a P + R controller in each α and β axis with harmonic compensation terms added, 'HC'.

The coordination between the DC–DC converter and the DC–AC inverter is simple. If the DC-link voltage is rising above the setpoint v_{dc} value then the DC–DC converter is delivering more power into the DC link than is being exported to the AC grid by the DC–AC inverter: the DC–AC inverter can export more real power by increasing the $i_{o,d}$ component, Figure 45.7, and the rate of change of v_{dc} will reduce and with further increases in $i_{o,d}$ will become negative moving v_{dc} back towards the setpoint value. If more power is being exported by the DC–AC inverter than is currently being injected to the DC link from the DC–DC converter, the DC-link voltage will fall below the setpoint value. The DC–AC inverter can then reduce the exported real power by reducing $i_{o,d}$ to allow the DC-link voltage to rise back to the setpoint value. In essence, the rate of change of DC-link voltage indicates power flow balance between the power sourced by the energy store via the DC–DC converter and the power delivered to the AC grid via the DC–AC inverter. This concept works

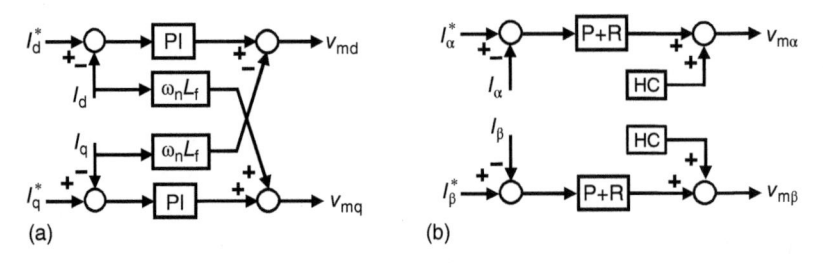

Figure 45.11 Current control options (a) dq frame based controller; (b) $\alpha\beta$ frame based controller.

for bidirectional power flow: the description above details energy flowing from the energy store to the grid.

The quality of the energy supplied to the AC grid must meet basic requirements and these will be set by a combination of grid codes and standards that may change depending on the rated apparent power and the voltage of the connection point. The grid codes and standards specify many performance indicators of the quality of the energy supplied by the grid-side converter, along with other important issues such as fault contributions, anti-islanding, ride-through conditions, and disconnection.

45.3.1 Fault Response

Under a fault condition on the grid, the grid voltage is likely to fall considerably. Therefore, to continue to develop the reference powers, P_{ref} and Q_{ref}, the inverter current must increase markedly. In such instances, the currents are likely to be excessive in the inverter. The inverter control scheme must therefore limit the current demand $i_{o,d}$ to avoid inverter failure due to overcurrent that would primarily lead to an overtemperature failure of the semiconductor switches. The inverter, therefore, lacks the fault current capability of electromagnetic machines which can often operate at 5–10 times the continuous current rating of the phase winding for short periods of time. This lack of overcurrent provision by the inverter presents an issue when designing suitable power system protection schemes as the difference between the current during rated operation and fault current is not as significant for inverter-connected resources. Reliance on overcurrent detection as a means of identifying faults and locating them is no longer certain and requires detailed analysis to devise a suitable protection scheme.

45.3.2 Single-phase Inverters

A single-phase inverter that is to be grid-connected typically uses an H-bridge circuit, Figure 45.12, with an AC filter on the output stage (not shown). The control scheme for this inverter will be like the three-phase system in that the current is injected into the grid and is controlled in magnitude and phase to generate the desired real and reactive power. An important difference between the three-phase and single-phase inverter is the power flow on the DC link. Under balanced

Figure 45.12 Basic single-phase H-bridge inverter circuit for DC–AC single-phase application. The DC-link current, I_{dc}, in a single-phase inverter has a DC value and a sinusoidal component at twice the grid frequency.

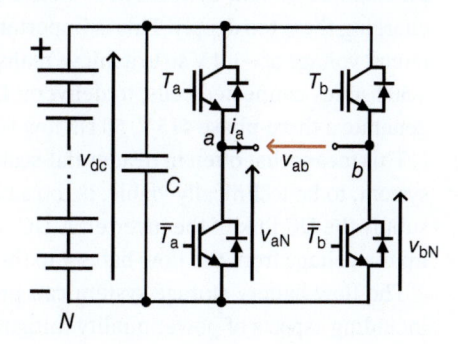

conditions, the three-phase power flow, when reflected onto the DC link ($V_{dc}I_{dc}$) and averaged over a PWM switching cycle, is a constant value. In a single-phase system, the flow of energy on the AC side is not constant over a fundamental cycle. For example, the output AC waveforms have four zero crossings per cycle as the voltage and the current both pass through zero twice. The 'instantaneous' power flows (power averaged over one PWM switching period) in the DC link of the single-phase system oscillates at twice the supply frequency with a DC offset. The DC offset is the delivery of real power and the AC oscillation is representative of the reactive power. The consequence of this bidirectional power flow in the DC link may cause issues if the DC source that supplies the energy to the DC link is not adequately buffered by a filter from the inverter's connection to the DC link. The filter must be designed to remove the large magnitude, low-frequency DC current. This implies that a relatively large filter component is required on the DC link.

45.3.3 Inverters for Flow Battery Energy Storage

Battery and other energy storage systems are sources and sinks of DC electrical energy. The inverter system, as shown previously in this section, can be configured to control both P and Q. The real power, P, can be bidirectional in the conventional inverter (both positive and negative) that allows the inverter system to charge ($-P$) and discharge ($+P$) the energy storage system. When charging the energy storage system, power is taken from the grid and when discharging the energy storage system, the stored energy in the storage medium is delivered back to the grid. The SOC of the energy storage system needs to be estimated continuously in order that the owner/operator can base decisions on how best to maximise financial returns in the markets targeted. From a technical perspective, SOC estimation is necessary to manage flow battery operations, minimising as far as possible operations that may accelerate aging factors or other processes contributing to degradation of the flow battery's membrane. SOC estimation in flow battery technologies is simple compared to traditional electrochemical batteries. The SOC can be measured quite accurately by sensing the open-circuit voltage of a small test cell that uses the electrolyte samples flowing into the main electrode system.

The flow-battery electrode system is often complex with series connection of multiple electrodes but avoids some of the challenging balancing requirements with, for example, Li-ion, to avoid over-stressing individual cells by overcharging or discharging them too deeply. This is important as the vanadium chemistry has an open circuit voltage of ~1.4 V so to achieve realistic DC-link voltages of, say 690 V, requires ~500 series-connected cells, to deliver a DC-link voltage suitable for an inverter to generate a three-phase 415 V, 50 Hz low voltage supply.

This means that often in residential-scale flow-battery storage systems, the storage system, to be technically viable, is too small to generate enough voltage to directly supply the DC link of the inverter. A DC–DC converter will then be required to step up the voltage from the flow battery to the DC link of the inverter.

The flow-battery storage system can provide limited grid system enhancements including aspects of power quality mitigation (through voltage restoration) and the capability of providing bidirectional power. This allows the system to act as both a generator and a load during different times of the day. This often results in less frequent curtailment of local renewable energy sources as there is additional flexibility in the demand-generation balance, and reduced voltage management requirements. One of the main interests in energy storage is to supplement variable renewables and provide replacement inertia. The storage element in combination with appropriate inverter controls provides the necessary ability to both fill-in gaps of generation at times of high demand, and absorb peaks of generation at minimum demand. The inverter control makes this possible and the inverter itself (without energy storage) is capable of delivering reactive power to support network voltage (Q can sometimes partly manage voltage problems), and perform active power filtering, remediating harmonic currents injected by local loads for example.

Other issues that require thought are maintaining control of the flow battery under grid faults that lead to a disconnection of the inverter from the grid. When an inverter disconnection occurs, the DC–DC converter, if used, must cease delivering energy to the DC link rapidly to avoid overcharging the DC-link capacitor. This must happen as the inverter is no longer able to inject power into the AC grid, hence no energy can be removed from the DC link by the inverter to manage v_{dc}.

The inverter control also must deal with a variety of grid disturbances including fast voltage sags, voltage phase angle jumps, and rapid frequency changes. The current state of inverter standards is only just beginning to acknowledge some of these challenges and how they impact the inverter controller.

45.4 Conclusions

This section has introduced the basic three-phase inverter circuit that is currently the workhorse of low voltage distributed generation for renewable generation technologies such as wind and photovoltaic. The three-phase inverter circuit is also utilised in energy storage systems that utilise a DC storage technique such as electrochemical and electrostatic storage (batteries and supercapacitors). The method of inverter synchronisation to the network has been explained and one example of an inverter control technique that controls the injection of real and reactive power onto (or off) the grid is described. This example uses pulse-width modulation and current controllers with an outer real and reactive power controller. Note that there are many other examples that have their own advantages and disadvantages.

Reference

1 C. A. Plet, M. Graovac, T. C. Green and R. Iravani, "Fault response of grid-connected inverter dominated networks," *IEEE PES General Meeting*, 2010, pp. 1-8, doi:https://doi.org/10.1109/PES.2010.5589981.

46

Flow-Battery System Topologies and Grid Connection

Thomas Lüth[1], Thorsten Seipp[2], and David Kienbaum[1]

[1] J.M. Voith SE & Co. KG | DSG, St. Pöltener Straße 43, 89522 Heidenheim, Germany
[2] Volterion GmbH & Co. KG, Carlo-Schmid-Allee 3, 44263 Dortmund, Germany

46.1 Introduction

A basic flow-battery system consists of a pair of tanks, a stack, and a power-conditioning system (PCS) for the grid connection. The energy storage capacity can be scaled by the amount of electrolyte in the tanks independently of the power rating, which is determined by the PCS and the stack. To achieve a higher power capability of a stack, either the cell/membrane area or the number of series-connected cells needs to be increased. However, the first is limited by an inhomogeneous electrolyte distribution inside the cells and the latter by increasing shunt currents. Also, restrictions regarding the manufacturing and handling of a stack exist, so it cannot be scaled up indefinitely. It is, therefore, necessary to connect multiple stacks together if a higher power capability is needed. Various options exist to design such a system of larger size, some of them also include multiple tank units or PCS. These so-called topologies show up different ways to connect the system components together and have specific advantages and disadvantages. In this chapter, the most common topologies of flow-battery systems are presented and discussed, starting with simple low-voltage systems for small-to-medium-sized battery systems up to high-voltage systems, which are better suited for large-scale energy storage.

Prior to that, the most relevant aspects of this comparison will be explained in the following sections.

46.1.1 Power-Conditioning System (PCS)

To connect a flow battery to an AC grid, a PCS is needed. It consists of bidirectional one- or three-phase inverters. To be able to feed into the grid, the inverter needs a DC-link voltage that is higher than the grid's peak voltage. In most cases, the battery voltage highly fluctuates and is below this value, so an additional bidirectional DC/DC converter is needed to adjust the voltage levels.

Flow Batteries: From Fundamentals to Applications, First Edition.
Edited by Christina Roth, Jens Noack, and Maria Skyllas-Kazacos.
© 2023 WILEY-VCH GmbH. Published 2023 by WILEY-VCH GmbH.

The efficiency of the inverter is of high importance for the total system efficiency, as the conversion losses occur both at charging and discharging. A simple half-bridge with an inductor can be used as a bidirectional DC/DC stage. However, its losses increase with a higher difference between both DC voltage levels. Therefore, another DC/DC topology, such as a dual-active bridge, is needed in case of higher voltage level differences, which usually requires a high-frequency transformer. This causes more effort and cost. However, it is useful in cases where galvanic isolation is desired.

If the battery voltage stays permanently above the grid's peak voltage during operation, the DC/DC stage might be omitted completely. This improves the efficiency and reduces the overall PCS cost. In this case, commonly available high-voltage inverter systems can be used. A pre-charge unit for the initial charging of the electrolyte up to the operational voltage level must be added though. Further information about PCS for batteries can be found in the chapter (see Chapter 45) and the literature [1–3].

46.1.2 Shunt Currents

Usually, cells in a flow-battery stack are connected electrically in series and hydraulically in parallel. Each half cell is then at a different voltage potential. The electrolyte gives a conductive connection between all the positive and negative half cells of a stack, respectively, which leads to parasitic shunt currents (see Chapter 4). If several stacks in the system are connected in series and are using the same tank and piping systems, additional shunt currents also occur between the individual stacks. As they are independent of the load current of the battery, the caused Coulombic efficiency loss [4] can have a significant effect, especially at part load. At standby, the self-discharge due to shunt currents generates heat, which might lead to critical electrolyte temperatures [5]. Additional issues regarding corrosion of the bipolar plates and felt as well as safety issues due to gas evolution can be a major problem [6]. Generally, shunt currents depend on the voltage differences between half cells, so they increase with more series-connected cells or stacks and higher SoC. They can be reduced by measures regarding the piping design. As an example, longer hydraulic paths through the pipes as well as piping inserts-like rotating wheels can be used. However, this comes at the cost of increased pump losses.

46.1.3 Reliability

Depending on the application, the reliability of the system has a different significance. While in some applications, such as uninterruptible power supplies or base radio station supplies, it might be critical, in other applications, the reliability aspect rather concerns running cost calculations regarding efforts in maintenance and the impact of the nonavailability of the system.

Apart from improvements of materials and components for larger durability, reliability on a system level can be increased by adding redundancy. This can reduce the number of single points of failure in the system. However, it must be considered that adding redundant components, such as stacks or pumps, might require also additional components, such as pipes or valves, creating new possible sources of failure and even new single points of failure [7].

46.1.4 SoC Band Limitation

Depending on the topology, only a certain part of the full SoC band of the electrolyte can be utilized. This lowers the actual energy storage capacity. Generally, the manufacturer of a flow battery specifies a maximum charging and minimum discharging voltage to prevent side reactions and deep discharge. The battery voltage consists of the open-circuit voltage (OCV), which depends on the SoC and the overpotential, determined by the internal stack resistance and the current (Eq. 46.1).

$$V_{\text{Bat}} = V_{\text{OCV}} \pm R_i \cdot I \qquad (46.1)$$

Due to the overpotential, the end-of-charge and end-of-discharge voltage is reached earlier than the theoretical limit, which would be the OCV at highest or lowest SoC.

If the current gets reduced at this point or the flow of electrolytes gets increased [8], a wider span of the SoC band can be utilized. However, it does not make sense to reduce the current indefinitely because of the self-discharge due to shunt currents and crossover. Furthermore, increasing the flow rate causes more auxiliary losses.

In a real system with multiple stacks, there is the problem that always one of the stacks reaches the charge termination the earliest, which might further limit the SoC band of the other stacks, depending on the topology. This can be caused by different internal resistances (R_i) or Coulombic efficiencies (CE).

46.1.5 Modularity/Flexibility

The flexible scaling of a flow battery regarding power and energy as one of the key advantages comes to play only if the battery components and the system topology actually allow it. To be fully able to scale a system to the customers' needs, either the system components, such as stacks, pumps, and tanks, must exist in various sizes, or the system design must be modular. The latter means, that it allows adding or removing multiples of the same component in a flexible way. From an economic point of view of a manufacturer, only this is feasible, so a high modularity of the used topology is advantageous.

46.1.6 System Size

Not every topology that exists, in theory, is possible or reasonable for all sizes in terms of power and capacity of real systems. Dependent on series or parallel connections of multiple stacks, there are different requirements for the current and voltage range of the battery-side PCS interface. With the inverters available on the market, these requirements cannot be always met, which leads to restrictions and disadvantages. However, in the following sections, it is assumed that there is an ideally suited PCS.

Furthermore, there are restrictions regarding other system components, which might not be available or difficult to manufacture in every size. As an example, a high-voltage system with a low total power rating is difficult to realize, as it would require many cells and stacks in series with a very small membrane area for each cell.

All in all, it is often a consideration of technical benefits versus effort and price, which differs depending on the system size. So, some topologies are rather suited for small, medium, or large-scale applications.

46.2 Topologies

Several possible system configurations from low-voltage to high-voltage systems are presented and described in the following sections.

46.2.1 Low-Voltage Parallel Connection (LV-P)

The most common and simple approach for a flow-battery system is the direct connection of one stack to the grid by an inverter. Therefore, a PCS compatible with low-DC voltages is needed.

To increase the power capability, multiple stacks can be connected electrically and hydraulically in parallel, as shown in Figure 46.1. This results in higher DC currents from the stacks to the inverter, so larger cable diameters are needed, or higher ohmic losses are caused. The amount of parallel-connected stacks is limited by the maximum DC current of the PCS.

Unequal internal stack resistances because of tolerances of the internal resistances lead to an inhomogeneous current distribution between the stacks. This makes

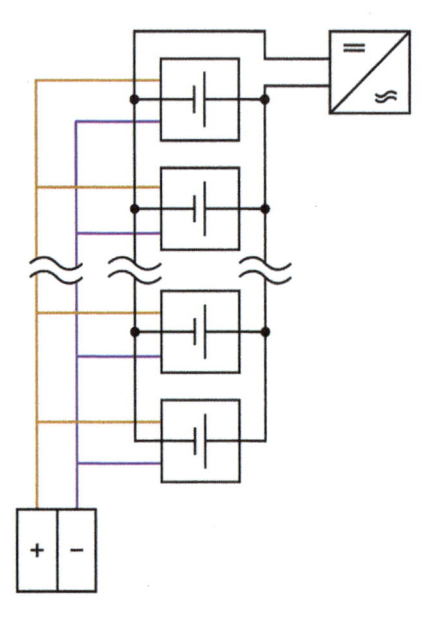

Figure 46.1 Topology LV-P: Parallel connection of multiple stacks to one PCS (DC/DC & DC/AC).

it necessary to measure the current of each individual stack to stay within the manufacturers' limits and ensure safe operation.

Therefore, the stack with the lowest internal resistance determines the current limit of the system and thus also the performance of the overall system. The stack with the highest internal resistance determines the utilization of the electrolytes via the maximum permissible stack voltage. In no-load operation, i.e. with $I_{\text{Inverter}} = 0$ A and the electrolyte supply switched on, equalizing currents can occur between the stacks connected in parallel. This happens due to the different internal resistances and Coulombic efficiencies, although they are supplied with electrolytes from the same reservoir and the same state of charge. However, this effect is self-limiting due to the voltage drop of the charge or discharge current occurring at the internal resistor. Regarding the charging efficiency and the utilization of the electrolyte, the different Coulombic efficiencies of the stacks can be neglected since the electrolyte in the tanks is mixed accordingly and the differences balance out.

In a low-voltage system with one or more parallel-connected stacks, only stack-internal shunt currents must be considered compared to systems with series-connected stacks. Therefore, the effects are relatively small and no special minimization measures on system level must be taken.

It can be concluded that this topology offers a certain but also limited modularity and scalability. However, the parallel connection of multiple stacks offers no redundancy and is susceptible to single failures. Due to its simple design and operational management, it is well-suited for small-scale applications.

46.2.2 Low-Voltage System with Several Inverters (LV-DC/AC)

To prevent the unequal current distribution at directly parallel-connected stacks, each stack can be connected to the grid with its own inverter. This can be advantageous, in particular, with increasing stack power and therefore increasing system power. The concept is depicted in Figure 46.2.

The length of the DC cabling can be reduced, which results in less ohmic losses. Instead, a more extensive AC cabling is necessary. The high number of inverters requires more control effort but also offers the opportunity to run every stack at its optimal operating point. A single stack reaching its voltage limit no longer terminates the charging/discharging process of the whole system and cuts down the total capacity. Instead, the current/power of each stack can be reduced individually when needed. This leads to good utilization of the electrolyte also in the boundary areas of the SoC. As with the previous topology, there are no restrictions caused by unequal Coulombic efficiencies.

If the stacks are also supplied individually by separate pumps or controllable valves, there is high flexibility regarding the part-load operation. Not only in this case but also in case of a stack failure, individual stacks can be shut down while the total system keeps running. Therefore, this topology can also provide high redundancy and reliability. Likewise, a highly modular design is possible, which makes it easy to adapt it to a specific application.

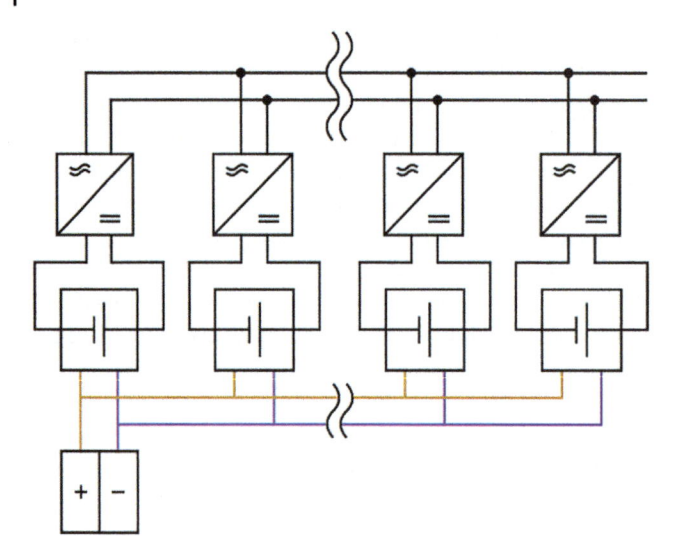

Figure 46.2 Topology LV-DC/AC: Each stack is connected to the AC grid by its own PCS (DC/DC & DC/AC).

In contrast, the usage of multiple low-voltage inverters is disadvantageous concerning the high cost and the fixed relation between inverter and stack. Both must be matched perfectly to maximize efficiency and reduce cost. Depending on the location of installation, different grid-connection rules and standards apply. For different markets, it can be necessary to adapt different inverters and it is more difficult to find off-the-shelf solutions. In general, the high cost can be less attractive for larger systems compared to other topologies.

46.2.3 Low-Voltage System with Several DC/DC Converters and One Central Inverter (LV-DC/DC)

The topology in Section 46.2.2 can be adapted by splitting the converter into a DC/DC stage and a DC/AC stage, as seen in Figure 46.3. This allows the stack-referenced DC/DC converter to be optimally matched to the stack. A high-voltage DC bus is then used to connect the stacks' DC/DC converters to a bidirectional central DC/AC inverter. This system can be more easily adapted to different grids than the topology in Section 46.2.2. Only the DC/AC inverter must be exchanged. The high-voltage bus potential is usually selected above the grid's peak voltage to be able to use a favorable inverter topology, so off-the-shelf solutions for the specific grid type can be used.

In comparison to the topology described in Section 46.2.2, this topology is almost as advantageous in reliability with the small difference that the DC/AC stage is not redundant anymore. In all the other not yet mentioned aspects, the same advantages and restrictions apply.

46.2.4 High-Voltage System with One Tank Pair (HV-1T)

As an alternative to the low-voltage topologies presented in Sections 46.2.1–46.2.3, an electrical series connection can be set up with several stacks, as shown

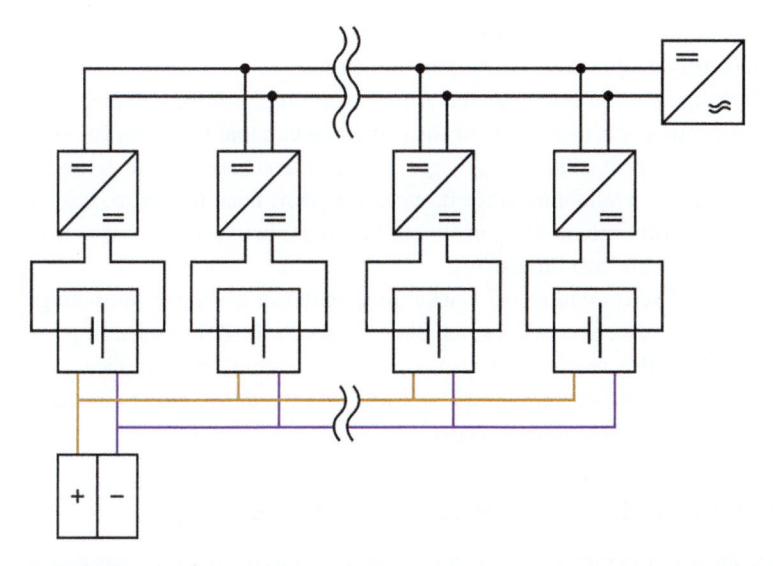

Figure 46.3 Topology LV-DC/DC: Each stack is connected to a high-voltage DC bus by its own DC/DC. A central DC/AC high-voltage inverter connects the bus to the AC grid.

in Figure 46.4. Due to the electrical series connection and hydraulic parallel connection of the stacks, the external shunt currents via the electrolyte distribution pipes add up to the stack-internal ones. This leads to the issues described in Section 46.1.2 and reduction measures must be taken, which are for instance described in refs. [9, 10].

By connecting the stacks in series, the string voltage can be matched to the input voltage range of commercially available high-voltage inverters. A more simple DC/DC stage can be used or it can be omitted completely, which also implies more efficiency and less PCS cost.

Voltage monitoring at the individual stacks connected in series is necessary because the stack with the highest internal resistance specifies the charge termination of the whole string, considering the stack voltage limits defined by the manufacturer. The total usable SoC band is then limited by the weakest stack.

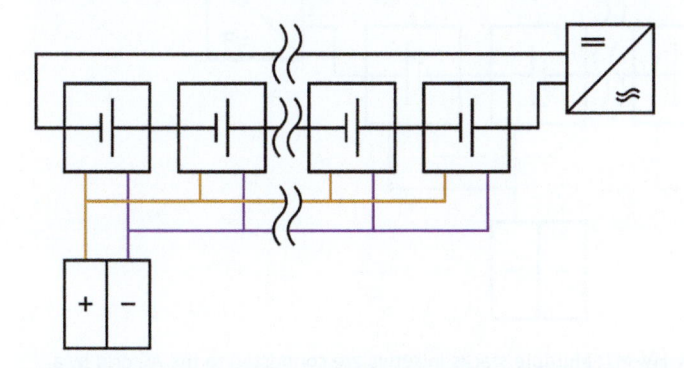

Figure 46.4 Topology HV-1T: Multiple stacks in series are connected to the AC grid by a high-voltage PCS. They share one common tank pair.

Variances in the Coulombic efficiencies of the individual stacks are not critical because the electrolyte intermixes in the common tanks.

The overall design can be implemented with less piping, pumps, and electrical connections in comparison to systems of the same power capability with the previous topologies. This brings cost advantages, but also disadvantages in terms of fault susceptibility and availability since if one component fails, the entire system fails. Also, the reduction measures against shunt currents between the stacks may be prone to errors and are cost intensive.

As the number of series-connected stacks must match a certain input voltage range of the inverter, this concept does not allow for much flexibility regarding the power capability. For this reason, it is also not suited for small-scale applications. However, it is possible to connect multiple of these systems on AC side to increase the power and create larger storage systems.

46.2.5 High-Voltage System with Multiple Tank Pairs (HV–MT)

The topology shown in Figure 46.5 is essentially identical to the topology presented in Section 46.2.4. The difference is that the series-connected stacks are not supplied by only one but by multiple tank pairs. Each tank pair is connected to one or more series-connected stacks.

This configuration has the advantage that the voltage driving the shunt current between the stacks is lower and thus losses and problematic effects can be reduced. In the special case of having a tank pair for each stack, inter-stack shunt currents can be avoided completely.

Regarding the stack voltage and the limited SoC band, the same restrictions apply as mentioned in Section 46.2.4. It should be noted that due to different Coulombic efficiencies of the stack groups (where a stack group can also consist of only one stack), a differently efficient charge or discharge must be assumed. This results in the stack group with the highest Coulombic efficiency charging the tank pair assigned to them faster than the stack groups with lower Coulombic efficiencies.

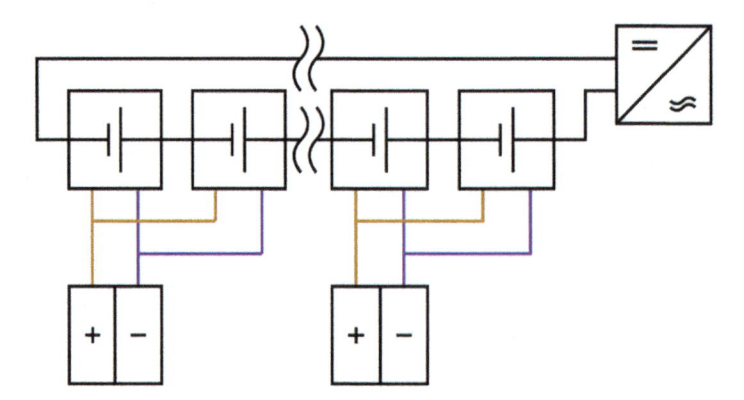

Figure 46.5 Topology HV-MT: Multiple stacks in series are connected to the AC grid by a high-voltage PCS. Multiple tank pairs exist and supply one or multiple stacks together.

When discharging, the stack group with the lowest Coulombic efficiency will discharge their associated tank pair the fastest. Unused capacity, thus, remains in each of the other tank pairs. This process is repeated for each cycle, regardless of whether it is a full or partial cycle. The states of charge in the various tanks, thus, slowly drift apart.

In a configuration with several pairs of tanks, care must also be taken to ensure thermal equilibrium between the tanks, since the conductivity of the electrolyte and thus the internal resistance of the stacks depends on the temperature, as shown in ref. [8]. A thermal imbalance has a negative effect on the usable capacity since the coldest module reaches the charge or discharge end voltage prematurely due to the voltage drop across the increased internal resistance. A fundamental problem of the series connection with individual tank pairs or tank groups is unevenly aged or imbalanced electrolytes with an overall oxidation state deviating from the mean value of 3.5 in a tank group. In this case, the capacity of the aged or not optimally adjusted tank pair is limiting for the entire system. This problem can be overcome by additional electrical switches, which can be used for bypassing the limiting tank group.

On the positive side, the series connection with tank groups allows high voltages to be achieved for the usage of correspondent PCS at acceptable shunt currents. However, operational management is complex, and the system is vulnerable in the event of a fault because it offers little redundancy. It is comparable in terms of modularity and sizing to Section 46.2.4.

46.2.6 Mixed Parallel-Series High-Voltage System (HV-MIX)

A mixture of parallel and series stack connections is described in ref. [11]. Figure 46.6 shows a simplified version based on the basic idea of the topology. Fluid connections have been omitted for clarity. Here, the individual battery stacks are arranged in a matrix, and each individual battery stack can be electrically disconnected from the part system using switches. Additionally, hydraulic decoupling is imaginable although not being described in ref. [11]. Each tank unit supplies multiple parallel-connected stacks, effectively avoiding shunt currents in between series-connected stacks. The clusters, dotted in Figure 46.6, are in turn electrically connected in series multiple times to reach a high-voltage level for the usage of an appropriate PCS.

The electrical parallel connection of several stacks leads to an even voltage distribution amongst them. As described for topology Section 46.2.1, unequal internal resistances then cause differences in the currents between these stacks. To detect a too large deviation, each battery stack should be equipped with a current measurement or as described in ref. [11] with an OCV/SoC measurement. The limitations due to unequal Coulombic efficiencies discussed in Section 46.2.5 also apply in this topology. On the one hand, the risk of unequal tank charge becomes less likely when multiple stacks are connected in parallel. On the other hand, this risk increases with the number of tank systems. Differences in the Coulombic efficiency of the individual stacks in one parallel connection are not critical since these are balanced

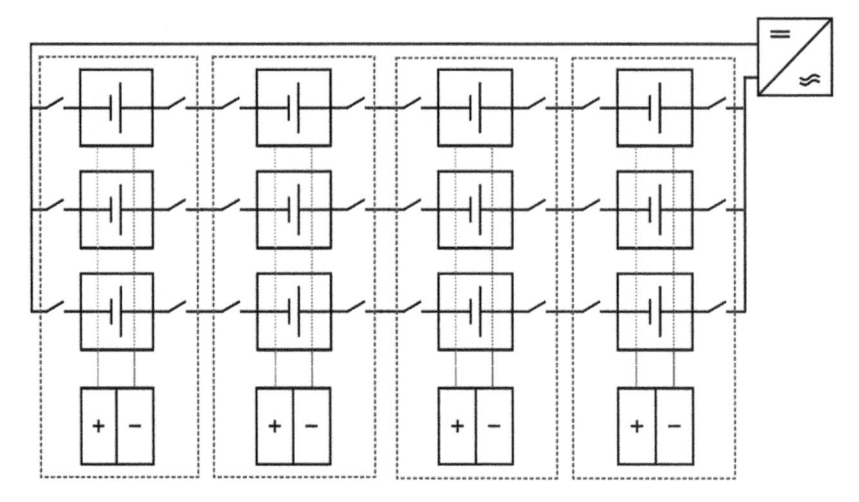

Figure 46.6 Topology HV-MIX: Combination of series and parallel connection of stacks. Each stack can be disconnected separately. Parallel-connected stacks share one tank pair. The fluid circuit is not depicted. Concept by Sumitomo. Source: Adapted from Refs. [11, 12].

on a potential level within the associated tank pair. The more stacks are connected in parallel, the less significant are slight variations in the Coulombic efficiency and the internal resistance. The mean value of the parallel-connected stacks is expected to be close to the mean value of the other in series-connected units then. As in any series connection of flow-battery stacks, voltage monitoring of the individual stacks is important to avoid destructive overvoltage. Generally, the installed switches offer the possibility to perform a balancing in case of SoC deviations.

With a typical stack size of 5–10 kW, this topology is only suitable for large systems. The high hardware costs for the electrical decoupling of the individual battery stacks and the high-cabling costs are critical, although the system is characterized by a high degree of redundancy on the stack level. If a stack fails, it can be easily compensated by the many parallel stacks.

46.3 Evaluation of the Topologies

Generally, any of the topologies presented in the previous sections are suitable for building a flow-battery system. All topologies have specific advantages and disadvantages. In Table 46.1, the advantages and disadvantages are evaluated qualitatively and the decision for the ratings will be explained here briefly.

The PCS efficiency and cost is generally rated as disadvantageous (−−) for low-voltage topologies and advantageous (++) for high-voltage topologies. The latter allows to omit the DC/DC stage or to use a simpler and more efficient topology. The LV-DC/DC topology has a small benefit, that the DC/DC can be fitted well to the flow battery, whereas standard DC/AC stages can be used for different grid applications and are thus rated only with a (−).

Table 46.1 Qualitative evaluation of the introduced topologies. Legend: +(+) = positive, O = neutral, and −(−) = negative.

Topology		PCS Efficiency/ Cost	Shunt current losses	SoC band Limitation by		Reliability	Modularity/ Flexibility	Preferable system size
				Ri	CE			
2.1.	LV-P	−−	+	O	++	−−	+	Small
2.2.	LV-DC/AC	−−	+	++	++	++	++	Small
2.3.	LV-DC/DC	−	+	++	++	+	++	Small
2.4.	HV-1T	++	−−	−	++	−−	−	Medium-large
2.5.	HV-MT	++	−/O/+	−	−−/−/O	−−	−	Medium-large
2.6.	HV-MIX	++	+	O	O	O	−−	Large

The low-voltage and the HV-MIX topologies are rated with a (+) in terms of shunt currents because these are limited to the stack-internal ones. For the HV-MT topology, it depends on the number of stacks in series that use a common tank system and is therefore rated between (−), (O), and (+). The latter applies when each stack is supplied by its own tank system. The HV-1T topology has the highest shunt current losses, as many series-connected stacks use the same electrolyte system (−−).

The SoC band limitation by R_i does not play a role for the low-voltage topologies with a DC/DC or DC/AC per stack (++). Because of the effect of uneven current distribution, it is rated (O) for the LV-P and HV-MIX topologies. The limitation is bigger for the HV-1T and HV-MT (−) topologies with several series-connected stacks, which cannot be disconnected by switches as is possible with HV-MIX (O).

In all systems where a common tank pair is used for all stacks, the Coulombic efficiency (CE) does not limit the SoC band (++). In contrast, this is relevant for topologies with multiple tank pairs, such as HV–MT. With many stacks per tank pair, the CE tolerances of the stacks compensate more (O) than in the case of few stacks (−) or only one stack per tank pair (−−). The latter would generally also apply for HV-MIX, but parallel-connected stacks compensate for tolerances, and disconnecting stacks by the usage of switches can help to use a larger SoC band. Therefore, it is rated (O).

The systems LV-P, HV-1T, and HV-MT have many single points of failure, so they offer lower reliability (−−) than the other topologies. The switches in HV-MIX (O) again allow to disconnect single stacks in case of failure, however, it is not protected against single failures, for example of the piping system. Systems with separate PCS, such as LV-DC/AC and LV-DC/DC, offer high redundancy. The latter has the common inverter as a single point of failure, so it offers a little less reliability (+) than the LV-DC/AC (++) topology. However, this comparison only applies when a single system, as depicted, is being used. In larger installations, multiples of these units can be connected on the AC side, thus increasing the overall system reliability.

In terms of modularity and flexibility, the systems LV-DC/AC and LV-DC/DC with separate PCS offer the most advantages (++), as the number of stacks can be chosen almost completely freely. The LV-P topology is a little more limited (+) because many stacks in parallel become difficult to handle. For HV topologies, a minimum number of stacks is always required to profit from the high-voltage PCS advantages. That makes it less modular and flexible. In comparison to HV-1T and HV-MT (−), the HV-MIX system needs even more stacks because of the additional parallel connection, so it is rated with a (−−) in this criterion.

46.4 Summary

In summary, smaller systems can be designed easily in low-voltage configurations. The advantages of low SoC band limitations and low system complexity bring fast adaption to different applications. High-voltage flow-battery systems are preferable for medium-to-large system sizes of the 3-digit kW to MW range. Low PCS cost and higher system efficiency bring benefits to the battery installations that pay off, especially in bigger projects. Furthermore, higher system complexity can be handled more easily in larger installations. Also, when several battery strings are installed in parallel on AC side to reach higher power and capacity, redundancy increases for high-voltage configurations as well. As the focus of flow-battery installations will probably advance toward larger projects, more and more installations with these concepts will surely be applied to the market in the future.

References

1 Wang, G., Konstantinou, G., Townsend, C. et al. (2016). A review of power electronics for grid connection of utility-scale battery energy storage systems. *IEEE Transactions on Sustainable Energy* 7: 1–1.
2 Xavier, L., Amorim, W., Cupertino, A. et al. (2019). Power converters for battery energy storage systems connected to medium voltage systems: a comprehensive review. *BMC Energy* 1: 7.
3 Stynski, S., Luo, W., Chub, A. et al. (2020). Utility-scale energy storage systems: converters and control. *IEEE Industrial Electronics Magazine* 14: 32–52.
4 Fink, H. and Remy, M. (2015). Shunt currents in vanadium flow batteries: measurement, modelling and implications for efficiency. *Journal of Power Sources* 284: 547–553.
5 Tang, A., McCann, J., Bao, J., and Skyllas-Kazacos, M. (2013). Investigation of the effect of shunt current on battery efficiency and stack temperature in vanadium redox flow battery. *Journal of Power Sources* 242: 349–356.
6 Darling, R., Shiau, H.-S., Weber, A., and Perry, M. (2017). The relationship between shunt currents and edge corrosion in flow batteries. *Journal of The Electrochemical Society* 164: E3081–E3091.
7 Reichelt, F. and Müller, K. (2020). Assessment of the reliability of vanadium-redox flow batteries. *Engineering Reports* 2: e12254.

8 König, S. (2017). *Model-based Design and Optimization of Vanadium Redox Flow Batteries,* Karlsruhe.
9 Horne, C.R., Sha, J.E., Lyle, W.D., and Mosso, R.J. (2011). Shunt current resistors for flow battery systems. Patent WO2012078786A2.
10 Sha, J. and Lin, B. (2014). Redox flow battery system configuration for minimizing shunt currents. Patent WO2014145788A1.
11 Kumamoto, T. (2015). Redox flow battery system and method for operating redox flow battery system. Patent US20170098849A1.
12 Sumitomo Electric Industries, Ltd. (2021). Datasheet: Redox Flow Battery Smart Energy Innovator. https://sumitomoelectric.com/sites/default/files/2021-04/download_documents/Redox_Flow_Battery_En.pdf (accessed 29 November 2021).

47

Vanadium FBESs installed by Sumitomo Electric Industries, Ltd

Toshio Shigematsu[1] and Toshikazu Shibata[2]

[1] Sumitomo Electric Industries, Ltd., Power Systems R&D Center, 1-1-3, Shimaya, Konohana-ku, Osaka 554-0024, Japan
[2] Sumitomo Electric Industries, Ltd., Energy Systems Division, 1-1-3, Shimaya, Konohana-ku, Osaka 554-0024, Japan

47.1 Historical Overview [1-3]

The development of energy storage batteries, including flow battery energy systems (FBESs), in Japan dates back to the 1970s. At that time, the difference between daytime and nighttime electric power demand was expanding with the popularization of air conditioners and the electric power load factor (=average load/maximum load) of electric power companies had declined year by year to less than 60%. To improve the load factor, large-scale energy storage technologies were expected to be developed, which would be able to store surplus electric power at night and deliver it in the daytime.

At that time, pumped hydropower plants, the only energy storage technology, accounted for about 10% of the total capacity of all power supply plants, but there were fewer sites for new construction due to environmental issues. The advanced energy storage battery was expected as a promising technology that would complement and replace pumped hydropower plants. In 1980, as a national project, the development of four new types of energy storage batteries, Fe/Cr, Zn/Br, Zn/Cl FBESs, and NaS batteries, was begun. In the mid-1980s, the development of FBESs and NaS batteries was also started jointly with electric power companies.

L.H. Thaller of NASA, USA proposed the principles of FBES in 1974 [4], and at almost the same time in Japan, Nozaki et al. of the National Institute of Advanced Industrial Science and Technology (AIST) started basic research as well. In 1975, Nozaki et al. reported feasibility study results including cost estimation through the conceptual design of real-scale FBES [5]. After that, Fe/Cr FBESs were developed to the stage of prototyping a 60 kW × 8 hours system in the above-mentioned national projects [6-10].

Flow Batteries: From Fundamentals to Applications, First Edition.
Edited by Christina Roth, Jens Noack, and Maria Skyllas-Kazacos.
© 2023 WILEY-VCH GmbH. Published 2023 by WILEY-VCH GmbH.

Around 1985, Vanadium FBES was invented by Professor M. Skyllas-Kazacos of the University of New South Wales Australia [11], dramatically improving the performance of FBESs. Beginning around 1990 in Japan, several companies have tested Vanadium FBESs of a few kW to 200 kW in national projects.

Sumitomo Electric began developing FBESs independently through joint research with Kansai Electric Power Co., Inc. (1985–2008). To begin, Fe/Cr FBES was developed and a trial demonstration test of a 60 kW × 8 hours system was carried out in 1989 [12–17]. Sumitomo Electric encountered some technical issues, such as the low-energy density of electrolytes and the capacity decay phenomenon due to hydrogen gas evolution in negative side reactions. Sumitomo Electric abandoned our effort to develop and commercialize Fe/Cr FBES and began to develop Vanadium FBES in 1992 [18]. A prototype test of a 450 kW × 2 hours Vanadium FBES was conducted in 1996. Subsequently, Sumitomo Electric commercialized Vanadium FBESs in 2001 and has delivered about 30 systems up to now, which range from small-scale systems for consumers to large-scale systems for power grids for both research and practical use. The delivery list is shown in Table 47.1.

In the following sections, the design concept and operation results of the typical Vanadium FBESs delivered by Sumitomo Electric are described along with development considerations made with regard to the energy demand context of the time.

Table 47.1 Sumitomo Electric's Vanadium FBESs delivery list [Total: 46 MW 159 MWh^{-1}].

	Customer	Application	Power, Energy	Completion
1	Utility	R&D	450 kW × 2 h	1996
2	Office building	Load leveling	100 kW × 8 h	2000
3	Utility	R&D	200 kW × 8 h	2000
4	NEDO project	Power stabilization of a wind turbine	170 kW × 6 h	2000
5	Construction Co.	R&D with PV	30 kW × 8 h	2001
6	Factory	UPS, peak cut	1.500 kW × 1 h (3000 kW × 1.5 sec)	2001
7	Utility	UPS, peak cut	250 kW × 2 h	2001
8	University	Load leveling, peak cut	500 kW × 10 h	2001
9	Research institute (Italy)	Load leveling	42 kW × 2 h	2001
10	Utility	R&D	100 kW × 1 h	2003
11	Office building	Load leveling	120 kW × 8 h	2003
12	Railroad Co.	UPS, Load leveling	30 kW × 3 h	2003
13	Office building	R&D	100 kW × 2 h	2003
14	Data center	UPS, emergency power supply	300 kW × 4 h	2003

Table 47.1 (Continued)

	Customer	Application	Power, Energy	Completion
15	Research institute	Load leveling	170 kW × 8 h	2004
16	Office building	Load leveling, Emergency power supply	100 kW × 8 h	2004
17	University	Load leveling, Emergency power supply	125 kW × 8 h	2004
18	Museum	Load leveling, Emergency power supply	120 kW × 8 h	2005
19	Utility	R&D with PV station	100 kW × 4 h	2005
20	NEDO project	Power stabilization of wind power	4000 kW × 1.5 h	2005
21	Sumitomo Electric	DC micro grid operation demonstration	2 kW × 5 h	2011
22	Sumitomo Electric	Demonstration with PV	1000 kW × 5 h	2012
23	Laboratory (Taiwan)	Micro grid demonstration	2 kW × 5 h	2012
24	Construction Co.	Micro grid demonstration	500 kW × 6 h	2015
25	Utility	Grid operation demonstration	15 000 kW × 4 h	2015
26	Sumitomo Electric	Behind the meter demonstration	500 kW × 4 h	2016
27	Utility (Taiwan)	Micro grid operation	125 kW × 6 h	2017
28	Utility (USA) (NEDO project)	Grid operation demonstration	2000 kW × 4 h	2017
29	Sumitomo Electric	Behind the meter demonstration	250 kW × 3 h	2017
30	Sumitomo Electric	Behind the meter demonstration	250 kW × 3 h	2018
31	Construction Co.	Peak cut/Islanding operation	250 kW × 3 h	2018
32	EPC (Belgium)	Behind the meter demonstration	500 kW × 3 h	2018
33	Government (Morocco)	Micro grid with PV	125 kW × 4 h	2019
34	Utility	Grid operation	17 000 kW × 3 h	2022

47.2 Typical Vanadium FBESs Delivered by Sumitomo Electric

47.2.1 1990s

At that time, the purpose of the development of advanced energy storage batteries was to level the electric power load of electric power companies and for that purpose, to build large-scale battery systems to serve the same function as pumped hydropower plants. Figure 47.1 shows the appearance of a 450 kW × 2 hours Vanadium FBES test facility installed at the Test Laboratory of Kansai Electric Power Co., Inc. in 1996. The system was designed and experimentally manufactured for the purpose of verification of basic elemental technologies assuming a large-scale system of MW class. The developed elemental technologies include real-scale size cell stacks, shunt current suppression measures, system configurations for a plurality of modules, and methods of long-term maintenance and management of the electrolytes.

The cell stack consisted of 4 sub-stacks and a sub-stack consisted of 15 cells. The electrode area in each cell was 5000 cm^2. These stacks were electrically connected in a 4 parallel/6 series configuration. Shunt currents control measures peculiar to flow batteries included sub-stacking construction during stacking many cells in series, making small and long piping for the electrolyte distribution to the sub-stacks, and modularization of tanks. The Vanadium FBES system performed at 82% energy efficiency (97% current efficiency and 85% voltage efficiency) with an average current density of 50 mA cm^{-2} [19].

The 450 kW Vanadium FBES was tested from December 1996 to June 2001. The fundamental technical issues identified during this term were durability of the materials such as positive electrodes, the reliability of electrolyte sealing technology for cell stacks, and electrical insulation technology for the whole battery systems.

Figure 47.1 The 450 kW × 2 hours Vanadium FBES installed at the test laboratory of Kansai Electric Power Co., Inc. (1996–2001).

47.2.2 2000s

In the 2000s, the electric power market was deregulated and electricity rates tended to decline. Up until that time, electric power companies had considered installing their own large-scale batteries in substations, but they changed to a policy of distributing small-capacity batteries to their consumers. By supplying the batteries to consumers for nighttime electricity at a low price, electric power companies gained a load-leveling effect that suppressed peak power in the daytime. Consumers benefitted from lower electricity rates applied at night and a decrease in contract peak power. Furthermore, new features were also added to increase the value of the battery such as momentary voltage drop compensation and emergency power supply.

47.2.2.1 A 500 kW×10 hours Vanadium FBES with Underground Tanks Installed at a University [2001–2011]

It was difficult to install Vanadium FBESs compactly to consumers (commercial buildings, factories, etc.) in urban areas where electric power demand was higher because the energy density was relatively low compared with other batteries. Sumitomo Electric developed rubber tanks that can be installed in underground spaces by focusing on the electrolytes, which occupy a relatively large volume but whose shape can be changed as needed. The rubber tank is foldable and can be carried in with other tank equipment through small openings in the basement. The electrolyte of Vanadium FBESs is a non-flammable aqueous solution, and there are few laws and regulations to restrict their use so they can be installed in commercial buildings. [20, 21].

Figure 47.2 shows a 500 kW × 10 hours Vanadium FBES installed through the use of an underground tank at a university campus. The electrolytes were stored in rubber tanks in a concrete frame installed underground and the cell stacks and necessary electrical equipment were installed on the first floor of the building. Load-leveling operation was carried out once a day to store inexpensive electricity at night and deliver the power in the daytime. An example of the operating results is shown in Figure 47.3.

The Vanadium FBES is a good example utilizing the design flexibility in which the cell stack and the tank can be installed independently. Although there was several percent decrease in energy efficiency and the resulting decrease in capacity, this system was actually operated for 10 years. In order to maintain the long-term performance during this period, various electrolyte management techniques were essential such as SOC monitoring, countermeasures for crossover phenomenon, and ion valence adjustment between positive and negative electrolytes.

47.2.2.2 A 1.5 MW × 1 hours Vanadium FBES with Momentary Voltage Drop Compensation Function in a Factory [2001–2007]

This system was an example of a Vanadium FBES installed in a semiconductor factory. In this area, there are frequent momentary voltage drops due to lightning strikes in winter, so countermeasures such as a UPS were necessary to protect the manufacturing line. The frequency of the lightning was extremely high and frequent

Figure 47.2 The 500 kW × 10 hours Vanadium FBES with the underground tanks at the university campus (2001–2011).

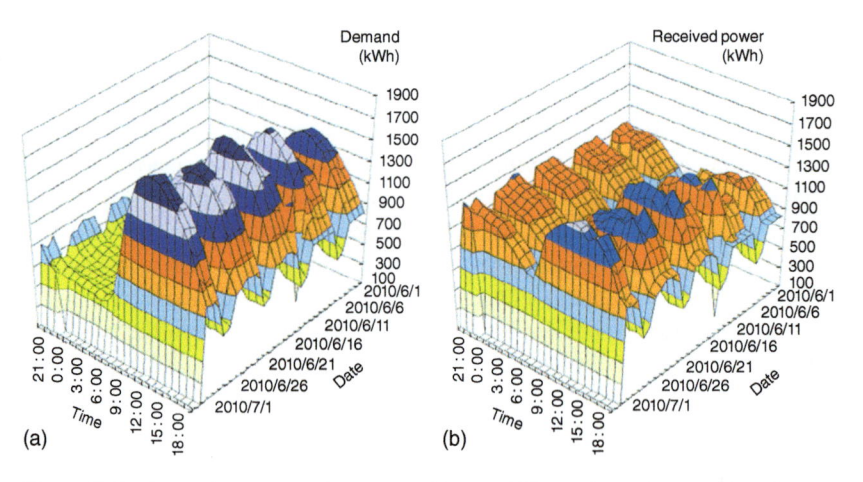

Figure 47.3 Operating test results of the 500 kW × 10 hours Vanadium FBES (a) Demand curves (b) Received power curves.

charge/discharge cycle operations were required for the battery, making it difficult to apply conventional lead-acid batteries due to the limitation of the charge/discharge cycle life.

A Vanadium FBES was adopted because there was no limitation of charge/discharge cycles in principle, and it was also suitable for irregular charge/discharge operations. Normally, the Vanadium FBES was connected to a commercial line and was operated with a peak cut operation of 1.5 MW × 1 hour per day for checking the soundness of the system. When a momentary voltage drop occurred, the system was designed to supply power of 3 MW for 1.5 seconds without momentary power interruption. As for the installation of the Vanadium FBES, the vacant space of the two-storied factory building was utilized. Cylindrical tanks made of polyethylene were installed on the first floor and cell stacks were installed on the second floor. Figure 47.4 shows the appearance of the installed Vanadium FBES. The battery capacity was designed to have the minimum required capacity of 1 hour. This system has actually compensated for the momentary voltage drops over 100 times during a five-year period.

This system is a good example of increasing the value of the Vanadium FBES for consumers by adding an instantaneous voltage drop compensation function as a new feature in addition to the peak cut function.

47.2.2.3 A 4 MW × 1.5 hour Vanadium FBES Installed in a Wind Farm [2003–2007]

Currently, the utilization of renewable energy sources, such as photovoltaic power generation and wind power generation, is being implemented as a measure against climate change. These naturally fluctuating power sources are characterized by power output that fluctuates according to the weather, and there was concern that the stability of power systems would be adversely affected when a large amount of

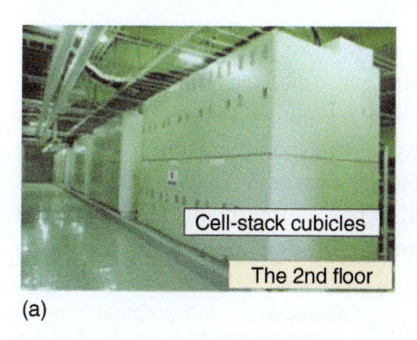

Figure 47.4 The 1.5 MW × 1 hour Vanadium FBES with momentary voltage drop compensation function at the factory (2001–2007). (a) Inside view of the 2nd floor. (b) Inside view of the 1st floor.

these power sources is introduced into the power grid. The purpose of the NEDO (New Energy and Industrial Technology Development Organization) project was to demonstrate that frequently fluctuating wind power output was able to be smoothed by applying batteries in combination with wind power generators.

Although the output fluctuation of wind power generation includes periods varying from seconds to hours, Vanadium FBESs can be flexibly and optimally designed by increasing or decreasing the amount of electrolyte according to the required specification. For short-period fluctuations, high-rated power output characteristics [22] can be utilized.

Figure 47.5 shows a conceptual configuration diagram of the entire system. The battery was installed at the interconnection point of the wind power generators and the smoothing procedure was carried out by the battery absorbing the fluctuating output. As a smoothing method, a smoothing target output value was set by removing the short-period components using a low-pass filter having a certain time constant with respect to the wind power output. The battery outputs the difference between this target value and the actual wind power output. A smoothed output which is the sum of this battery output and the actual wind power output was sent to the power grid.

Demonstration tests of the NEDO project had been conducted for two terms. The first project titled "A survey on the possibility of installing batteries with wind power generators" was conducted as the first term in 2000. Three types of batteries (Vanadium FBES, NaS, and lead-acid batteries) were installed with the existing wind power generators at the respective sites [23]. The Vanadium FBES project was entrusted to the Institute of Advanced Energy Science and Technology and a Vanadium FBES with a rated output of 170 kW × 6 hours (maximum output 275 kW) was installed in conjunction with one unit (AC275 kW output) of the wind power generators at the Horikappu Wind Power Plant of Hokkaido Electric Power Co., Inc.

Several demonstration tests were conducted. For example, smoothing operation tests in which the smoothing time constant was changed (8 minutes, 1 hour, 8 hours, etc.), and patterned output operation tests in which the battery was operated and controlled so that the total output matched a predetermined output pattern. In addition, several improved technologies were tested and their usefulness verified (as practical technologies) with the goal of improving economic and system efficiency. For example, (i) utilization of high-rated output characteristics

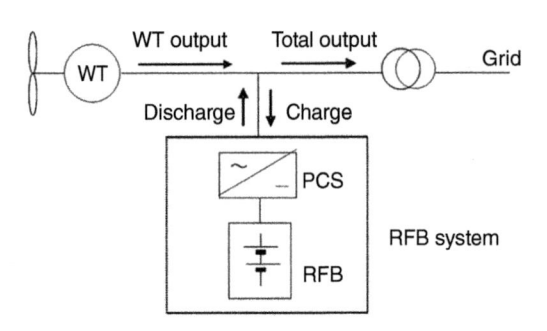

Figure 47.5 A conceptual configuration diagram of the entire system with a FBES installed in a wind power plant.

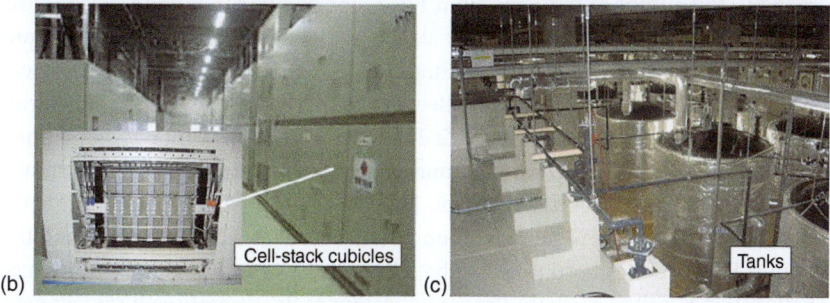

Figure 47.6 The 4 MW × 1.5 hours Vanadium FBES at the Tomamae winvilla power station in Hokkaido (2003–2007). (a) The building for the Vanadium FBES. (b) Cell-stack cubicles. (c) Electrolyte Tanks.

(1.5 times the rated output for short time), (ii) variable time constant control, and (iii) electrolyte flow rate control. These techniques were applied to the following demonstration tests at the wind farm [24].

In 2003, a NEDO Project titled "Technology Development for Stabilization of Wind Power Systems" was conducted as the second term [25]. The project was entrusted to Electric Power Development Company (J-Power) and the demonstration tests were conducted at the Tomamae Winvilla Power Station in Hokkaido (19 wind power generators, total output 30 600 kW) with a newly installed Vanadium FBES (rated output AC4 MW × 1.5 hour: maximum output 6 MW).

This location has a lot of snow in winter and the system component equipment was all installed in a building, including the power conditioning systems (PCSs), battery control boards, battery cubicles that house cell stacks, electrolyte tanks, and piping. The exterior of the building and the appearance of the Vanadium FBES are shown in Figure 47.6. The Vanadium FBES was composed of four banks and the cell stacks (DC rated output 45 kW per cell stack) in one bank were electrically connected in a 6 parallel/4 series configuration and the rated output of one bank was 1 MW. Each bank consisted of four modules, each module consisted of six cell stacks, positive and negative electrolyte tanks (each 15 m^3), and one heat exchanger.

The purpose of this project was to demonstrate that the short-term output fluctuations (from a few seconds to several tens of minutes) were able to be smoothed.

The smoothing control method was similar to the case of the aforementioned single wind power generator.

In actual operation, the various system control technologies, including methods, described above were utilized. For example, the remaining capacity of the battery may reach its lower limit due to the energy losses during charging/discharging operations or the charged capacity may reach its upper limit because charging energy may exceed discharging energy in some cases. In order to avoid these situations, ancillary charging/discharging output control was conducted while performing smoothing control operations to maintain the remaining capacity level appropriately at all times. This control was called battery capacity feedback control.

The verification test results of the Vanadium FBES are shown in Figure 47.7. The chart shows the output of the Vanadium FBES, the ancillary charging/discharging output, and the monitor cell voltage as well as the wind power generation output when the smoothing operation with a time constant of 10 minutes was performed. The monitor cell voltage indicates the electromotive force of the battery and corresponds to the state of charge. It was verified that the wind farm output was smoothed as expected and the SOC level was also maintained within the predetermined range by the battery capacity feedback control. In addition, the following control methods were also carried out. When the wind power output suddenly changes and the Vanadium FBES cannot keep up with the wind power output fluctuations because the battery has a limited capacity, the time constant of the smoothing filter was temporarily reduced to keep performing the smoothing operation as much as possible. This operation was called the time constant variable control. The bank control tests were also conducted to operate with the optimum number of banks according to the output required for the battery system from the viewpoint of improving the system efficiency. These system operation control technologies have been utilized as basic control technologies for subsequent large-scale Vanadium FBESs [26].

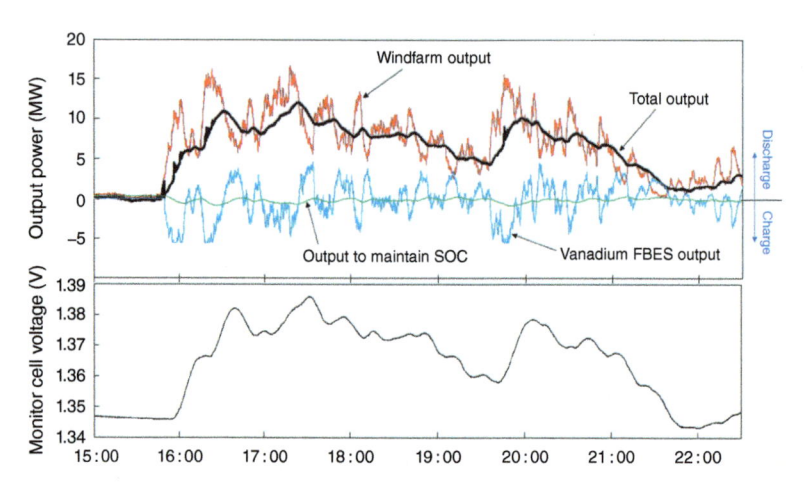

Figure 47.7 Verification test results of the smoothing operation with the Vanadium FBES (time constant: 10 minutes).

47.2.3 Recent Vanadium FBESs Since 2010

As a measure against climate change, we have entered an era in which the introduction of renewable energy sources is actively promoted on a global scale, with the aim of realizing a carbon-neutral and decarbonized society. Concerned about the impact of output fluctuations peculiar to renewable energy sources on power systems, energy storage batteries have become more important as a measure of stabilizing power systems. Sumitomo Electric has also promoted the development of Vanadium FBESs that can meet these requirements by using elemental technologies and system operation technologies that have been developed previously.

47.2.3.1 A Vanadium FBES for Optimal Energy Management in a Factory with Photovoltaic Power Generation (in Operation Since 2012)

In 2012, Sumitomo Electric installed a 1 MW × 5 hours Vanadium FBES at its Yokohama Works, connected it with a 100- kW concentrated photovoltaic power generation system, gas generators, and other systems, and began demonstration tests of optimal energy management according to the power demand of the plant [27]. Figure 47.8 shows the appearance of the entire test equipment. The equipment shown in the figure was as of 2012 and some of it has since been replaced with more advanced equipment.

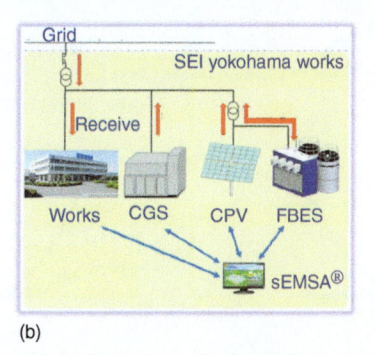

Figure 47.8 The 1 MW × 5 hours Vanadium FBES connected with the 100-kW concentrated PV system, etc. at Sumitomo Electric's Yokohama Works (as of 2012). (a) Exterior of the Vanadium FBES and CPV system. (b) Diagram of the entire system.

The 32 cell stacks are housed in 8 battery cubicles and the electrolytes are stored in 16 cylindrical polyethylene tanks (25 m^3 per unit). The rated output of the PCS is 1 MW, and it consists of one 500 kW unit and two 250 kW units. The Vanadium FBES is installed inside a dike outdoors and normal rainwater is drained by drainage pumps. In the unlikely event that the electrolyte leaks, measures have been taken to prevent the electrolyte from leaking out of the dike.

The operation of the equipment is managed by the energy management system (EMS). The EMS manages and optimizes the operation of each equipment to minimize the total energy cost, taking into consideration the energy efficiency characteristics and fuel cost of each equipment, such as batteries and generators, and factory power demand. In basic peak cut operation, the Vanadium FBES is charged during off-peak hours such as at night, and the peak power during the daytime is reduced along with the operation of the other generators (see Figure 47.9). Recently, there have been cases where EMS controls and operates the Vanadium FBES in response to demand response (DR) signals from the electric power companies. Furthermore, verification tests of supply and demand adjustment operations are being conducted with the Vanadium FBES as one of the energy resources of VPPs (virtual power plants).

The purpose of this Vanadium FBES is to verify the long-term stability of the newly developed technology as well as to verify the performance. Figure 47.10 shows the transition of the current efficiency and the cell resistance of the Vanadium FBES, which are indicators of long-term performance. Operation conditions typically consist of a current density of around 0.15 A cm^{-2} and an SOC upper limit of approximately 80%, with electrolyte management operations such as partial remixing of the positive and negative electrolytes being performed on a daily basis. It has been verified that it can be operated stably over the long term.

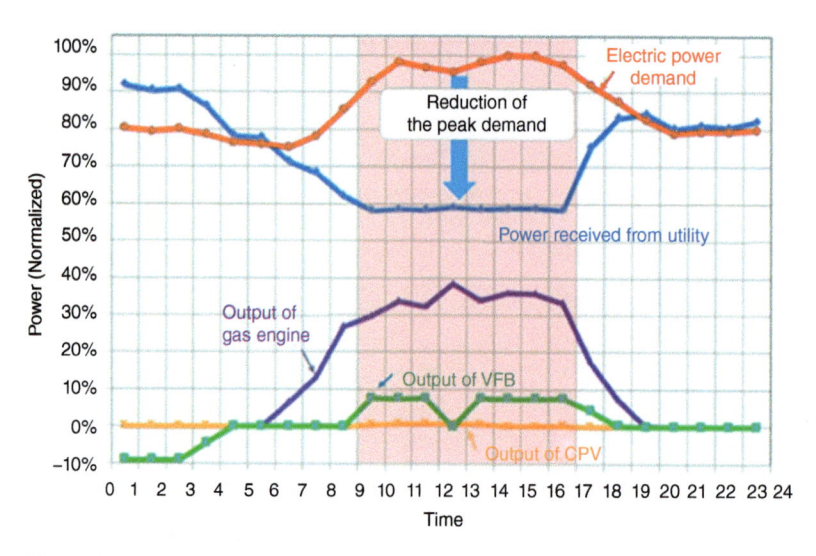

Figure 47.9 Basic peak cut operation test results with the Vanadium FBES, the CPV and the gas generators controlled by EMS.

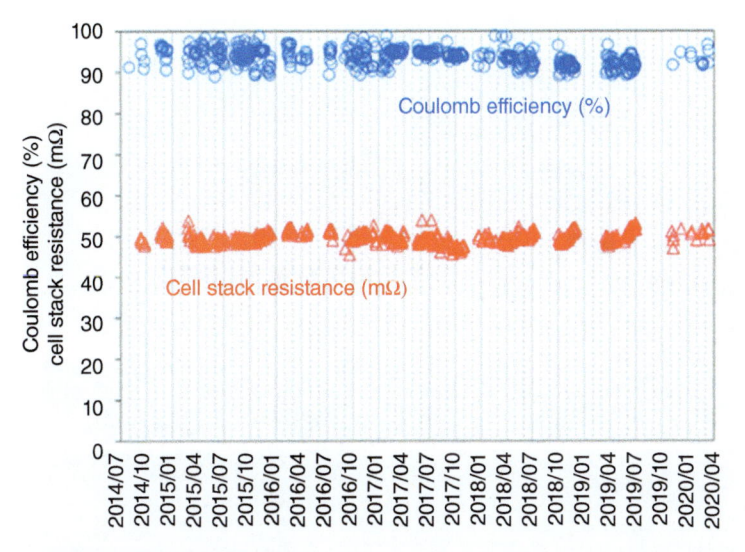

Figure 47.10 Transitions of the Coulomb efficiency and the cell stack resistance of the Vanadium FBES.

47.2.3.2 Operation Example of Grid Control (2015–2018 Demonstration, 2019 ~ Practical Operation)

A 60 MWh Vanadium FBES was installed in the Minami-Hayakita substation of Hokkaido Electric Power Co., Inc. (HEPCO), and operation started in December 2015, to demonstrate the application of the storage battery as one of the dispatchable power source. This system has been operated to demonstrate various types of control modes, such as load frequency control, compensation for wind power and PV output fluctuation, and control for countermeasures against overgeneration [28–32]. Figure 47.11 shows the exterior of the battery building and the appearance of the system. The system has a nominal rated output power of 15 MW, a maximum output power of 30 MW, and an energy capacity of 60 MWh. In operation, the maximum current density of a cell stack is 0.35 A cm^{-2}. The system consists of 13 banks that have 5 MWh of energy capacity (1250 kW × 4 hours). The system is operated by the central load dispatching center of HEPCO. This project was promoted by the Ministry of Economy, Trade, and Industry (METI).

As for short periodic fluctuation mitigation control, governor-free equivalent control and auto-frequency control have been evaluated. In governor-free equivalent control, the battery output is controlled by the power conditioner of the Vanadium FBES according to a frequency deviation detected at the substation. In auto-frequency control, area control error (ACE) is calculated by using the frequency deviation detected at the central load dispatching center and is allocated between the battery system and other plants. Figure 47.12 shows an example of the comparison of frequency deviation between the period with governor-free equivalent control (12 : 30 ~ 14 : 00) and that without the control (14 : 00 ~ 14 : 30). In the actual grid, the effectiveness can be seen from the reduction of spike-like

Figure 47.11 The 60 MWh Vanadium FBES installed in Hokkaido, Japan (a) Battery System Building. (b) Battery Cubicles (2nd floor). (c) Tanks, Pumps and PCSs.

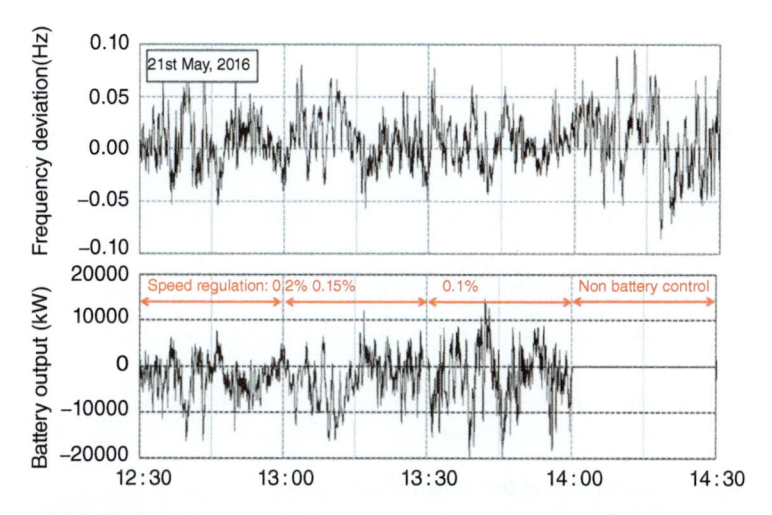

Figure 47.12 Test results of governor-free equivalent control operation.

frequency deviation, although the effectiveness of the control was weak due to the small battery output against the system capacity of the Hokkaido area.

In order to evaluate aging characteristics, energy capacity and system efficiency were measured once a year during the demonstration period. Aging characteristics were evaluated for each bank, taking into consideration energy losses, including conversion loss and energy consumption of the auxiliary subsystems. The measurement results are shown in Figure 47.13 and Figure 47.14. Energy capacities exceeding rated capacities were confirmed in all measurements. And it was judged that the calculated decrease rate from a statistical point of view taking into consideration the temperature and the operating condition of the auxiliary power, was not significantly affecting the system. System efficiency was kept at 70% which is its initially designed value. No significant change is seen in capacity and efficiency; the cell is operating as predicted towards a designed lifetime of 20 years.

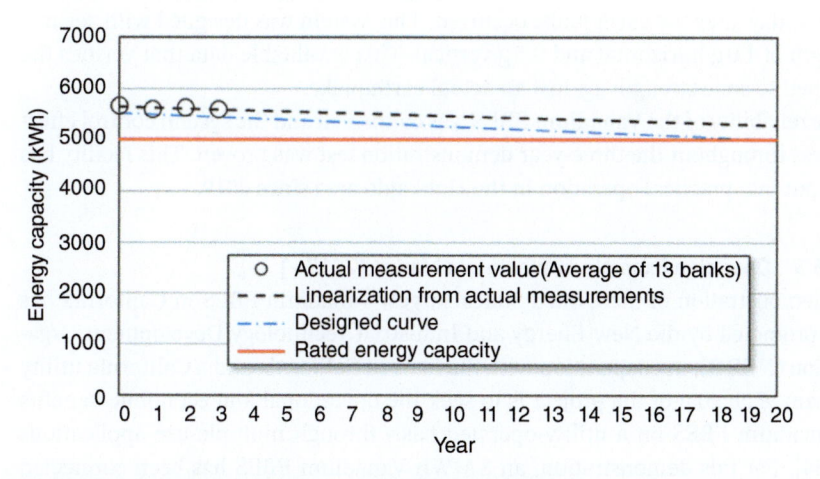

Figure 47.13 Aging characteristics of the energy capacity.

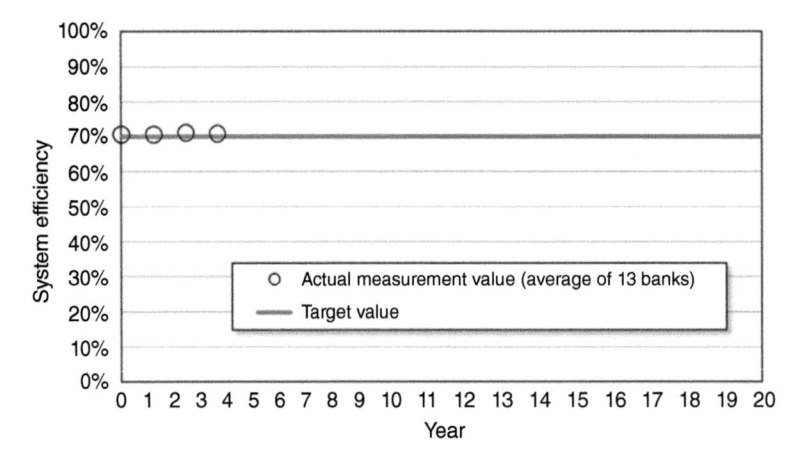

Figure 47.14 Aging characteristics of the system efficiency.

Availability was also evaluated. Availability is defined as Eq. (47.1).

$$\text{Availability} = \frac{\sum_{i=1}^{13}\{T_{\text{Total}}(i) - T_{\text{SMDT}}(i) - T_{\text{USDT}}(i)\}}{\sum_{i=1}^{13}\{T_{\text{Total}}(i) - T_{\text{SMDT}}(i)\}} \qquad (47.1)$$

$T_{\text{Total}}(i)$: Total timeframe of "Bank i"
$T_{\text{SMDT}}(i)$: Scheduled Maintenance Down Time of "Bank i"
$T_{\text{USDT}}(i)$: Unscheduled Down Time of "Bank i"

In this definition, availability of 99.3% was confirmed for the three-year demonstration period. The low occurrence of unexpected serious trouble and shutdown due to temperature restrictions contributed to keeping high availability.

During the demonstration period, a large earthquake of magnitude 6.9 occurred in this area. The Vanadium FBES site was only 16 km away from the epicenter, but there was no damage or breakdown of the battery system and it was possible to re-operate the next day after the earthquake occurred. This system was designed with seismic strength of 1.0 g horizontal and 0.5 g vertical. This is valuable data that verifies the designed seismic strength against an actual earthquake.

The reliability of the Vanadium FBES was confirmed and the system control effectiveness throughout the three-year demonstration test was proven. This facility has been put into practical operation in the Hokkaido area since 2019.

47.2.3.3 Operation Example of Wholesale Market [2017~]

The demonstration of the United States' largest Vanadium FBES in California has been promoted by the New Energy and Industrial Technology Development Organization (NEDO), in cooperation with the State of California and a California utility company. The goal of the project is to seek the operational and economic benefits of Vanadium FBES on a utility-operated basis through multiple-use applications [33, 34]. For this demonstration, an 8 MWh Vanadium FBES has been connected to the distribution grid. The system is shown in Figure 47.15.

Figure 47.15 The 8 MWh Vanadium FBES installed in California.

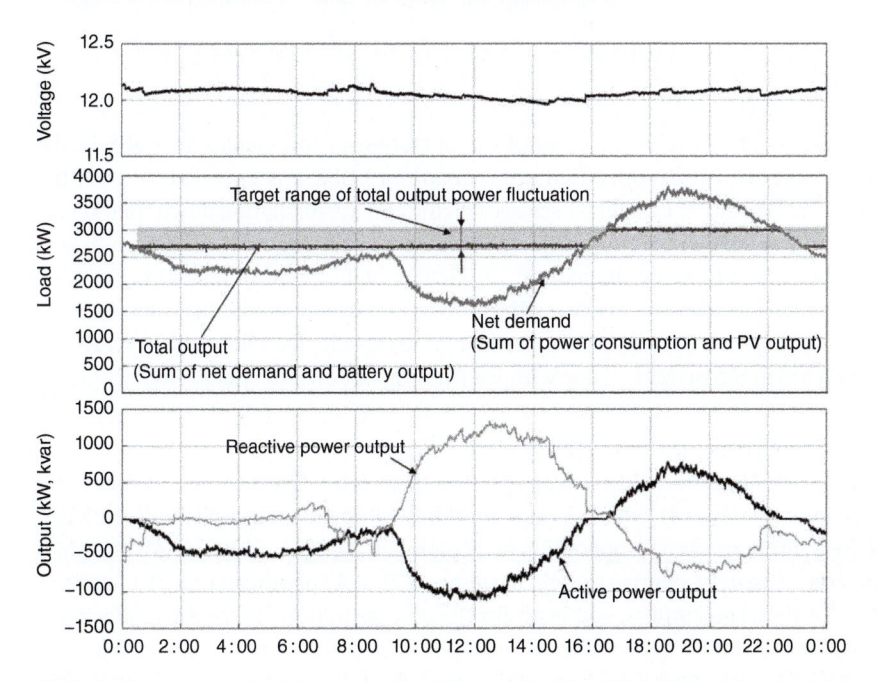

Figure 47.16 Active and reactive power output of the Vanadium FBES in the distribution grid.

Figure 47.16 shows an example in which the Vanadium FBES was operated in the distribution grid. The output of photovoltaics connected to the distribution grid caused fluctuations in NET demand (sum of consumer load and PV output) and voltage fluctuations. The Vanadium FBES outputs active power so as to suppress fluctuations in NET demand. In addition, the Vanadium FBES outputs reactive power in order to suppress the change of voltage to within a certain range. As shown in Figure 47.16, it was confirmed that fluctuations in NET demand and voltage could be suppressed by these controls.

This Vanadium FBES was certified as one of CAISO's power resources and participates in the energy market and ancillary service market. These markets consist of the "day-ahead market" and "real-time market." The former is a market that predicts the electricity demand of the next day. The latter is for adjusting the deviation from the prediction.

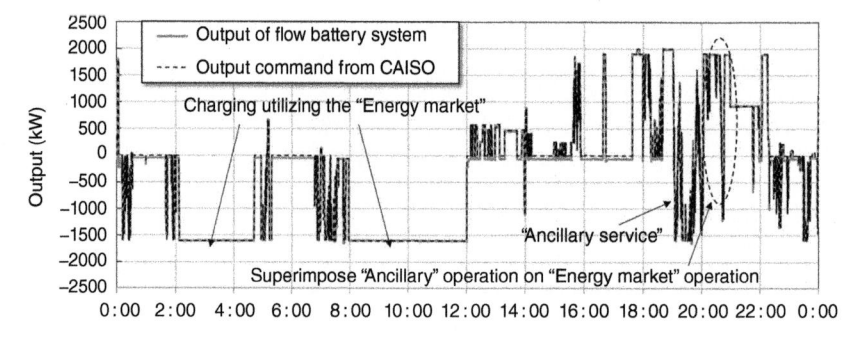

Figure 47.17 Superimposing power output for ancillary service and energy market in CAISO.

Figure 47.17 shows an example of the operation of the Vanadium FBES in the CAISO market. In this operation, the Vanadium FBES was operated for both the energy market and the ancillary service (regulation up, regulation down) market for frequency adjustment simultaneously. As shown in the figure, it can be confirmed that the output from the Vanadium FBES accurately follows the command value from CAISO. In addition, the Vanadium FBES could be operated by using the wide state of charge (SOC) range, and by utilizing the capacity for 4 hours, it is possible to participate in the market all day without running out of battery capacity.

47.2.3.4 Microgrid Operation [2018~]

A 1.7 MWh Vanadium FBES was installed into John Cockerill (JC)'s microgrid demonstration system which is named "Micro Réseau Intégré Seraing (MiRIS)" in Belgium, and started operation in September 2019[34, 35]. The MiRIS plant consists of battery energy storage systems, 1.71 MW photovoltaic (PV) systems, 500 kVA diesel generators, and loads for the demonstration of off-grid applications. A Vanadium FBES demonstration was performed both on-grid and off-grid at the MiRIS plant. The Vanadium FBES used for this demonstration is shown in Figure 47.18. The system is fully containerized and its configuration contributes to footprint reduction as well as lower transportation costs and less installation work, in comparison to that of a conventional plant system, which consists of separate tanks, electrolyte circulation, auxiliary system, and cell stacks.

During on-grid operation, the Vanadium FBES was successfully charged or discharged from or to the grid according to set-point value with fast response time and sufficient power quality, which enables peak shaving and demand response applications. During off-grid operation, the Vanadium FBES acted as a voltage source and maintained the frequency and voltage of the MiRIS microgrid.

Figure 47.19 shows the voltage and frequency trends when the output of the Vanadium FBES is changed. The Vanadium FBES takes the role of a voltage source in off-grid mode. As a result, it was confirmed that voltage fluctuation was kept within 2%, which is sufficient quality, despite the sudden change in load consumption and PV power. Assuming the Vanadium FBES operated in parallel with generators in off-grid mode, frequency control with the droop method (synthetic inertia control)

Figure 47.18 The 1.7 MWh Vanadium FBES installed in Belgium.

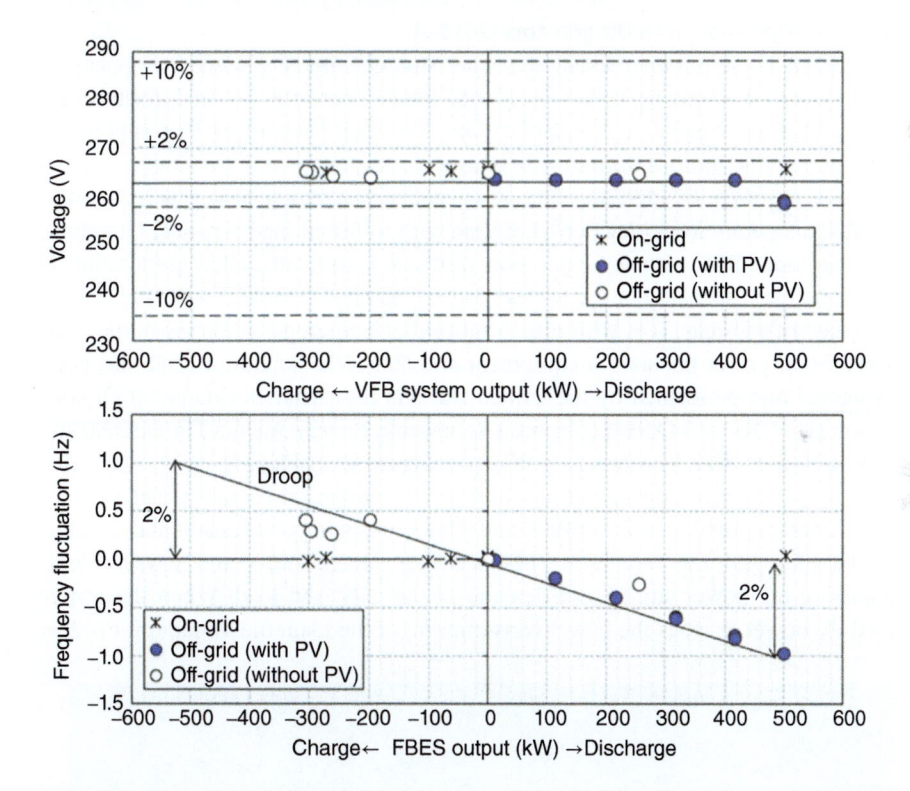

Figure 47.19 Voltage control and frequency control with droop method.

is applied. As a result, it is confirmed that the frequency trend for the Vanadium FBES output corresponded to a droop gain (the ratio of the frequency deviation of the microgrid to the rated frequency when the VRBF system outputs at the rated power) of 2%.

The transition from grid interconnection to microgrid without power outage is called a "seamless transition." Figure 47.20 shows an example of the voltage

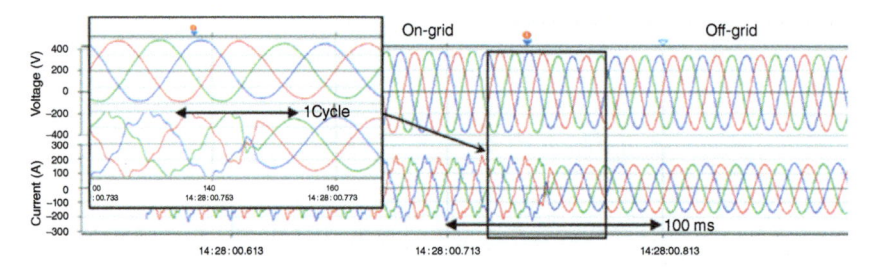

Figure 47.20 A seamless transition from on-grid operation to off-grid operation.

waveform during the transition to off-grid. As this figure shows, the Vanadium FBES achieved a seamless transition from on-grid to off-grid operation without disturbance of the voltage waveform.

47.2.3.5 Operation in an Off-grid Area [2019~]

Microgrids using photovoltaic systems and storage battery systems are appearing as a means of supplying power in off-grid areas where the power infrastructure is not well developed. In this application, charging/discharging that suppresses long-period fluctuations from sunrise to sunset and short-period fluctuations due to clouds is required. In addition, since it is installed in a remote area, it is required to be able to operate for a long period of time with minimal maintenance. Therefore, a Vanadium FBES with few operation restrictions and no cycle deterioration is very suitable. Especially in the African region, about 600 million people belong to non-electrified areas. It is with this in mind that a microgrid demonstration with 1 MW solar power generation equipment and 500 kWh Vanadium FBES has been promoted and performed by the United Nations Industrial Development Organization (UNIDO) at MASEN (Morocco Renewable Energy Agency) in Ouarzazate, Morocco since 2019. The system is shown in Figure 47.21 [36, 37].

Figure 47.22 shows the operating waveform on a microgrid consisting of a photovoltaic system, the Vanadium FBES, and a load consisting of a resistor and a motor. In the microgrid, the Vanadium FBES acts as a voltage source and plays a role in adjusting the power supply and demand of the CPV and load. When the generated PV power exceeds the power consumption of the load, the Vanadium FBES is

Figure 47.21 The 500 kWh Vanadium FBES installed in Morocco.

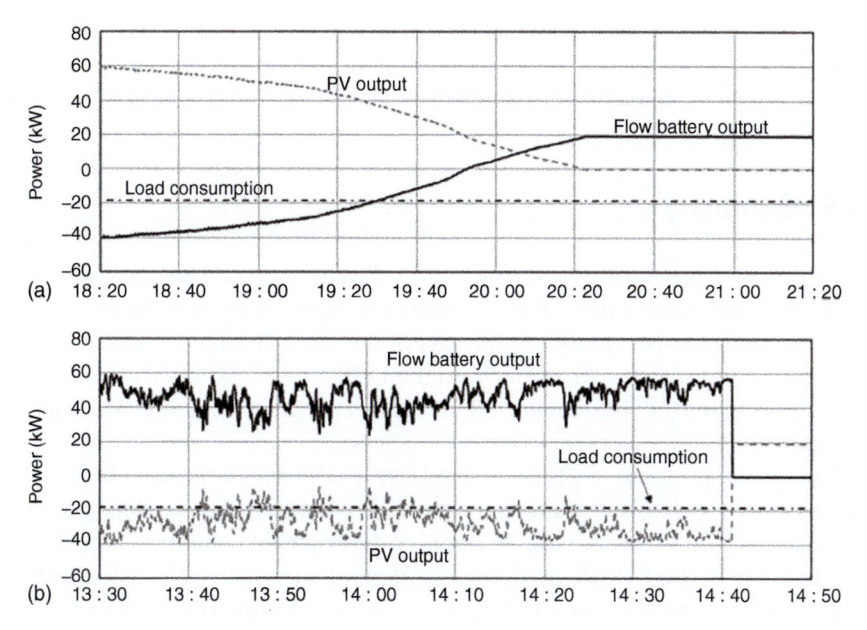

Figure 47.22 Test results of islanding mode operation (a) Sunny Day. (b) Cloudy Day.

charged with surplus power. When the power consumption of the load exceeds the generated power, power is discharged from the Vanadium FBES to keep the power quality constant. As shown in Figure 47.22, in fine weather, the decrease in power generation over a long period due to the sun setting is compensated by the discharge of the Vanadium FBES. In cloudy weather, although the generated power of the PV fluctuates for a short period, it was confirmed that the output fluctuations are compensated for stably with a sufficiently fast response speed, and the total power of the PV and Vanadium FBES is maintained at 20 kW of load.

This system was installed in desert areas and is required to operate in high temperatures and high solar radiation conditions. In response to this requirement, the design of the air-cooled heat exchanger was optimized, and as a result, as shown in Figure 47.23, the electrolyte temperature could be maintained almost constant

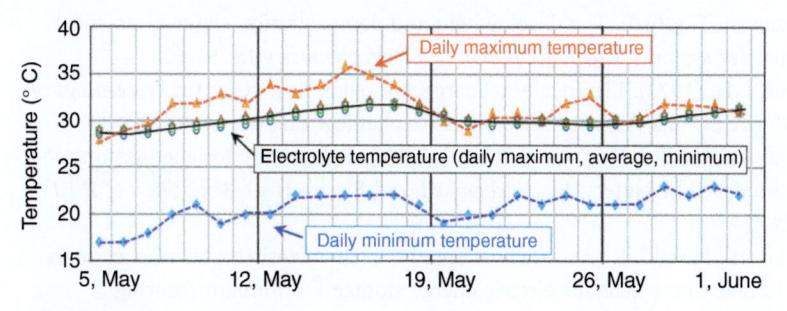

Figure 47.23 A correlation between air temperature and electrolyte temperature during 1-month operation test.

during continuous operation for one month. It shows that stable continuous operation can be performed without using an air conditioner, which causes a decrease in system efficiency.

47.3 Summary

About 50 years have already passed since L. H. Thaller proposed the basic principle of FBES and development of FBESs started in Japan. Sumitomo Electric now has a development history of about 35 years in this field. During this period, it is noteworthy that the invention of the Vanadium FBES by M. Skyllas-Kazacos dramatically improved the performance of FBESs and made great progress toward their practical use.

Although the principle of FBESs is simple, it is extremely difficult to build up a large-capacity system and guarantee performance that can withstand long-term practical use. To do so, a wide range of advanced development is required, including basic research on redox reactions, development of cell materials such as electrodes and membranes, development of cell stack structures, and system development that integrates them.

Currently, Vanadium FBESs have already reached the level of practical use in electric power systems, but further improvement in performance and cost reduction are required for full-scale introduction. In order to meet these demands, it is necessary to promote further R&D activities and verification tests based on the various research results, knowledge, and experiences described in this book. We hope that Sumitomo Electric's findings described in this chapter will contribute to the future development of FBESs.

References

1 Shigematsu, T. (2011). Redox flow battery for energy storage. *SEI Technical Review* 73: 4–13.
2 Shigematsu, T. (2019). Recent development trends of redox flow batteries. *SEI Technical Review* 89: 5–11.
3 Shigematsu, T. (2019). The development and demonstration status of practical flow battery systems. *Current Opinion in Electrochemistry* 18: 55–60.
4 Thaller, L.H. (1974). Electrically rechargeable redox flow cells. In: *Proceedings of the 9th IECEC* (ed. American Society of Mechanical Engineers), 924–928.
5 Nozaki, K., Kaneko, H., Ozawa, T. et al. (1975). Potentiality for energy storage with a redox flow battery. *Electrochemical and Electro-Heat Work Shop of IEEJ CH-75-3*.
6 Kaneko, H., Nozaki, K., and Ozawa, T. (1977). Characteristics of redox systems used in redox flow cells for electric energy storage I: criteria, measuring methods and results of chromium, titanium, and iron systems. *Bulletin of the Electrotechnical Laboratory* 41 (11): 877–887.

7 Nozaki, K., Kaneko, H., and Ozawa, T. (1979). Prospect of electric energy storage by secondary batteries. *Circulars of the Electrotechnical Laboratory* 201: 51–60.

8 Nozaki, K., Kaneko, H., Negishi, A. et al. (1984). Performance of ETL new 1kW redox flow cell system. In: *19th IECEC*, 844–849.

9 Hamamoto, O., Takabatake, M., Yoshitake, M. et al. (1985). Research and development of 10 kW class redox flow battery. In: *20th IECEC*, vol. 2, 98–104. Society of Automotive Engineers, Inc. SAE P-164.

10 Kanari, K., Nozaki, K., Kaneko, H. et al. (1990). Numerical analysis on shunt current in flow batteries. *Proceeding of 25th IECEC* 3: 326–331.

11 (a) Sum, E., Rychcik, M., and Skyllas-Kazacos, M. (1985). Investigation of the V(V)/V(IV) system for the positive half-cell of a redox battery. *Journal of Power Sources* 16: 85–95; (b) M. Skyllas-Kazacos and RobinsR.G. (1986). All-vanadium redox battery. US Patent No. 4, 786,567; Australian Patent 575247; Japan Patent 2724817.

12 Shimizu, M., Mori, N., Kuno, M. et al. (1986). Test results of a 1 kW RF battery. *Kansai-section Joint Convention of Institutes of Electrical Engineering* G3-36 G123.

13 Shimizu, M., Mori, N., Kuno, M. et al. (1988). Development of redox flow battery. In: *Proceeding of 172th ECS Meeting 88-11*, 249–256.

14 Sakamoto, T., Mori, N., Kuno, M. et al. (1989). Development of redox flow battery. *SEI Technical Review* 28: 180–184.

15 Tanaka, T., Sakamoto, T., Mori, N. et al. (1990). Development of a redox flow battery. *SEI Technical Review* 137: 191–195.

16 Tanaka, T., Sakamoto, T., Mori, N. et al. (1991). Development of a 60kW class redox flow battery system. *The third international conference batteries for utility energy storage* (18–22 March Japan).

17 Shigematsu, T., ItoT., Fujitani, K. et al. (1993). Development of a redox flow battery system. *The fourth international conference batteries for utility energy storage* (September Germany).

18 Tokuda, N., Kumamoto, T., Shigematsu, T. et al. (1995). Development of a redox flow battery. *SEI Technical Review* 40: 77–82.

19 Tokuda, N., Kumamoto, T., Shigematsu, T. et al. (1998). Development of a redox flow battery. *SEI Technical Review* 45: 88–94.

20 Tokuda, N., Kanno, T., Hara, T. et al. (2000). Development of a redox flow battery. *SEI Technical Review* 50: 88–94.

21 Tokuda, N., Motoi, K., Hara, T. et al. (2001). Development of a redox-flow battery for installation inside office buildings. *SEI Technical Review* 52: 32–37.

22 Sasaki, T., Enomoto, K., Shigematsu, T. et al. (2002). Evaluation study about redox flow battery response and its modeling. *Meeting Abstracts of CEPSI*: 186–191. (Fukuoka Japan).

23 Investigation for Introducing Battery Energy Storage System to a Wind Power Generation (2002). NEDO-NP-0004 (Japan).

24 Shigematsu, T., Kumamoto, T., Deguchi, H. et al. (2002). Applications of a vanadium redox-flow battery to maintain power quality. *IEEE Transmission and Distribution Conference and Exhibition 2002: Asia Pacific* OR-25 (6–10 October Japan).

25 Koshimizu, G., Numata, T., Yoshimoto, K. et al. (2007). Subaru project analysis of field test results for stabilization of 30.6 MW wind farm with energy storage *EESAT*. (23–26 September San Francisco).

26 Development of Technologies for Stabilization of Wind Power in Power Systems (2008). NEDO Report. 20090000000094 (Japan).

27 Nakahata, H., Ayai, N., Shibata, T. et al. (2013). Development of smart grid demonstration systems. *SEI Technical Review* 76: 8–13.

28 Shinya, K., Matsumura, Y., Matsumoto, T. et al. (2018). Demonstration project of large-scale storage battery system at Minami-Hayakita substation – Overview of the demonstration project. *Proceedings of Grand Renewable Energy 2018 international Conference*, PACIFICO Yokohama, (Yokohama, JAPAN).

29 Shinya, K., Matsumura, Y., Matsumoto, T. et al. (2018). Demonstration project of large-scale storage battery system at Minami-Hayakita substation – Verification results of frequency fluctuation control. *Proceedings of Grand Renewable Energy 2018 international Conference*, PACIFICO Yokohama, (Yokohama, JAPAN).

30 Shibata, T., Sano, T., Yano, K. et al. (2018). Demonstration project of large-scale storage battery system at Minami-Hayakita substation – Evaluation of the 60 MWh vanadium flow battery system performance. *Proceedings of Grand Renewable Energy 2018 international Conference*, PACIFICO Yokohama, (Yokohama, JAPAN).

31 Shibata, T., Hayashi, S., Yano, K. et al. (2019). Performance evaluation of a 60 MWh vanadium flow battery system over three years of operation. *Proceedings of International Flow Battery Forum*, Le Centre de Congrès de Lyon (Lyon, France).

32 Project report of NEPC (2019). http://www.nepc.or.jp/topics/pdf/190308/190308_5.pdf.

33 Nagaoka, Y., Fukumoto, S., Hirata, Y. et al. (2018). Pilot project of the VFB for multiple-use application. *Proceedings of International Flow Battery Forum*.

34 Ooka, T., Kitano, R., Nagaoka, Y. et al. (2021). Demonstration of the wholesale market operation using the flow battery system in California. *Proceedings of 2021 Annual Meeting of IEEJ (The Institute of Electrical Engineers of Japan)*, On-line. No. 7-31.

35 Fukumoto, S., Moriguchi, M., Sano, T. et al. (2021). The demonstration and operation of a vanadium flow battery system for microgrid application. *Proceedings of International Flow Battery Forum Online conference*, Kyoto International Conference Center (Kyoto, Japan).

36 Moriguchi, M., Fukumoto, S., Nagano, H. et al. (2020). Redox flow battery system for the effective use of surplus electric power. *Journal of IEIEJ (The Institute of Electrical Installation Engineers of Japan)* 40: 8.

37 Shibata, T. (2019). Flow battery system application for microgrid. *Powering African Innovations TICAD7*.

48

Industrial Applications of Flow Batteries

CellCube: History, Progress, and Outlook of Commercial VRFB Deployment

Pavel Mardilovich and Martin Harrer

Enerox GmbH, IZ NÖ-Süd Str. 3 Obj M36, 2355 Wiener Neudorf, Austria

Lichtenegg, Lower Austria. A large wind turbine is rising above a field by the side of a two-lane road winding its way down a valley. Around it are several solar panels, and a collection of smaller experimental wind turbines, some with vertical blades, other with shrouds. This is Lichtenegg Energy Research Park, opened by EVN in 2011. Next to the visitor's center, situated in the shadow of a utility building is a small white container, a Cellstrom FB10-100 flow battery. Installed at the Park's inception in 2011, it is quietly humming along, pumping electrolyte, storing and releasing the wind energy daily. One of Enerox's first FB10 installations, it has been continuously running with no stack replacement or electrolyte adjustment for the first 10 years of its service.[1]

48.1 Company History: How Funktionswerkstoffe Forschungs und Entwicklung GmbH Became Enerox, and What Is CellCube

The origin of Enerox dates back to the founding of Funktionswerkstoffe Forschungs- und Entwicklungs GmbH (FWG) by DDr. Martha Schreiber in Eisenstadt; Austria in 2000. The company started off as a research company investigating electroceramics and, a little later, lithium-ion battery components and heat exchangers based on a "lost core" casting technology. One of the electrochemists at the company, Dr. Christoph Haag, has done his postdoctoral work with vanadium flow batteries at the University of New South Wales, in Australia, in the research group of the Prof. Maria Skyllas-Kazakos, who is widely recognized as the founder of the vanadium flow battery technology in use globally today. Given the novelty of VRFBs at the

---

[1] As of writing of this section in 2021, two of the ten stacks in the Lichtenegg battery were replaced. One stack exhibited increased resistance, and the whole two stack string was replaced. The second stack tested normal, and both are being evaluated by the R&D in the context of the materials aging study.

Flow Batteries: From Fundamentals to Applications, First Edition.
Edited by Christina Roth, Jens Noack, and Maria Skyllas-Kazacos.
© 2023 WILEY-VCH GmbH. Published 2023 by WILEY-VCH GmbH.

time, this alone made Eisenstadt the epicenter of flow battery expertise in Austria. In 2001, Dr. Haag's Austrian Ph.D. advisor, Prof. Besenhard, referred FWG for an energy storage project with ASFINAG, the Austrian motorway operator. The goal was a reliable long-term battery to power service areas. It was decided that the most suitable battery technology for the project was vanadium flow technology, and in 2001, FWG started its flow battery business.

With little to build on, some of the earlier solutions were quite unusual. For example, the stack was a collection of individual cells connected in series, rather than the bipolar-plate design used across the industry today. But three months after beginning the project, with much trial and error, a proof of concept was ready, with a functional 250 W stack, a fluid circuit, and a control unit to keep it all running. A year later, as the "The Return of the King" hit the movie theaters and the first SARS-Cov-1 coronavirus was being contained, three men were laboring on a side of a road in Austria (Figure 48.1), putting the finishing touches on nothing more, nothing less than a first vanadium redox flow battery (VRFB) in Europe.

This demonstration system with 1 kW power and 50 kWh storage capacity was operated as an off-grid system fed by wind and solar and has powered a huge signal panel and was the proving ground for the basic battery concept. Note that this early installation of the VRFB technology used a very high energy-to-power ratio – a preview of one of the key capabilities of flow batteries.

The running prototype ended up one of several such prototypes installed, with subsequent installations including Italy and India. At this point, the optimal battery design was far from defined, and the installation featured many project-specific modifications. On the one hand, that increased the engineering costs, but on the other, it offered a range of testing environments to explore the early battery concepts. These batteries ran as part of research or pilot projects and did not function for too long – the first installation with the ASFINAG was decommissioned only after two years – but during this time, they produced learnings that prepared the ground for the next stage in the CellCube evolution.

Some of these findings might seem trivial today. For example, we know today that the graphite felt needs to be activated to guarantee efficiency and sustain long-term performance. At the time, no industry catering to the flow batteries existed, and many materials, such as electrodes and membranes, had to be sourced from other applications. For example, the electrode felts, while chemically robust, were hydrophobic, performed poorly, and lead to a fast increase in stack resistance. Crucial contribution in developing efficient electrodes, as well as other functional components was the work of the FGW research team, led by Dr. Adam Whitehead. Their innovative solutions have enabled the functionality of the early installations as well as the many improvements of the subsequent battery designs. In fact, rigorous R&D continues to be a significant component of CellCube's competitive advantage into the present day.

Another early, design-specific revelation was that one should not place pumps, stacks, or other serviceable components inside the secondary containment: in one instance a cracked flange led to the root cause being inconveniently submerged under a meter of vanadium electrolyte. As in many early technologies, there also

Figure 48.1 First battery: installation in Leobersdorf and its maintenance. Note the paper towels – indispensable item for any flow battery startup.

were some bizarre design issues that one simply does not think of, such as when a wasp decided to build a nest in a smoke detector and triggered a fire alarm.

In 2006, FWG was bought by Solon labs (later renamed to Younicos). Right about this time, the company has changed its name to Cellstrom GmbH. At this point, it might be worthwhile to clarify the naming. The company itself has gone through

Figure 48.2 Clockwise from top left: the first lab prototype, the stack used in the first field demonstration system, the first stack using bipolar plates and based on elastomeric frame, and the stack design for the first series product (FB10-100).

four names: FWG, Cellstrom, Gildemeister Energy Storage, and as of 2018 Enerox GmbH. CellCube is a product brand for all Enerox VRFBs that was established in 2012 and had since been used interchangeably with the Enerox name to refer to the company itself.

So.

With new investment and building on the knowledge collected from earlier prototypes, the Cellstrom team developed a new stack architecture to build on a bipolar plate concept (Figure 48.2). The project leader for the product development, Martin Harrer, had previously worked on carbon dioxide scrubbers based around a filter press design using elastomeric frames, a solution that he incorporated into the new design. It proved to be very effective in securing the battery against leakages, and this stack design remained the core of the Cellstrom (and afterward CellCube) batteries for the next 15 years, undergoing several performance-enhancing revisions, but with little substantial modification.

Based on this new stack design the first containerized flow battery was developed in 2008. The battery, FB10-100 (the name stands for "Flow Battery" 10 kW and 100 kWh), was capable of 10 kW for 10 hours and offered a standalone turnkey solution that could be delivered anywhere in the world. It featured two-level electrolyte/power plant architecture, with two serviceable compartments, one for the hydraulics and stacks, and one for inverters and electronics. One of the driving design principles, which carries forward today, was reliability-oriented field serviceability, where any repairs could be carried out with limited infrastructure even in remote and isolated locations.

Shortly afterward, the design was upgraded to allow for higher power requirements. Namely, FB10/20/30 battery with capacities from 40 to 130 kWh enabled addressing varying project and application needs with minimal hardware changes, cashing in on the flow batteries' ability to scale power and energy independence. For the next decade, the FB10 and FB10/20/30 batteries formed the core of Cellstrom product range, accounting for over 100 installations worldwide across the range of applications and climates spanning from sunny Abu Dhabi, UAE to frosty Ulyanovsk, Russia (Figure 48.3).

For vanadium flow batteries, the economics favors large scale, and soon the work began in drafting a larger VRFB concept. This direction received a strong impetus in 2010, as the company was bought by DMG, a world-leading manufacturer of tooling machines, bringing in additional capital and production know-how. This work has led to the development of the FB200 battery. This was the first-ever containerized design where the power and energy were housed in separate modular units. This format allowed 2- to 8-hour storage duration, giving the product energy and power

Figure 48.3 Installations across a wide range of climates, clockwise: Russia, Romania, Vietnam, and UAE.

Figure 48.4 Early FB10-100 stack production and assembly.

versatility. The first installation was a 2-hour CellCube FB200-400 installed at the Gildemeister HQ in 2012 to provide peak-load shaving and CO_2 footprint reduction goals. In 2013, an 8-hour FB200-1600 battery was installed on the island of Pellworm. This was the first >1 MWh installation of a CellCube, and it offered islanding, base load, and grid support services.

Over the next years, the company focused on manufacturing and delivering the existing battery designs (Figure 48.4). Technological development focused mainly on cost optimization, including transition to standard 40-foot container for larger batteries, modifications to DC/DC architecture, and tuning the electrolyte-to-stack ratio.

One challenge worth mentioning has been copper contamination. It was discovered that either initial contamination, or one introduced through stack failure, led to increased side reactions and electrolyte overoxidation. Operating a battery in such a state inevitably led to a cascade of charge unbalancing, oxidative damage, and further contamination. Once copper was introduced to the system, it needed replacement of failed stacks and expensive electrolyte exchange to return the battery to operation. A two-prong solution to this was the introduction of more rigorous quality control measures on the incoming electrolyte and the development of new patented copper-free current collectors. These two introductions have greatly decreased the costs associated with system operation and maintenance for future projects.

More recently in our history, 2015 promised to begin like any other year with more R&D being undertaken and more uses for the VRFB being assessed and studied. The parent DMG was merging with Mori Seiki to form DMG Mori, electrochemical technology development was going strong, and the batteries were flying off the assembly line. But then, somewhere in the merger process, the newly formed DMG Mori decided to consolidate their business around the core competencies in tool fabrication, and the business of vanadium flow battery R&D and manufacturing was sold off, seemingly as an afterthought to the merger process.

The new owner stepped up the R&D spending, talked a big game, and planned for large projects. In one case, one thousand batteries – over ten times the batteries installed to date – were planned to be delivered for an African project. The funding was to come in part from the EU and with a significant amount from the Chinese Belt and Road Initiative. What went on behind closed doors is anybody's guess, but after half a year no money materialized, and Gildemeister energy storage filed for insolvency. The only good thing to come out of this ambitious comedy of errors was the development of the new generation of CellCube batteries: Release 4, which centered around a large-scale containerized solution, building on lessons from the previous FB200/250 systems.

In 2018, Stina Resources bought the assets and the name Enerox GmbH was adopted, while Stina formed an umbrella CellCube Energy Storage Systems Inc. The commercial operations resumed, and the company was able to focus on improving the new Release 4 FB250-1000 battery design and preparing it for commercial production. In 2020, Enerox secured a significant multi-million USD investment from Bushveld Minerals, a London-listed vanadium miner and processor. Bushveld Minerals, to its own credit, was a start-up, and today supplies 50% of the world's primary vanadium. As an organization, it has been operationalizing its own forward vertical integration and has supported its own value chain strategy through this investment. This acquisition now gave Enerox the momentum to ramp up production of its FB250 flagship design, forge an ambitious R&D team, and take a growth-oriented market position.

48.2 Elephant in the Room: Can Flow Batteries Deliver the Duration They Promise?

In recent years, there has been a boom in founding of new flow battery companies, each aiming to capitalize on innovative electrochemical technologies. This goes hand in hand with a healthy growth of interest from investors, which is giving solid substance to the many energy storage startups. Over the last decade, the flow battery field has also benefitted from suppliers looking to offer flow battery-tailored components where in the past this was a neglected niche. For example, as mentioned before, in the early days the graphite felts, typically used for VRFB electrodes, had to be borrowed from other applications, and the suppliers offered only the basic graphite materials, which had to be activated in-house using proprietary methods. Now multiple suppliers are vying for the flow battery market share, developing and offering felts with optimized activation characteristics and ever-increasing quality.

While the market signals are encouraging, one concern is highlighted by potential participants and that is that far fewer flow batteries have been installed in comparison to the better-established nickel-metal hydride or lithium-ion technologies. This frequently (and understandably) invites questions about the track record. For all the promises of 20-year lifetime and 20 000+ cycles, where is the math?

This question becomes much more important when one moves past the product design and installation conversation and begins to look at the costs associated with running and maintaining the system. Availability of the energy storage capacity and the cost of maintaining it are key factors that determine whether a battery project will pencil out economically. And with any recent technology, there is often limited historical data to support this analysis.

Fortunately, for VRFB, this has been changing. As more flow battery installations come online, the scope of data that one can use to make performance projections is growing. With over 100 installations in the field, some built over a decade ago, Enerox has been developing an increasingly robust knowledge base of past and current project performances. This encompasses different climates, operating regimes, supplier inputs, and associated maintenance histories.

Based on this track record, we have analyzed statistics on maintenance events, including frequency, repair cost, and availability impact. The amount of data lets Enerox approach these questions with good granularity, evaluating risks on a component and subcomponent level, and establishing correlations with supplier, batch, or even production team. For example, Figure 48.5 offers a snapshot of stack replacement events. Here we can see that overall, about 30% of the stacks have been replaced over the operating time of all small batteries (10–30 kW units). However, if we partition this by battery generation, we can see that the second generation FB30-130 shows a considerably lower replacement rate of 8% vs. 59% for an earlier generation FB10-100, reflective of refinements in stack design, production, and quality control measures. Such analyses add tremendous value on many levels.

One key advantage is that now Enerox can offer its customers data-based performance warranties, production and availability guarantees, and cost-optimized service plans. As the size of installations grows (typically now at the MW size) and the projects move away from the experimental beginnings of a decade ago and enter the world of commercial reality, it is increasingly important for the end-user to have a business plan that works. This we refer to as the "use-case/business-case efficiency frontier." Being able to say how an energy storage asset will perform and how much that will cost is crucial, both at the euro per kWh and at the levelized

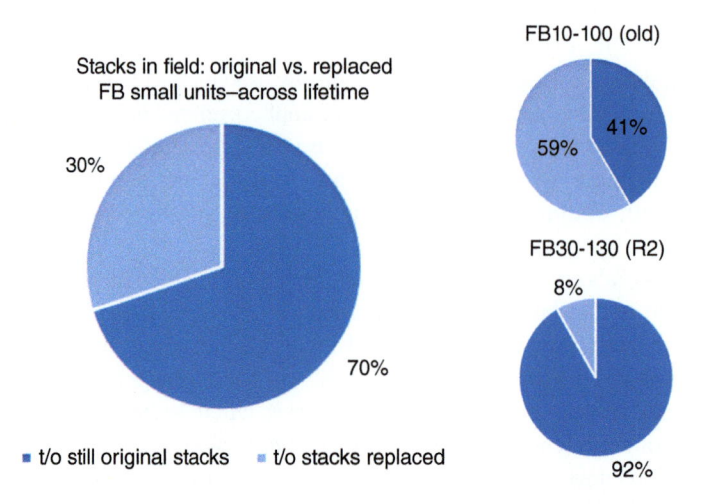

Figure 48.5 Break down of stack issues by battery generation: the growing pains can be seen in the high ratio of the stacks that needed to be replaced in the first generation of FB10-100 batteries. Compare that to the next generation FB30-130 installations.

cost of energy storage (LCOES) level. Enerox is, therefore, in an excellent position to support this analysis with hard numbers, and this level of certainty is a key component of Enerox's competitive offering.

On the engineering side of this, such analysis of past performances is very helpful in identifying weaker points of the product. And this lets us learn not only what fails more often, but what is the cost impact, and possible correlations with other failure events. Such learnings are invaluable when prioritizing development goals. For example, stack replacement is a significant maintenance event. These failures frequently co-occur with electrolyte over-oxidation. The physics behind this correlation is well-known: an over-oxidized electrolyte puts additional oxidative stress on the stack internals, leading to corrosion, loss of hermeticity, and contamination of the electrolyte via dissolution of metallic stack components. These contaminants, depending on their nature, can either contaminate other stacks or stimulate further oxidation of the electrolyte. In the end, the repairs include anything from replacement of a single stack and a quick electrolyte re-service to replacement of all the stacks on a hydraulic circuit and cleaning or replacing the entire electrolyte volume. Given the potential cost of the latter, preventing such events is a high priority. To date, Enerox has implemented manufacturing and quality control measures for electrolyte and the stack, and are in the process of implementing an in situ sensing and re-balancing system to detect and avert such issues automatically.

Not the least of CellCube arguments is the laboratory-level performance of cycling test stacks. In such environments, various parameters can be accurately monitored, and degradation tracked with more precision than is accessible in a field operation. In one study, (Figure 48.6) several stacks have been continuously

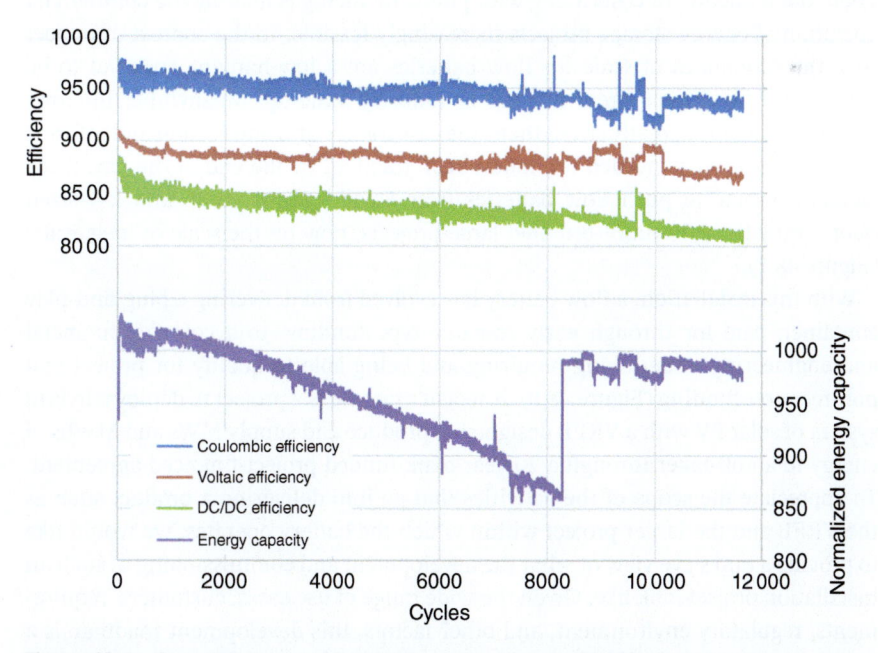

Figure 48.6 Over three years this stack has cycled nearly 12 000 times. In a small system air oxidation is more of an issue, so in this case, one rebalancing was necessary.

cycled over the full state of charge range over the last three years. Over this period, the test performed nearly 12 000 cycles with a minimal and predictable drop in performance. Of course, this cannot be scaled to calendar aging for all of the battery components – for example, the pumps on this test rigs ran only for the duration of the experiment – however, as far as the electrochemical power stack is concerned, 12 000 cycles are equivalent to 30 years of daily charge-discharge cycles.

48.3 Anatomy of A Project: What Does It Take to Put in a MW-Plus Battery in the Field?

Energy storage can serve many different applications from frequency response to grid connection and transmission infrastructure upgrade deferral to the ubiquitous energy arbitrage. Of all the possible applications, though, none drove the public awareness of the need for energy storage as much as its use to back up solar and wind power to provide reliable, on-demand energy. Over the last decade, the notion of large-scale energy storage evolved from an engineering curiosity to a crucial, need-fulfilling component of electrical grids essential in their future-proofing and decarbonization.

Now in 2021, there is a confluence of factors creating fertile environment for large-scale battery growth. Moving forward, these factors will affect organizational strategy and innovation in many sectors. At the forefront, the established need for energy storage is driving policy changes and creating a favorable environment for energy storage and, specifically, long-lived, long-duration flow battery deployment. Then, the reduction of costs along with public financing is making the commercial operation of energy storage projects increasingly feasible. In the context of project cost, the economies of scale for flow batteries are a low-hanging fruit yet to be picked, and so drive a great interest in industry scale-up. Meanwhile, Enerox's existing experience greatly strengthens the case for key grid application and informs the financial models needed for commercial traction. At the end of the day, these factors combine to push flow batteries from small-scale research and kW-sized pilot installations, to profit-oriented large projects, now on the scale of megawatts (Figure 48.7).

With this installation, a flow battery has evolved from delivering a plug-and-play container, paid for through early research-type funding, to a complex financial and engineering undertaking requiring and being able to qualify for project and non-recourse funding (Figure 48.8). A recent example is a project to deploy a hybrid system of solar PV with a VRFB designed to produce and supply MWs and MWhs of energy to an off-taker through a 20-year bank-funded project-financed agreement. To appreciate the scope of the activities that go into delivering a product such as the VRFB and the larger project within which the battery operates, we would like to provide a bird's eye view of what the development and commissioning of such an installation project look like. Given the wide range of use cases, customers' requirements, regulatory environment, and other factors, this development roadmap is a process tailored to specific conditions and will be composed differently each time.

Figure 48.7 Early days of battery delivery and today's assembly of FB500-2000 on an in-house testing site.

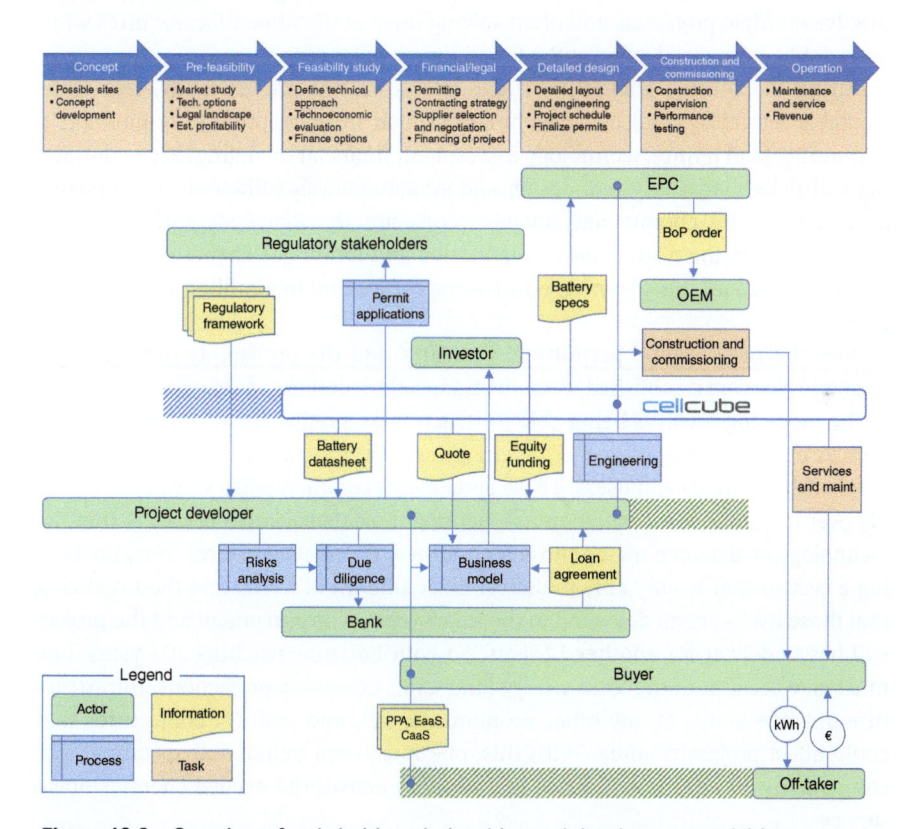

Figure 48.8 Overview of stakeholder relationships and development activities.

The anatomy of a project presented here serves more as an example and illustration of the need to view project development of this nature as a complex system rather than a prescriptive instruction.

The key word here is "commercial," so addressing this aspect is probably a good place to start. Unlike experimental installations of a decade ago, the new projects are large, they cost a lot of money, and they need to make financial sense. The first step, before the first shovel is sunk into the ground, involves a substantial amount of work in evaluating the solution feasibility, which must have a clear understanding of why, how, and what the VRFB is expected to do. The use case range of energy storage is very broad, which puts it in contrast to many other technologies, with which it will couple or will replace. For example, PV, as a use case, does one thing and one thing only: produce energy. So too does wind and, by and large, so do coal-fired power plants. Gas-fired turbines are capable of doing a few things, and in practice are used to deliver on a handful of use-cases in the field. Batteries – and VRFB specifically – are capable of several dozen use cases. The complexity of using energy storage to solve multiple problems, and often solving them at the same time, requires without a doubt substantial effort at the feasibility and planning stages.

During the project origination stage, efforts toward pre-feasibility, feasibility, bankable feasibility, project development (which involves project "origination"), permitting, land tenure, technology assessment, financial modeling, legal contracting and risks analysis, system design and system analysis (different to technology assessment), and commercial considerations are the many steps that go into planning before the next phase of negotiation and funding. Depending on the size of the installation, this planning effort alone can run up to a million USD for each project.

Once the risks around permitting, licensing, and the probability of reaching a financial close are established, then the financial modeling can take place.

The modeling involves being able to structure and refine the revenue contracts, such as power purchase agreements/energy as a service/capacity as a service agreement with off-takers and users. These agreements commercialize services and supply over long time horizons, therefore, joining the dots back to a certainty that the technology choices can match those long-term agreements is critical. Imagine buying a system that is only able to deliver a set amount of kWhs and then realizing that those kWhs are all delivered in the first 8 years of an agreement and the project still has to deliver for another 12 years. So with horizons reaching 20+ years, one must take account of the local energy landscape, economic projections, infrastructure upgrade plans, or any other economic, social, and political conjectures that could affect project revenue. With this, one must then define and specify all the components that must be in place to effectively deliver the needed energy storage services.

There is a range of applications for energy storage (Table 48.1) and depending on the target applications and services the battery system must be designed and sized appropriately. Flow batteries are particularly well suited for services that require continuous daily cycling and duration of multiple hours. On top of that, they are also capable and are often used to provide short- and medium-term services, such as primary and secondary response, albeit at lower cost-efficiency than some of the

Table 48.1 Various energy storage services that can be monetized (information from Schmidt et al. [1]).

Role	Application	Description
System Operation	1. Energy arbitrage	Purchase power in low-price and sell in high-price periods on wholesale retail market
	2. Primary response	Correct continuous and sudden frequency and voltage changes across the network
	3. Secondary response	Correct anticipated and unexpected imbalances between load and generation
	4. Tertiary response	Replace primary and secondary response during prolonged system stress
	5. Peaker replacement	Ensure availability of sufficient generation capacity during peak demand periods
	6. Black start	Restore power plant operations after network outage without external power supply
	7. Seasonal storage	Compensate long-term supply disruption or seasonal variability in supply demand
Network Operation	8. T&D investment deferral	Defer network infrastructure upgrades caused by peak power flow exceeding existing capacity
	9. Congestion management	Avoid re-dispatch and local price differences due to risk of overloading existing infrastructure
Consumption	10. Bill management	Optimize power purchase, minimize demand charges, and maximize PV self-consumption
	11. Power quality	Protect on-site load against short-duration power loss or variations in voltage frequency
	12. Power reliability	Cover temporal lack of variable supply and provide power during blackouts

alternative technologies. So, at times it might make sense to place flow batteries in tandem with high-power short-duration technologies, such as LIBS, flywheels, or supercapacitors to support high bursts or power, if the application calls for that.

Once the energy storage requirements are defined and the overall system is aspirationally designed, the technology suppliers offer information necessary for the techno-economic analysis and engineering of the solution. This includes things such as hardware costs, requirements for grid interface, communication and control protocol standard, siting requirements, delivery schedule, and maintenance outlays. Typically, at this point, an engineering and procurement company (EPC) is engaged in developing the framework of designing the solution in anticipation of carrying out future construction.

Another layer to the project is the permits for site control, planning permission, grid connection agreements, and other legal compliance issues. Not least of these are short- and long-term safety considerations, environmental impact assessment, and at times outreach efforts to target public acceptance.

A crucial issue for energy storage project development is risk integration. Will all the components work the right way? In this particular aspect, Enerox's prior experience and track record place it in a strong position to guarantee outcomes and inform long-term cost and performance projections. At this point in the development of the flow battery industry, the reliability of the technology for such projects is typically underwritten by specialized insurers. This offers a layer of certainty to the project.

The revenue and costs associated with all the above aspects are then folded into the business model that underpins the commercial viability of the whole project. The bankable "gift-wrapped" project is then presented to investors for equity funding and to the bank for debt borrowed against future project revenue. Once all the pieces are in place, and the financing is secured, the construction can begin.

Many factors go into installing a project (Figure 48.9), so each project is different, but in a way, each project is also the same. The first step is the scope review and a breakdown of tasks and allocation of responsible parties (Table 48.2). Here, every

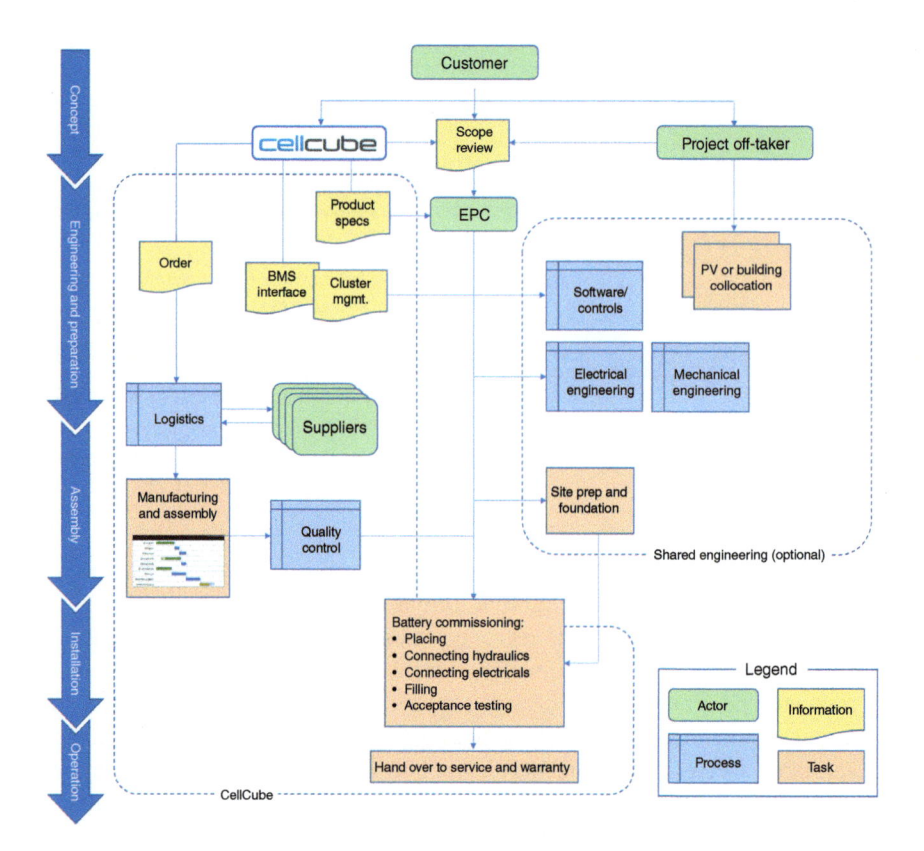

Figure 48.9 An overview of a CellCube installation.

Table 48.2 An illustrative excerpt from a scope review document listing some of the issues to consider.

	Enerox	Customer
Overall project management		
Define site control and BESS facility operation requirements		
Existing site documentation: supply any and all relevant engineering and construction drawings of existing site such as a site plot plan, single line diagram and network/SCADA communications diagram etc.		
Provide BESS equipment weights and dimensions necessary for foundation design		
Site civil design (foundations, grading)		
BESS engineered equipment package design including;		
Overall BESS system and equipment design up to high side of the isolation transformer		
Detailed design for the interconnection of all equipment up to high side of the isolation transformer, including cable schedule		
Design BESS structure including seismic restraints for battery rack frames		
Design BESS HVAC System		
Design BESS fire suppression system		
Design foundations, transformer oil containment, grading		
Design medium voltage system and equipment, including short circuit, ground fault and overcurrent protection		
Install auxiliary power supply		
Design aux power system and equipment		
Define SCADA and facility operation requirements and specifications		
Design DC conduits for power and communication cabling from "Battery System" to inverters		
Design AC conduits for AC cabling from Inverters to transformers including medium voltage collection system; also for auxiliary power connection point		
Design electrical grounding system and including ground fault detection system and lightning protection		
Design outdoor lighting system		
Provide site final design drawings for approval		
Specify revenue meters		
Specify 2nd source for aux. power supply		
Design substation/GSU to point of connection		
Design protection relay systems		

(Continued)

Table 48.2 (Continued)

	Enerox	Customer
Compliance engineering per codes and standards		
Overall safety program – site		
Safety program – BESS system		
Provide PE stamped documents		
Provide final site construction drawing package		
Builders all risk insurance for entire project		
Site preparation studies: seismic studies, grounding studies, soil analysis, geotechnical survey, site survey		
Sub-surface remediation		
Environmental control plan(s)		
Quality assurance plan		
Foundation, grading work, containment, as required		
Temporary facilities (construction trailer, toilets)		
Temporary equipment storage area for all equipment, including power for refrigerated containers (at site)		
Power and water supply during installation		
Submitting and acquiring of all relevant interconnection agreements, local permitting including building/grading permits		
Construction waste disposal		

step necessary for the successful completion of the project is considered within a detailed scope review with the customer. This step outlines the technical responsibilities of each party from start to finish of the project.

CellCube FB250 battery series weighs 80 tons per electrolyte unit and 100 tons for each electrolyte unit with a power unit on top. For this reason, an appropriate foundation must be prepared to ensure problem-free operation. Typically, this work lies within the customers' scope and is contracted to a specialized engineering firm, which plans the site, designs the batteries foundation with consideration of the batteries point loads, engineers the electrical conduit layout, determines where the batteries and auxiliary equipment is positioned and handles the construction aspects of site preparation. To accomplish the site preparation, Enerox participates in coordination meetings and provides all relevant technical information needed for the design. In many cases, the batteries are installed as part of a larger project, such as a photovoltaic installation or a new construction project. In all situations, it must be scoped and agreed as to who in the group of stakeholders must consider, and finalize, all necessary permits for their construction site prior to the shipment of their CellCube.

In parallel, the batteries to be installed are put into production. For timely completion of the project, requires meticulous planning of all the steps involved in production and delivery, starting from the lead time for ordering every single component, to assembly timeline of the necessary number of systems and factory throughput, all the way to system assembly time and shipping. On top of that incoming quality control inspections must be undertaken to ensure the delivered component is in line with Enerox's internal specifications. The same goes for the outgoing inspections. All batteries leaving the manufacturing facility must pass a series of non-destructive tests to ensure a seamless installation and functionality at their final location. All in all, timely and reliable order fulfillment requires a robust and flexible logistical network.

Regardless of the project structure and subcontracting of tasks, installation and commissioning of the battery invariably involve a CellCube installation crew. After the components are delivered to the construction site, our installation team reviews our on-site safety manual and identifies any major safety risks. The batteries are placed on the prepared foundations, the hydraulic systems are connected, and pressure-checked for hermeticity. Given specifics of local codes of each project, electrical terminations are made by locally qualified electricians. Once all is in place and connected, the batteries are filled with electrolytes. At this point, one must be mindful of possible environmental impact, chemical safety measures, and local regulations governing the transportation and storage of large quantities of electrolytes. Once filled, the batteries will run through the startup procedure and commissioning can begin. Upon completion of commissioning, the customer is then invited to witness the site acceptance test or final acceptance test. This test is undertaken on all projects and is used to verify that the battery operates per the contractual agreements. As the customer signs off on this test, this starts both the warranty period and commercial operations. At this point, the project is complete, and the responsibility for continued O&M is handed over to the services team or to the nominated contractor.

48.4 I'm on a Boat!: Innovation at CellCube

After several years of optimization, FB250-1000 Release 4 represents the flagship CellCube product designed to meet a range of challenges in providing energy storage at the flexible MW scale installation. It can offer tunable duration in the 4–8 hours range, it can be coupled together using DC or AC bus, depending on the system requirements, and its modular nature and use of the off-the-shelf components greatly improve the economics. However, it is too early for Enerox's R&D to rest on its laurels, and the next generation Release 5, is already in development, building upon leaning from earlier products, aiming for higher efficiencies and energy densities, longer service life, and further cost reduction. But in addition to the technological progress on the core product, there are a few rather exciting research leaps that Enerox is pursuing.

One of these is the application of flow battery technology to maritime transport. River shipping in Europe is a large industry, and as much of the transportation infrastructure, it comes with significant CO_2 footprint. Unlike rail, it is not easy to electrify.

Figure 48.10 An artist's illustration of the VRFB-powered barge refueling at a charging electrolyte pontoon. Source: Enerox GmbH.

LIBs can be a solution, but fire risk in confined spaces is a concern, meanwhile, the time required to recharge most batteries would introduce undesired shipping delays.

The innovative concept that CellCube is developing together with a global household name in shipping in Europe is to electrify cargo vessels using VRFBs. One feature that makes VRFBs particularly promising in this application is that the energy is stored in the electrolyte, and so the vessel can be rechanged by simply "refueling" it with charged electrolyte at a filling station along the transportation route while offloading the "spent" electrolyte (Figure 48.10). As the vessel continues on, the electrolyte left in the station will be recharged for the next barge. This arrangement allows very quick recharge of the transport, bypassing normal charging rates and the delay these inevitably introduce. Additionally, the electrolyte recharging at the filling station can be done using renewable energy sources or grid energy overproduction, thus offering additional benefits of enabling renewables and stabilizing the grid.

48.5 Sunny Upside: An Egg Pun or a Realistic Outlook?

With a proven track record, solid products, established logistical network, and favorable market tailwinds, CellCube is confident as it faces the future.

To meet the pipeline of the current and future projects, the company is well on the way to ramp its production capacity to 30 MW per year in 2022, with further expansion planned. The growth brings with it new challenges. Not the least of these is the ability of the many suppliers to keep up and deliver the necessary volumes while maintaining quality. Understandably supply chain management is a critical component of the scale-up activity. This means analyzing criticality of various inputs, increasing resiliency via multiple sourcing, and continuous supplier assessment in key areas. In the case of vanadium electrolyte – unarguably the most crucial component of the battery – CellCube is building close relationships

Figure 48.11 The Lichtenegg battery. Source: Enerox GmbH.

with vanadium producers and electrolyte manufacturers to guarantee its ability to deliver the megawatt-hours necessary.

The growth in flow battery production and installation presents not only challenges but also opportunities. Growing battery manufacturing also requires a growing ecosystem of producers and service providers, and this growing market proffers growth opportunities for existing RFB suppliers as well as opens the door for new entrants.

An important point to make here is that a larger and more robust flow battery supplier ecosystem increases resiliency of the flow battery industry overall and improves economies of scale for all participants. Together with increasing visibility of flow batteries as a viable component of renewable energy package; this enables these batteries to have a strong impact in the energy storage market and positions them to play a key role in the energy infrastructure of the future.

Ten years after its installation, the Lichtenegg battery (Figure 48.11) is still humming along with 99% of its initial capacity remaining. At 10 kW and connected only to a test grid, it does not make much of an impact on greenhouse gas emission or key grid services, but its resilience and the vanadium-based flow battery technology that made – and continues to make – it possible is the foundation on which CellCube is setting itself up to journey into the sunrise.

Acknowledgements

The facts presented in this article were made possible by the work of the many team members of FWG, then CellStrom, then Enerox over the last 20 years. Without their hard work, creativity, and dedication at every level, little of what is presented here would have been a reality. In addition, we would like to extend special gratitude to our VP of Business Development Peter Oldacre, Project Engineer Ryan Hunt, and CEO Alexander Shoenfeldt for their contributions to structuring the discussion on topics of project development, commercial installation, and company vision.

Reference

1 Schmidt, O., Melchior, S., Hawkes, A., and Staffell, I. (2019). Projecting the future levelized cost of electricity storage technologies. *Joule* 3 (1): 81–100. ISSN 2542-4351, https://doi.org/10.1016/j.joule.2018.12.008.

49

Applications of VFB in Rongke Power

Huamin Zhang[1,2]

[1] *Rongke Power, China*
[2] *Dalian Institute of Chemical Physics, Chinese Academy of Sciences, China*

Vanadium flow battery (VFB) technology is the only flow-battery technology that has been commercialized for a large-scale energy storage application. Since 1980s, Professor M. Skyllas-Kazacos's research team, from the University of New South Wales (UNSW), Australia, has accomplished a lot of research in the field of VFB technology and made important contributions to the development of VFB technology [1–4]. Based on several years of the research experience of proton-exchange membrane on fuel-cell technology, Professor Huamin Zhang's research team, from the Dalian Institute of Chemical Physics (DICP), Chinese Academy of Sciences (CAS), started research on the key technology and the core materials of flow batteries since 2000, such as electrolyte, bipolar plate, ion-conductive membrane, stack structure design and integration, system integration, and control management of VFB system. A set of production and mass production technologies have been developed, for instance, electrolyte with high reactivity, high stability [5], carbon plastic composite bipolar plate with high conductivity, low cost [6], and non-fluoride ion-conduction membrane with high conductivity, high-ion selectivity, high stability, and low cost [7–9]. In addition, there have been obvious breakthroughs in the design and integration of high-performance stacks.

In 2006, a joint laboratory was established in cooperation with Bolong (Dalian) Industrial Investment Co., Ltd., to develop engineering and industrialization technologies of VFB. On top of that, Dalian Rongke Power Co., Ltd. (RKP), whose major business is to achieve industrialization of VFB, was jointly established by Bolong (Dalian) Industrial Investment Co., Ltd., DICP CAS, and professor Huamin Zhang in 2008. It has achieved mass production of electrolyte, carbon plastic composite bipolar plate, non-fluoride ion-conduction membrane of VFB, high-power-density stack, and large-scale energy storage systems. A manufacturing factory with battery production capacity of 300 MW per year was subsequently established, and more than 30 demonstration projects have been deployed, including 5 MW/10 MWh, 10 MW/40 MWh, and 200 MW/800 MWh VFB energy storage systems.

Flow Batteries: From Fundamentals to Applications, First Edition.
Edited by Christina Roth, Jens Noack, and Maria Skyllas-Kazacos.
© 2023 WILEY-VCH GmbH. Published 2023 by WILEY-VCH GmbH.

49.1 Development and Application of Core Materials for VFB

49.1.1 Electrolyte

Electrolyte is the active material of VFB for energy storage. The energy storage capacity of flow-battery system is determined by the concentration and volume of the active material in electrolyte. Types and concentrations of additives and impurity elements in electrolytes not only affect the energy storage capacity of VFB system, but also affect the reactivity, stability, and durability of VFB stack and energy storage system directly. Vanadium ions with different valences acting as the active material are stored in positive and negative electrolyte of VFB. Sulfuric acid solution is usually used as the supporting electrolyte, and this electrolyte is called sulfuric acid electrolyte. To improve the solubility and stability of vanadium ions, another kind of electrolyte with a certain concentration of hydrochloric acid in sulfuric acid has been developed [10], which is called the mixed acid electrolyte. Electrolytes with a certain concentration of phosphate and other compounds as the stabilizer have developed as well. The main challenges of the production technology of high activity and high stability electrolyte are as follows:

(1) The stabilization technology of bivalent vanadium (V^{2+}) in negative electrolyte. Due to V^{2+} in the negative electrolyte can be oxidized extremely easily by air, the valence state of electrolyte between positive and negative side will be in imbalance, which causes degradation of energy storage capacity. So, pumping inert gas into the negative electrolyte tank to ensure the stability of V^{2+} of negative electrolyte during the running of VFB system is one of the most important factors that enable VFB system to be at stable operation for a long time. In addition, in order to ensure all the vanadium ions with different valences can be dissolved without precipitation, the concentration of vanadium ions in the sulfuric acid system electrolyte is usually controlled at $1.7\,mol\,l^{-1}$.

(2) The stabilization technology of pentavalent vanadium (VO_2^+) in positive electrolyte. When the electrolyte temperature rises, the stability of VO_2^+ in the positive electrolyte gets worse, accompanied by V_2O_5 precipitation. This not only affects the normal operation of the VFB, but also blocks the electrolyte flow and accordingly damages the stacks. In order to improve the high-temperature stability of VO_2^+, Skyllas-Kazacos and co-workers proposed the use of additives as staibilizing agents [4b] and selected ammonium and phosphate compounds as well as a range of anti-scaling agents to inhibit precipitation in both the positive and negative half-cell solutions over a broader temperature range in solutions of up to 3 M vanadium concentration [11]. A kind of electrolyte with appropriate kinds of additives along with reasonable concentrations that can operate stably at 45 °C has also been developed by Zhang's research team [5, 12].

(3) The online and offline recovery technology of electrolyte with valence state imbalance. After the long-term operation of VFB system, the valence states and the concentrations between positive and negative electrolyte will be an imbalance, which is caused by the gradual side reactions on the positive and

negative electrode or the accumulated cross behavior of the electrolyte through the ion-exchange membrane for years, which will lead to degradation of the energy storage capacity. Therefore, a fast and accurate analytical method for different valence vanadium ions and an online and offline recovery technology of the imbalance electrolyte have been exploited by Zhang's research team, which can greatly improve the reliability and lifetime of the energy storage system [13].

After years of research, our team has collected knowledge of the impurity elements that affect the reaction activity and stability of electrolyte and have developed effective purification technology of the impurity elements and the production technology of electrolyte. Other than that, we have optimized the additives that can effectively improve the stability of electrolyte. Furthermore, we have developed industrialized production equipment and have established a manufacturing factory with an energy storage capacity of 1000 MWh year^{-1}. The electrolyte produced has been successfully applied to more than 30 energy storage stations built by RKP. And a lot of electrolyte has been sold to several VFB manufacturers in Japan, the United States, South Korea, Germany, and other countries so far.

49.1.2 Bipolar Plate

Similar to the structure of the proton-exchange membrane of fuel cell, the VFB stack is composed of several or dozens of single cells following the assembly way of filter press. There is an ion-exchange membrane in the middle of the single cell. Two electrodes are put on both sides of the membrane, connecting with the bipolar plates. The bipolar plate is used to connect all single cells in the stack and to isolate the positive and negative electrolyte solutions between adjacent single cells. It also collects the current generated by the electrode reaction on both sides of the bipolar plate. In addition, the bipolar plate serves as a rigid support to the electrode, since the electrode material in the stack should be compressed when assembling stacks.

Graphite and carbon plastic composites are the main bipolar plate materials used in VFB. Graphite plate is a brittle material with high price. Its impact strength and bending strength are very low, so it is easy to fracture during the assembly process of stack. At the same time, due to the brittleness of the graphite bipolar plate, the pressing force of the stack is limited, which is easy to cause electrolyte leakage of the stack [14]. On the other hand, the thickness of graphite bipolar plate should be high enough thin, which increases the volume and weight of the VFB stack. All of these restrict the application of graphite bipolar plate in the VFB, especially in the large-scale energy storage application.

Stemming from the needs of industrialization, Prof. Zhang's research team [15] has carried out the development of carbon plastic composite bipolar plate material with high conductivity, high mechanical strength, and low cost at the early stage of research. Carbon plastic composite bipolar plates were first developed and used for the Zn/Br battery in the 1980s [16] and novel conducting plastic formulations were also developed and implemented in VFB stacks at UNSW in the 1990s [17].

Carbon plastic composite bipolar plate is made by molding, injection molding, and other methods after mixing the polymer and conductive filler. In the carbon plastic composite bipolar plate, the polymer constitutes a toughening network. The polymers used are usually polyethylene, polypropylene, polyvinyl chloride, etc. The mechanical strength of the composite can be improved by adding carbon fiber, glass fiber, polyester fiber, or other short fibers into the polymer. The electrical conductivity is provided by the conductive network formed by the conductive carbon material, and the materials that can be used as conductive fillers include carbon fiber, graphite powder, and carbon black. Carbon plastic composite bipolar plate has good corrosion resistance and toughness, simple production process, and low cost.

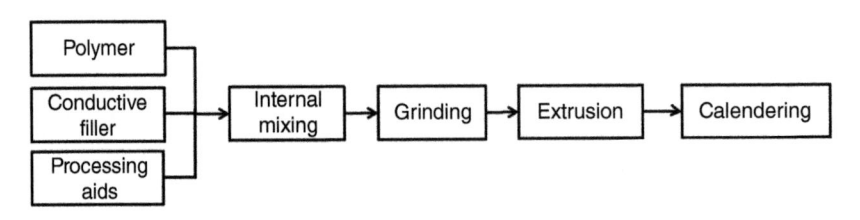

The high conductivity and high toughness of carbon plastic composite bipolar plate material are contradictory. It is difficult to produce carbon plastic composite bipolar plates with high conductivity, large area, high mechanical strength, and high toughness. Through the systematic study of the relationship between the selection of raw materials, ratio, mixing method, and the electrical conductivity, and mechanical strength of composite materials, Prof. Zhang's team [6] developed the batch production technology of high-efficiency production of carbon plastic composite bipolar plate with good mechanical properties and toughness, electrical conductivity, and large area, which adopts extrusion-calendering technology to make the thickness and width of the carbon plastic composite bipolar plate adjustable. The produced composite bipolar plate has uniform thickness and high surface smoothness, and its width and thickness can reach 800 and 1 mm, respectively. The bulk conductivity of the bipolar plate is above 15 S cm^{-1}, and the bending strength is higher than 30 MPa. At present, the annual production capacity of 100 000 m^2 has been formed. So far, more than 40 VFB energy storage power stations implemented by the RKP–DICP cooperation team have used the carbon plastic composite bipolar plate developed by Prof. Zhang's team (Figure 49.1).

49.1.3 Ion-Conducting Membrane

In VFB, the ion-exchange (conducting) membrane serves to prevent the cross-mixing of active materials in the positive and negative electrolyte, respectively, while conducting ions to form an internal electric circuit. In VFB stacks, membranes are used in a harsh environment of strongly oxidizing (VO$_2^+$) ions with strong acidic, high potential voltage, and high current. As a result, ideal ion-conducting membranes should have such features [18]:

Figure 49.1 The production process of carbon plastic composite bipolar plate.

49.1.3.1 Excellent Ion Conductivity

As for sulfuric acid-based VFBs, V^{2+}, V^{3+}, VO^{2+}, and VO_2^+ ions act as active materials, while sulfuric acid is used as the supporting electrolyte. To reduce the battery internal resistance and ohmic polarization, along with to improve the battery voltage efficiency, ion-conducting membranes should have excellent ion conductivity.

49.1.3.2 High ion Selectivity

In VFBs, the cross-mixing of vanadium ions with different valence states will lead to self-discharging and eventually contribute to lower Coulombic efficiency of batteries. Simultaneously, the charges and hydrous protons of vanadium ions with different valence states are different. Thus, their diffusion rates across the membrane are also different. The low ion selectivity of membranes will cause the imbalance of vanadium concentration and electrolyte volume of the positive and negative sides during long-term battery operation, which will result in battery capacity degradation. Consequently, the ideal ion-conducting membranes for VFBs should possess excellent ion conductivity and high ion selectivity simultaneously, to avoid the crossover of the positive and negative electrolyte, the battery capacity degradation and self-discharging.

49.1.3.3 Outstanding Mechanical and Chemical Stability

In the strong oxidizing and acidic operation, along with high-voltage conditions of VFBs, ion-conducting membranes should have excellent mechanical, chemical, and electrochemical stability to ensure the reliability and lifespan of energy storage systems.

49.1.3.4 Low Cost for Large-Scale Commercial Application

The economic efficiency of energy storage stations appeals to ion-conducting membranes with low price, to achieve widely large-scale utilization.

Perfluorinated sulfonated ion-exchange membranes exhibit great ion conductivity and high chemical stability in VFBs. However, the ion selectivity of perfluorinated sulfonated membranes is relatively low, and their price is very high, restricting the commercialization of VFBs seriously [9]. Non-perfluorinated ion-exchange membranes possess high selectivity and low cost, however, low chemical stability. In order to solve these defects of perfluorinated sulfonated and non-perfluorinated ion-exchange membranes to promote the industrialization of VFBs, Prof. Huamin Zhang's research team found that the introduction of ion-exchange groups into non-perfluorinated polymers is the essential reason why they can easily degrade in VFBs [7]. This conclusion is based on the experience of Prof. Zhang's research team in the fields of proton-exchange membrane fuel cells and ion-exchange membranes used in VFBs. According to the traditional ion-exchange mechanism, ion-exchange membranes must have ion-exchange groups. Nonetheless, to introduce ion-exchange groups will lead to the easy degradation of non-perfluorinated ion-exchange membranes. As a result, Zhang's team broke through the restriction of ion-exchange mechanism and innovatively proposed the concept of "ion-sieving conducting" mechanism without ion-exchange group [19]. Then, a series of non-perfluorinated porous ion-conducting membranes with various pores were successfully prepared. Based on pore size exclusion effect, the high selectivity of vanadium ions to protons was achieved. The issue of low chemical stability of non-perfluorinated ion-exchange membranes caused by introducing ion exchange groups was accordingly solved fundamentally. The choice range of membrane materials for VFBs was also extended, which developed a completely novel idea to explore high-performance and low-cost membrane materials for VFBs. Afterward, Prof. Zhang's team successfully confirmed the concept of ion-sieving conducting mechanism [20]. Based on the optimization of pore structures of porous ion-conducting membranes, non-perfluorinated ion-conducting membranes have been successfully developed with high selectivity and high conductivity.

As mentioned above, non-perfluorinated ion-conducting membranes have outstanding features of excellent electrochemical performance, high chemical and mechanical stability. When non-perfluorinated ion-conducting membranes were used in VFBs, the as-assembled VFB could stably run for more than 13 000 cycles, without obvious efficiency fade, exhibiting great stability and reliability (Figure 49.2) [10].

Based on this result, Zhang's team developed scale-up technology and processing equipment for mass production. Consequently, the batch production of high-performance non-perfluorinated ion-conducting membranes has been realized (Figure 49.3).

Moreover, Rongke Power (RKP) assembled eight 25 kW VFB stacks utilizing non-perfluorinated ion-conducting membranes, carbon-plastic composite bipolar plates, vanadium electrolytes, and carbon felt-based electrodes developed by DICP. Then a 200 kW VFB energy storage system was integrated using these eight 25 kW stacks. This 200 kW VFB energy storage system demonstrated an outstanding performance with Coulombic efficiency of 96.5%, voltage efficiency of 80.0%, and energy efficiency of 77.1% at 80 mA cm^{-2}.

Therefore, the improvement of non-perfluorinated ion-conducting membranes can effectively reduce the ohmic polarization at high-current density, thus

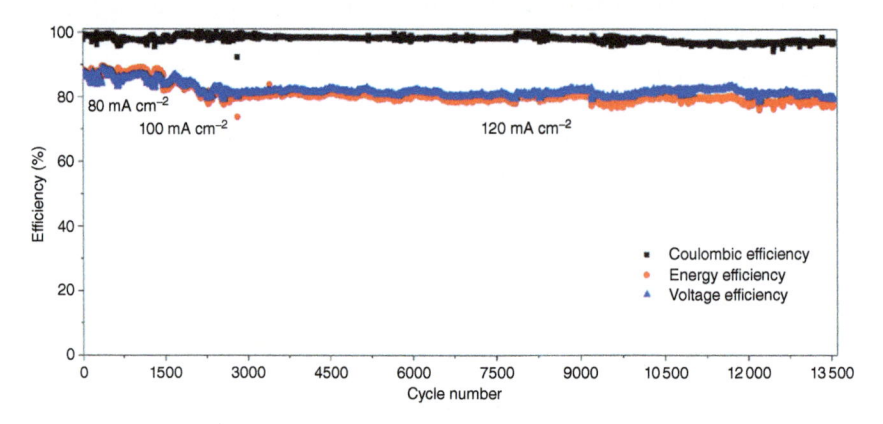

Figure 49.2 The durability of the non-perfluorinated ion-conducting membrane developed by DICP at 80-100-120 mA cm^{-2}.

Figure 49.3 The equipment for the mass production of non-perfluorinated ion-conducting membranes developed by DICP.

improving the power density of the batteries. The utilization of high-performance and low-cost non-perfluorinated ion-conducting membranes in VFBs can subsequently reduce the production cost of VFB stacks.

49.1.4 Key Technologies of Stack and Energy Storage System of VFB in RKP–DICP Team

The stack is the key component of VFB energy storage system. The battery's performances, such as the rated current density, distribution of electrolyte, energy efficiency, electrolyte utilization, and capacity retention will determine the reliability, stability, cost, and service efficiency of the system. One of the most effective ways to achieve the cost reduction is to improve the rated power density and the electrolyte distribution uniformity of VFBs under the premise of the constant energy efficiency, by which, the material consumption of a VFB stack can be reduced significantly. In the last decade, the RKP–DICP cooperating team (RKP–DICP team) has devoted continuously to stack research and development. In combination with the promotion strategies of the electrical conductivity of bipolar plates, carbon felt-based electrodes and non-fluoride ion-conduction membranes, the reaction activity of electrodes, and the selectivity of the membranes, the rated current density has been increased evidently, and the stack cost has been reduced effectively as a result.

The internal resistance of a VFB consists of three parts, i.e. activation polarization, ohmic polarization, and concentration polarization. The VFB usually employs the improved carbon felt as the electrode, which is highly reactive for the electrochemical reaction. Hence, a great research effort has been made by the RKP–DICP team to reduce the ohmic polarization and concentration polarization for achieving high power density VFB stacks.

By combination of numerical simulations and experiments, through stack material innovation and stack structure design innovation, we have decreased the ohmic

Key technical indicators	2012	2019	2025	2030
Rated current density	80 mA cm^{-2}	150 mA cm^{-2}	300 mA cm^{-2}	400 mA cm^{-2}
Cost (for 1 MW/5 MWh scale)	700 \$ kWh^{-1}	500 \$ kWh^{-1}	350 \$ kWh^{-1}	260 \$ kWh^{-1}
Lifetime	>10 years	>15 years	>20 years	>20 years

Figure 49.4 The development roadmap of the VFB energy storage system.

polarization and improved the electrolyte distribution, and established the design and optimization strategy for obtaining high power density VFB stacks. The rated current density has been increased from 80 to 260 mA cm^{-2} at constant energy efficiency of >80%, which means the power density has been increased over three times and subsequently reduces the materials consumption significantly. Figure 49.4 shows the developing roadmap of the VFB energy storage system, including rated current density, cost, and lifetime, which is based on a 1 MW/5 MWh VFB system, and the material cost of V_2O_5, stabilizing at US\$ 10 000–12000 ton^{-1}.

As shown in Figure 49.4, the rated current density of a 32 kW VFB stack in 2012 was 80 mA cm^{-2}, and the related system cost was about US\$ 900 kWh^{-1} for a 5-hour system. When the current density increased to 150 mA cm^{-2} in 2018, the system cost has been reduced to US\$ 500 kWh^{-1}. As the VFB technology progresses, the current density shall be improved to 300 and 400 mA cm^{-2} in 2025 and 2030, respectively, which indicates that the system cost shall be reduced to US\$ 350 and US\$ 260 kWh^{-1} respectively for a 5-hour system. Meanwhile, the lifetime shall be increased from 10 years to 15 years and 20 years, respectively. For the improved VFB system, the electrolyte cost shall account for >70% of the total cost. Though there is a trace of side reaction occurring during the charge–discharge process, resulting in some vanadium ions crossover through the membrane and accordingly the imbalance of vanadium ions of the two sides, the online or off-line recovery strategy of electrolytes has been successfully developed, which is fairly effective to improve the economic value of the electrolyte. After 10–20 years of continuous operation, the VFB system, including the stacks and electrolytes, can be efficiently recycled. Therefore, the VFB technology is the preferred technology for large-scale energy storage fields thanks to the advantages of high safety, high performance-price ratio, and environmental friendliness throughout the lifetime.

For the urgent need for energy structural adjustment along with the widespread of renewable energies in China, RKP established an equipment manufacturing plant in 2016, where several production lines for high-performance non-fluoride ion-conduction membrane, bipolar composite plate, and 300 MW stack have been developed for the industrialization of VFBs. Figure 49.5 shows a picture of the stack production department of the factory.

A micro-grid power system, consisting of a 1.57 MW solar battery system and a 300 kW/1.2 MWh VFB energy storage system, has been installed on the rooftop,

Figure 49.5 RKP 300 MW year^{-1} VFB equipment manufacturing plant.

which offers electricity for the factory by itself, and sells a surplus over Dalian power grid.

49.1.5 Applications of VFB Technology

Energy storage system (ESS) applies to all aspects of power generation, transmission & distribution, and utilization. On the grid side, ESS can provide a variety of services such as peak and frequency regulation, backup power, black start, and demand response for grid operation, becoming an important method to improve the flexibility, economy, and security of traditional power system. On the power generation side, ESS can significantly improve the capacity of renewable energy integration, considered as the key technology to promote the replacement of the main energy source from fossil energy to renewable energy. On the customer side, ESS can achieve diverse functions such as supporting the distributed power and micro-smart grid construction, serving as the backup power, and improving demand-side response, applying to the C&I micro-grids, backup power, electric vehicle charging station, 5G base station, etc.

Aiming at two major application fields of power grid peak regulation and renewable energy integration, RKP has designed and launched three product lines of VPower, TPower, and ReFlex, as shown in Figure 49.6.

49.1.5.1 VPower

The electrolyte tanks are placed outside the container. Power and capacity can be freely combined according to customer requirements. Containers and electrolyte tanks can be flexibly installed such as tiling or stacking, greatly reducing the system footprint. VPower products are suitable for the construction of high power (several megawatts to hundreds of megawatts) and a large capacity (long-term energy storage, ≥4 hours) energy storage system. Cell stacks, pipes, and other components

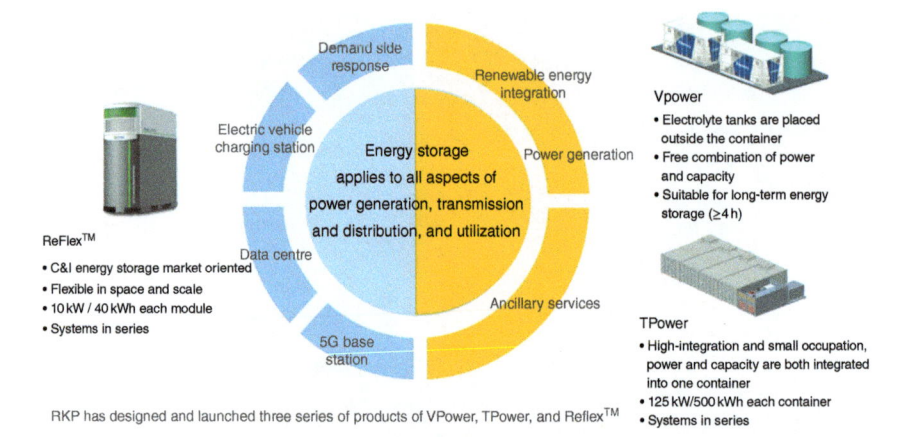

Figure 49.6 RKP designed and launched three product lines of VPower, TPower, and ReFlex. Source: Reflex Winkelmann GmbH.

are prefabricated in a 20-foot container. The power rating of a standard battery module is 250 kW. Two standard battery modules are connected to a 500 KW PCS in series to form a 500 kW energy storage unit. Multiple groups of energy storage units are connected to the bus in parallel to build a larger energy storage system. VPower products have been applied to the Dalian 200 MW/800 MWh VFB Energy Storage Station.

49.1.5.2 TPower

A highly integrated product, the power units, capacity units, and control units are all integrated into one 20-foot prefabricated container. TPower products can be used outdoors, having the prominent advantages of high integration level, high energy density, and small footprint. TPower products are suitable for constructing energy storage systems with power ranging from 100 kW to 100 MW and an energy storage duration of two or four hours. A standard battery module is 125 kW/500 kWh. Four standard battery modules are connected to a PCS in series to form a 500 kW energy storage unit. Multiple groups of energy storage units are connected to the bus in parallel to build a larger energy storage system. The containers can be tightly arranged without fire segregation, greatly reducing the footprint. The 10 MW/40 MWh VFB energy storage system adopts RKP's TPower products, covering an area of about 2600 m^2, including the battery system, PCS, and transformer.

49.1.5.3 ReFlex

Aiming at the diverse user side market, RKP and the American team jointly designed and launched the ReFlex products, with a small and compact form factor. A standard battery module is 10 kW/40 kWh. 5 ~ 21 standard battery modules can be connected to a PCS in series flexibly to form an energy storage unit. Multiple groups of energy storage units are connected to the bus in parallel to build an energy storage system required by customers. The ReFlex products have been applied to the PV+ESS micro-grid project at Heron Island.

By January 2021, more than 30 VFB energy storage station projects had been deployed by RKPDICP team, including 100 MW/400 MWh, 10 MW/40 MWh, and 5 MW/10 MWh with applications covering the grid side, transmission, and distribution side and user side, such as peaking power station, wind farm power output smoothing energy storage station, and microgrid energy storage station.

In order to avoid the intermittent and unstable characteristics of wind power generation when connected to the grid, to reduce the wind power generation curtailment when electricity usage is low, and to improve the economic payback of wind farms, the RKP–DICP team provided a set of 5 MW/10 MWh VFB system to Guodian Longyuan at Faku Woniushi in Liaoning Province (Figure 49.7). The project was connected to the grid in December 2012 and was accepted by State Grid and Guodian Longyuan in May 2013. The functions of this energy storage station are planned power generation tracking, wind power output smoothing, transient real power control for emergency response, and transient voltage emergency support. This energy storage station consists of VFB system, power inverter, transformer, and site monitoring system. This 5 MW/10 MWh VFB consists of fifteen 352 kW/700 kWh VFB units which are connected to the 35 kV busbar of the wind farm and operated together with the 50 MW wind farm. Until February 2021, this system has been in stable operation for more than eight years with a large amount of practical operation data and has gained a lot of experience in engineering construction, operation, and maintenance.

Controlled by the energy management system, this 5 MW/10 MWh VFB system has achieved the following functions:

(1) Under the dispatch instruction of the energy management system, the energy storage system can quickly switch between charging and discharging and adjust the power output, which smooths the power output of wind farm in order to reduce the adverse effects on the grid voltage and frequency. Smooth power output can be achieved this way.
(2) The wind farm can effectively improve the ability of tracking plan power output by storing or releasing the power in real-time from VFB system.
(3) When the power grid dispatching center issues the instruction to the wind farm to limit the power output of the wind farm, the energy storage power station will charge the energy storage system via the control by the energy management system (EMS), and after the power limit command has been removed, the energy storage system discharges to the power grid, reducing the wind power curtailment, improving the utilization of wind resources and increasing the economic benefits of the wind farm.
(4) The energy storage power station accepts the dispatch instruction from the power grid dispatch center in Liaoning Province and participates in the peak shaving of the power grid, which effectively improves wind power penetration.
(5) The overload capacity of the energy storage system can reach 150% of the rated output power (5 MW) that the output power can reach 7.5 MW, and has the long-term operation capability when the power output is overloaded to 120%, which can be used to meet the requirements of the transient real power emergency response and transient voltage emergency support function at the wind farm and the network.

Figure 49.7 Guodian Longyuan Woniushi wind farm 5 MW/10 MWh VFB energy storage station. (a) The exterior view of the VFB energy storage power station of Guodian Longyuan Woniushi wind farm. (b) 5 MW/10 MWh VFB energy storage power station battery stack area. (c) 5 MW/10 MWh VFB energy storage power station electrolyte area. Source: Prof Huamin Zhang in DICP/EnergySuperStore.

(6) The remote-communication, remote-measurement, remote-commissioning, and remote-control functions are achieved between the energy storage station and the dispatch center of Liaoning Province. According to the operation demand of the power grid, the energy storage system can be dispatched in real-time, and participate in the tasks of power grid frequency regulation and load balancing.

This energy storage power station is currently the world's longest-running megawatt-scale VFB energy storage power station, which fully verified the reliability and stability of VFB system. From the operational point of view, the energy storage power station can not only be operated onsite, but also can directly accept the instruction from the dispatch center of Liaoning Province which demonstrates a stable operating capability and rapid response capability, and fully proves that the VFB system can meet wind power fluctuation control, planned power generation capability and meet the functional requirements of grid services.

Having the successful operation of the above-mentioned 5 MW/10 MWh VFB energy storage station, in 2017, RKP was approved by the State Energy Administration to build a 200 MW/800 MWh energy storage demonstration project. The project will be implemented in two phases, each at 100 MW/400 MWh.

As shown in Figure 49.8, the energy storage power station is modular in design. The 500 kW/2 MWh module, which consists of sixteen 32 kW battery stacks, is the basic energy storage module. Such module can achieve solo charge and discharge control. A set of 25 MW/100 MWh energy storage unit consists of 50 basic energy storage modules, each with a monitoring system. Another 4 sets of 25 MW/100 MWh energy storage units are composed of 100 MW/400 MWh VFB energy storage system. Finally, the two 100 MW/400 MWh sub-systems are combined into a 200 MW/800 MWh VFB peaking power station. This is currently the world's largest VFB energy storage power station.

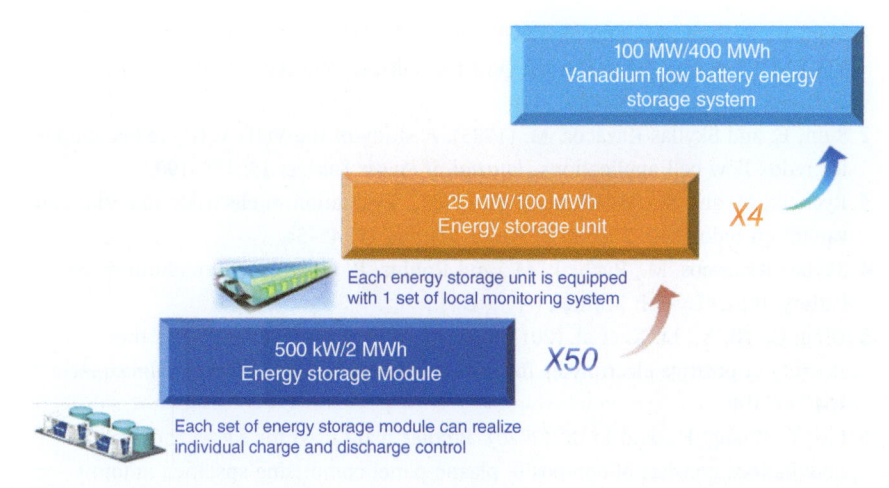

Figure 49.8 Modular structure of 200 MW/800 MWh energy storage power station. (Phase 1: 100 MW/400 MWh).

Figure 49.9 Bird's eye view of Dalian 200 MW/800 MWh VFB energy storage station.

Affected by the abnormal high prices of international vanadium in 2018–2019 and the COVID-19 in 2020, the first phase of the project (100 MW/400 MWh) was postponed and has completed the grid connection commissioning in August in 2022. The project shall be collected to the grid formally in the late October, in 2022. The main functions of the energy storage power station are grid-level peak shaving, emergency backup power supply, and black start. The project will play an important role in the value verification, industrialization, and policy formulation of VFB energy storage batteries. The figure below is a bird's eye view of Dalian 200 MW/800 MWh VFB energy storage peaking power station (Figure 49.9).

References

1. Sum, E., Rychcik, M., and Skyllas-kazacos, M. (1985). Investigation of the V(V)/V(IV) system for use in the positive half-cell of a redox battery. *Journal of Power Sources* 16: 85–95.
2. Sum, E. and Skyllas-Kazacos, M. (1985). A study of the V(II)/V(III) redox couple for redox flow cell applications. *Journal of Power Sources* 15: 179–190.
3. Rychcik, M. and Skyllas-Kazacos, M. (1987). Evaluation of electrode materials for vanadium redox cell. *Journal of Power Sources* 19: 45–54.
4. Skyllas-Kkazacos, M., Rychcik, M., and Robins, R. (1988). All vanadium redox battery. In: Unisearch Limited (AU), US.
5. Ding, C., Ni, X., Li, X. et al. (2015). Effects of phosphate additives on the stability of positive electrolytes for vanadium flow batteries. *Electrochimica Acta* 164: 307314.
6. Liu, T., Zhang, H., and Li, X. (2020). Bipolar plate used for e.g. vanadium flow battery, consists of composite plastic panel comprising specified amount of polypropylene having preset melt index, conductive filler and lubricant, CN107565146-A.

7 Yuan, Z., Duan, Y., Zhang, H. et al. (2016). Advanced porous membranes with ultra-high selectivity and stability for vanadium flow batteries. *Energy & Environmental Science* 9: 441–447.

8 Lu, W., Yuan, Z., Zhao, Y. et al. (2017). Porous membranes in secondary battery technologies. *Chemical Society Reviews* 46: 2199–2236.

9 Shi, M., Dai, Q., Li, F. et al. (2020). Membranes with well-defined selective layer regulated by controlled solvent diffusion for high power density flow battery. *Advanced Energy Materials* 10: 2001382.

10 Li, L., Kim, S., Wang, W. et al. (2011). A stable vanadium redox-flow battery with high energy density for large scale energy storage. *Advanced Energy Materials* 1: 394–400.

11 Cao, L., Skyllas-Kazacos, M., Menictas, C., and Noack, J. (2018). A review of electrolyte additives and impurities in vanadium redox flow batteries. *Journal of Energy Chemistry* 27: 1269–1291.

12 Skyllas-Kazacos, M. (1996). High energy density vanadium electrolyte solutions, methods of preparation thereof and all-vanadium redox cells and batteries containing high energy density vanadium electrolyte solutions. WO9635239A1.

13 Zhang, H., Lu, W., and Li, X. (2019). Progress and perspectives of flow battery technologies. *Electrochemical Energy Reviews* 2: 492–506.

14 Qian, P., Zhang, H., Chen, J. et al. (2008). A novel electrodebipolar plate assembly for vanadium redox flow battery applications. *Journal of Power Sources* 175: 613–620.

15 Zhang, H., Liu, Z., Han, X. (2010), et al. Carbon-plastic electricity-conductive bipolar board for liquid energy-storing battery and manufacture thereof. China Patent No. CN101308923B.

16 Tomazic, G. and Skyllas-Kazacos, M. (2014). Flow battery technologies. In: *Electrochemical Energy Storage for Renewable Sources and Grid Balancing* (ed. P.T. Moseley and J. Garche). Elsevier.

17 Zhong, S., Kazacos, M., Kazacos, M.S., and Haddadi-Asl, V. (1997). Flexible, conducting plastic electrode and process for its preparation. US Patent No. 5,665,212.

18 Li, X., Zhang, H., Mai, Z. et al. (2011). Ion exchange membranes for vanadium redox flow battery (VRB) applications. *Energy & Environmental Science* 4: 1147–1160.

19 Zhang, H., Zhang, H., Li, X. et al. (2011). Nanofiltration (NF) membranes: the next generation separators for all vanadium redox flow batteries (VRBs)? *Energy & Environmental Science* 4: 1676–1679.

20 Yuan, Z., Zhu, X., Li, M. et al. (2016). A highly ion-selective zeolite flake layer on porous membranes for flow battery applications. *Angewandte Chemie International Edition* 55: 3058–3062.

50

Metal-Free Flow Batteries Based on TEMPO

Tobias Janoschka and Olaf Conrad

JenaBatteries GmbH, Otto-Schott-Strasse 15, 07745 Jena, Germany

50.1 Introduction

Since the concept of metal-free flow batteries was introduced in the late 2000s, numerous cell chemistries that employ organic redox species in both posolyte and negolyte were studied in the laboratory scale. Two of the most promising candidates for scale-up and industrialization are nitroxide-free radicals, in particular, 2,2,6,6-tetramethylpiperidine-N-oxyl (TEMPO) derivatives for the posolyte and 4,4′-bipyridinium ions (viologen) for the negolyte. This chapter provides a deeper insight into the battery-related chemistry of TEMPO and viologen systems, their challenges and advantages. In the first part of the chapter, redox reactions, physicochemical properties, and synthetic procedures of selected derivatives are presented.

The second part reviews the battery-cycling behavior. Starting out with the general performance of laboratory-scale cells, the scale-up of a battery system to kWh size is addressed. The world's first metal-free flow battery was integrated into a smart grid with renewable energy sources (wind, PV, and biomass), and variable power consumption units were introduced as an example. In the final part, the raw material pipeline of organic compounds that are employed in flow batteries (i.e. a new raw material base) is discussed briefly in the context of conventional metal-based batteries.

50.2 Properties, Synthetic Procedures, and Redox Reactions

50.2.1 TEMPO

Organic radicals of the general structure $R_2N–O^{\bullet}$ are commonly known as nitroxide radicals (IUPAC: aminoxyl radical) [1]. The first example of a nitroxide was discovered by Fremy in 1845 (Scheme 50.1). He prepared potassium nitrosodisulfonate, a compound that forms bright violet radicals in aqueous

Flow Batteries: From Fundamentals to Applications, First Edition.
Edited by Christina Roth, Jens Noack, and Maria Skyllas-Kazacos.
© 2023 WILEY-VCH GmbH. Published 2023 by WILEY-VCH GmbH.

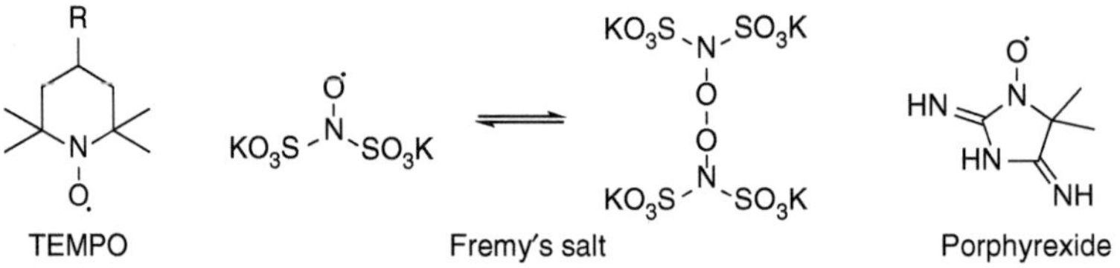

Scheme 50.1 Schematic representation of TEMPO derivatives, potassium nitrosodisulfonate (Fremy's salt), and porphyrexide.

solutions and ocher-colored dimers upon crystallization [2]. Later on, heterocyclic porphyrexide was discovered to be the first known stable organic nitroxide radicals [3]. Followed by numerous aryl- and alkyl-substituted species being stable in solution or even in their isolated form, the excellent stability of a number of nitroxides made them promising candidates for industrial application in catalysis and energy storage. Most prominently, the persistent radical TEMPO is employed in the catalytic oxidation of alcohols and has been described as active material in several publications on flow batteries [4–6].

The majority of persistent nitroxide radicals owe their exceptional stability to mesomeric and steric effects. It is further promoted by the relatively high activation barrier of potential degradation reactions. The high delocalization energy of the strong three-electron nitrogen–oxygen bond contributes to low-energy content of the radical. Dimerization of two radicals is hindered because a week O—O bond would need to be formed, not compensating for the net energy loss of the removal of electron delocalization. By implementing the right substitution pattern, the stability of nitroxide radicals can be optimized. In particular, N,N-alkyl-substituted nitroxides, such as TEMPO, are only stable if the α-carbon does not carry a hydrogen atom. Otherwise, they disproportionate into nitrones and hydroxylamines. Exceptions include bicyclic radicals (Bredt's rule) as well as electronically stabilized nitroxide radicals, such as "SG-1," which is used in nitroxide-mediated polymerization [7, 8].

Using easily available bulk chemicals, TEMPO derivatives are prepared on an industrial scale (Scheme 50.2). In the first step, the universal precursor triacetonamine is synthesized from acetone and ammonia. Several reaction routes have been described in the last 120 years, such as the direct aldol condensation reaction or procedures using phorone and acetonin as intermediate. Afterward, the carbonyl group can be used for subsequential functionalization of the molecule, forming alcohols, esters, amines, ethers, etc. Regarding the application of TEMPO derivatives in batteries, this functionalization step is the key to tune the molecule's solubility. Finally, the amine is oxidized yielding the nitroxide radical that displays a bright

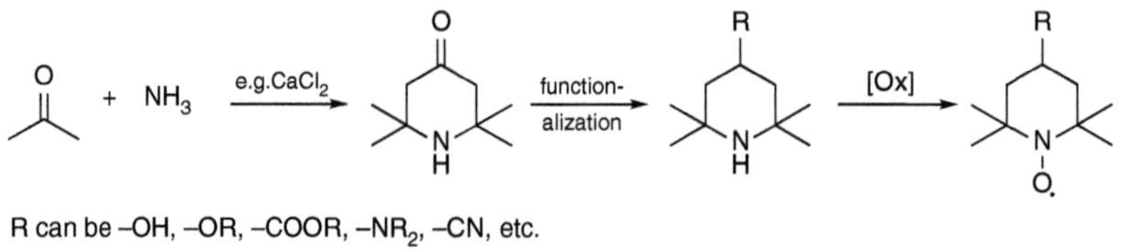

R can be –OH, –OR, –COOR, –NR$_2$, –CN, etc.

Scheme 50.2 Schematic representation of the synthesis of TEMPO derivatives.

(orange) color. Peroxides, such as peracetic acid or hydrogen peroxide/sodium tungstate, are commonly employed oxidation agents [9].

TEMPO has become the most investigated nitroxide for battery applications. In 2002 Nakahara et al. first reported a polymer battery using a polymethacrylate bearing the persistent radical as cathode material (PTMA-NO) in combination with a lithium anode [10]. This type of solid-state battery has since become known as "organic radical battery" (ORB) [11–17].

With a TEMPO and N-methylphthalimide cell that employed acetonitrile as solvent, the concept of ORB was translated into the flow-battery context by Li et al. in 2011 [18]. Polymers bearing TEMPO and viologen groups (see Section 50.3) were introduced for flow batteries in 2015, using easier-to-handle aqueous solutions and size-exclusion membranes [19]. Shortly afterward, two aqueous flow batteries based on low-molecular-weight TEMPOL (R=OH) [20] and TEMPTMA (R=N(CH$_3$)$_3$Cl) [21] in combination with methyl viologen were reported. Numerous other studies applied TEMPO derivatives in laboratory-scale flow cells and demonstrated the versatility of the redox molecule [22–27].

The nitroxide radical used in these studies can undergo reversible one-electron oxidation to form an oxoammonium cation or reduction to an aminoxyl anion (Scheme 50.3, Figure 50.1). Because the reoxidation of the aminoxyl anion is kinetically hindered, only the oxoammonium cation is used in battery application. While many organic molecules experience great structural rearrangement upon oxidation or resonance effects slow down the redox reaction at conjugated π-systems, TEMPO benefits from the localization of electron density at the N–O center and only slight structural changes. This implies a high electron-transfer rate constant (10^{-2} to 10^{-3} cm s^{-1} in aqueous solution depending on supporting electrolyte) and, therefore, fast redox kinetics. Subject to the substituent in 4-position, the redox potential in aqueous solution varies from 580 mV for R=COOH up to 790 mV (Ag/AgCl) for R=N(CH$_3$)$_3$Cl [21, 28].

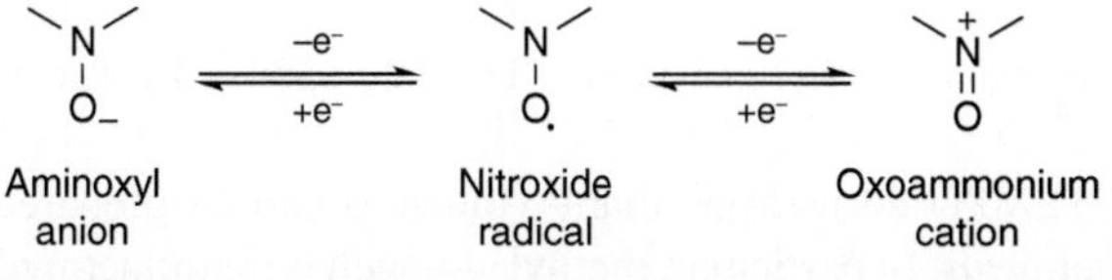

| Aminoxyl anion | Nitroxide radical | Oxoammonium cation |

Scheme 50.3 Schematic representation of the redox couples of nitroxide radical derivatives.

A challenge in building a full-flow-battery system based on nitroxides lies in finding a suitable counter electrode. Viologens are identified as versatile and reliable materials. They will be discussed in detail in Section 50.2.2 [19–22, 29–32].

However, similar to vanadium batteries, an all-TEMPO flow battery is desirable. It would allow for straightforward rebalancing of an aged electrolyte, where negolyte and posolyte are intermixed due to membrane failure and crossover effect. Due to the irreversible nature of the nitroxide/aminoxyl anion reaction in most solvents, molecules combining two different redox-active centers were studied, e.g. TEMPO combined with phenazine [27], indigo carmine [33], or viologen groups [34]. Eventually, Wylie et al. achieved an all-TEMPO system by introducing

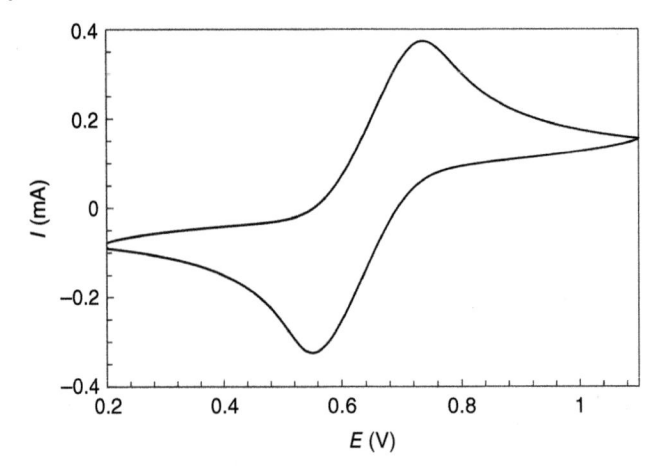

Figure 50.1 Cyclic voltammogram of TEMPOL in 0.1 M NaCl$_{aq}$ using an Ag/AgCl reference and a glassy carbon working electrode.

an ionic liquid as solvent in 2020 [35]. While the all-TEMPO battery's performance and Coulombic efficiency (60%) still need to be improved, the concept lays the foundation for future research in the field of TEMPO flow batteries.

50.2.2 Viologen

For the substance class of N,N'-disubstituted 4,4'-bipyridine derivatives, the name *viologens* was coined in the 1930s by Michaelis and Hill, who introduced the molecules as a new class of oxidation–reduction indicators [36]. The most prominent representative is N,N'-dimethyl-4,4'-bipyridinium chloride (methyl viologen), which displays a vibrant violet color in its reduced state. Besides being used as indicator in biology, viologens are known for their application as herbicide in agriculture, as photocatalysts in the production of hydrogen, and as electrochromic layer in windows. For a comprehensive overview of this diverse class of compounds, the authors recommend the excellent book "The Viologens" by P.M.S. Monk [37–39].

Like the aforementioned TEMPO derivatives, many viologens can be prepared from easily available bulk chemicals. In particular, methyl viologen is manufactured on an industrial scale of more than 100 000 tons per year [40]. Two major synthetic routes are commonly employed, both in laboratory and industrial production (Scheme 50.4). One is based on 4,4'-bipyridine as universal precursor, which is obtained from a reductive coupling reaction of pyridine with sodium in liquid ammonia. 4,4'-bipyridine is reacted with alkyl halides, e.g. methyl iodide, to form the cationic viologen salts. Ion-exchange reactions may be used afterward, to tune the substance properties. The second route follows an inverse process. Pyridine is quaternized with an alkyl halide first and subsequently, the pyridinium salts are coupled in the presence of cyanide ions [37].

Bivalent viologen cations Viol^{2+} can undergo two one-electron reduction steps forming the radical cation Viol$^{+\bullet}$ and the neutral Viol^0, respectively (Scheme 50.5).

Scheme 50.4 Schematic representation of two commonly used routes for the synthesis of methyl viologen.

Scheme 50.5 Schematic representation of the redox couples of viologen derivatives.

A wide range of redox potentials can be accessed by introducing electron-donating or withdrawing groups, e.g. as *N*-substituent, and by choice of the counter ion. For example, the first reduction step occurs at −650 mV for R′ = methyl/Cl⁻, −850 mV for R′ = OCH$_3$/I⁻, and −350 mV for R′ = CH$_2$CN/I⁻ (aqueous solution, Ag/AgCl) [41]. In aqueous solution, only the first reduction step is reversible. The second step leads to precipitation of the nonpolar species and side reactions. Hence, most water-based flow-battery systems employ the Viol^{2+}/Viol$^{+\bullet}$ redox pair (Figure 50.2). At 2.4 mol l^{-1}, the solubility of methyl viologen in water is excellent,

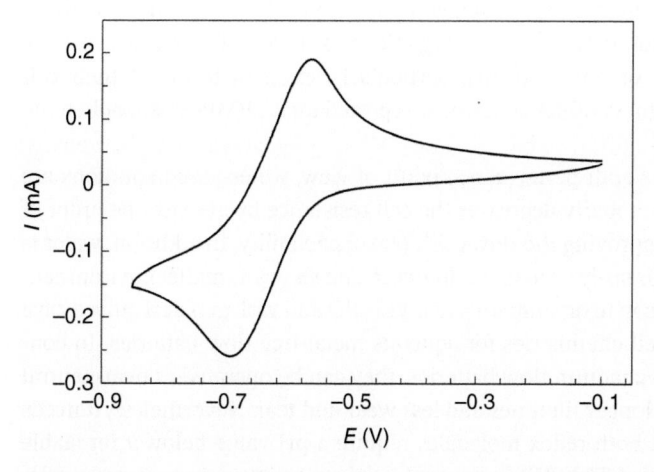

Figure 50.2 Cyclic voltammogram of the first reduction step of methyl viologen (Viol^{2+}/Viol$^{+\bullet}$) in 0.1 M NaCl$_{aq}$ using an Ag/AgCl reference and a glassy carbon working electrode.

allowing for stable battery operation at a concentration of 2 mol l^{-1}. To unlock the additional energy storage capacity of the second reduction, Viol$^{+\bullet}$/Viol^0, solubility enhancing functionalization can be introduced. Trimethylammonium moieties show promising results [42]. Luo et al. presented a π-conjugation-extended viologen, that not only introduced trimethylammonium groups for improved solubility but also introduced a thiazolo[5,4-d]thiazole spacer for a smaller potential gap between the first and the second reduction step and, hence, a more stable cell voltage [31].

Most viologens employed in aqueous flow batteries exhibit fast redox kinetics with electron-transfer rate constants of about 10^{-3} cm s^{-1}, making them an ideal counter electrode for TEMPO [21, 37, 41, 43].

50.3 TEMPO Flow Batteries – Selected Examples

50.3.1 Fundamental Research

TEMPO has been used in full-flow-battery-type test cells that employ two liquid-storage solutions, i.e. posolyte and negolyte, and in hybrid systems combining a solid electrode (e.g. zinc, lithium) with a liquid one [5]. By utilizing organic solvents or water, two fundamental strategies for building these flow-battery test cells have been developed successfully. Organic solvents allow for straightforward application of commercially available TEMPO derivatives for the posolyte and suitable hydrophobic redox molecules for the negolyte. For example, an acetonitrile-based supporting electrolyte was used for the TEMPO/N-methylphthalimide cell by Li et al. [18]. It revealed a cell voltage of 1.6 V and was operated for 20 consecutive charging/discharging cycles. However, due to low ion mobility and rather high resistance, the cell's current capability was limited to 0.35 mA cm^{-2}. The same challenges were encountered in a mixed-component posolyte made from TEMPO and ferrocene in acetonitrile and other organic cells alike [44]. Yet, flow-battery chemistries based on organic solvents, particularly, excel in terms of their cell voltage. One of the highest values achieved is reported for a TEMPO/benzophenone cell (2.4 V) [45].

However, from an overall performance point of view, water-based solutions are preferred. Higher ion mobility decreases the cell resistance by at least one order of magnitude, thereby improving the flow cell's power capability. In addition, water is not flammable (safety), easily available at low cost, and shows a smaller environmental burden in comparison to organic solvents. TEMPO and viologen derivatives have become established cell chemistries for aqueous metal-free flow batteries. In contrast to conventional vanadium flow batteries, they can be operated at near-neutral pH values, promising longer lifetimes and less wear and tear. Nevertheless, it needs to be pointed out that both redox molecules require a pH value below 7 for stable operation. In the case of TEMPO, the oxoammonium cation may undergo nucleophilic attack of hydroxyl ions at elevated pH. Similarly, the methyl groups of methyl viologen may be cleaved off, resulting in methanol formation. As is common for many redox molecules showing a negative standard potential, viologens should be handled in an oxygen-free atmosphere, once they are in their reduced state (Viol$^{+\bullet}$) in order to avoid unwanted reoxidation.

Although water is the preferred solvent, a challenge lies in identifying stable and highly soluble TEMPO derivatives possessing a redox potential within the electrochemical stability window of water. Commercially available redox catalyst TEMPOL (R=OH) has good solubility in water (2.1 mol l^{-1}), has a suitable potential, and can be combined with methyl viologen to build an organic (metal-free) flow cell. For full-cell operation, aqueous sodium chloride solution was employed as supporting electrolyte, limiting the solubility of TEMPOL to 0.5 mol l^{-1} and the theoretical energy density to 8 Wh l^{-1} at a cell voltage of 1.25 V. After 100 consecutive charging/discharging cycles at 40 mA cm^{-2}, the cell revealed a capacity decay of ~20%. This decay can be attributed to side reactions, such as autoxidation of the OH group because TEMPO is a renowned oxidation catalyst [20].

By introducing a trimethylammonium group in 4-position, the solubility and stability of TEMPO were significantly improved. The resulting redox molecule TEMPTMA revealed a solubility of 2.3 mol l^{-1} in aqueous sodium chloride solution. Using a 2 mol l^{-1} solution (38 Wh l^{-1}), an extensive cycling test at 80 mA cm^{-2} was carried out and no capacity decay was observed after 100 cycles. In stress tests, a peak current density of up to 200 mA cm^{-2} was demonstrated [21].

Another robust cell is based on TEMPO and viologen derivatives both using a 3-(trimethylammonium)propoxy group as solubility enhancing substituent. The resulting TMAP–TEMPO demonstrates an impressive solubility in water of 4.6 mol l^{-1} and the viologen of 2 mol l^{-1}. Because the cationic groups enhance the Coulomb repulsion (electrostatic force) and reduce the risk of bimolecular side reactions, excellent capacity retention of 99.993% per cycle over 1 000 cycles at 40 mA cm^{-2} and practical concentrations of 0.1 mol l^{-1} can be achieved (in total 6% loss). Peak current densities of up to 300 mA cm^{-2} were reported [22].

Besides the actual redox molecules, other components of a redox flow cell, such as membrane and electrode, need to be optimized for even better cell performance. Unfortunately, research on electrode materials for TEMPO flow batteries is limited to only a few examples. In one instance, a carbon paper electrode was improved with nitrogen-doped three-dimensional reduced graphene oxide [46]. The functionalization was tested in a TEMPOL/methyl viologen cell, demonstrating improved wettability, charge-transfer kinetics, and activation polarization. In another example, nanofiber-functionalized graphite felts were prepared using a microwave-assisted technique [47]. In a test cell employing TEMPO and viologen polymers, the redox kinetics and the current rating of the cell could be improved. This effect is attributed to a 50-fold increase in surface area, while oxygen functional groups on the graphite surface are expected to show no significant influence on the cell performance. Both examples indicate that high-surface-area electrodes, in particular, such with functionalization placed close to the membrane, promise improved battery characteristics.

Membranes that separate the positive and the negative cell compartment are known to define the cell resistance and, hence, limit the overall performance.

Organic solvent-based cells usually rely on polyolefin membranes, e.g. polyethylene, as they are sufficiently stable in the hydrophobic solvents. However, crossover of the redox molecules remains high with this membrane type, lowering the overall lifetime of the cell [48, 49].

Where possible, ion-exchange membranes are applied. They selectively allow the counter ions to pass while retaining TEMPO, viologen, and other relevant redox molecules. The best lifetime can be achieved if ionic TEMPO derivatives in aqueous electrolyte solutions are used because their retention is significantly higher than observed for neutral molecules [18, 20–22].

Dialysis membranes, which separate both half-cells and electrolytes by a size-exclusion effect, can be used in polymer flow batteries, due to larger molecule size. Up to 10 000 cycles can be realized with such a set-up [19, 26].

In most experiments, commercial membranes are used as they are readily available. However, in-depth membrane research directed at enhancing cell performance needs to be expanded in the future. Small et al. analyzed five commercial anion-exchange membranes for a TEMPOL/methyl viologen cell [50]. In another study, a cationic TEMPO derivative and methyl viologen were used as reference systems [51]. Both studies correlated the polymer structure with membrane properties and transport phenomena. They identified water uptake (TEMPO crossover) and ion-exchange capacity (power capability and conductivity) as two crucial but opposing parameters that can be tuned by intelligent membrane design.

To ultimately overcome the limitations introduced by the membrane, the concept of membrane-free cells was introduced. Marcilla and coworkers presented a system based on two immiscible solutions, one containing TEMPOL in water and the other one p-benzoquinone in a mixture of $PYR_{14}TFS$ (ionic liquid) and propylene carbonate. It could be operated for 300 cycles at $2\,mA\,cm^{-1}$. Due to the crossover of active species at the solution interface the Coulombic efficiency was limited to 80% (self-discharge). However, the research paves the way for new cell concepts to be developed in the future [52].

50.3.2 Scale-up

Because the TEMPO-based flow-battery chemistry has been a success on laboratory scale, the compound was chosen by JenaBatteries (Germany) for scaling up the technology and for developing an industrial product. In 2019, the company presented the world's first metal-free flow battery integrated into a smart grid.[1] At the Application Centre for Renewable Resources (ACRRES) in the Netherlands, a battery system was installed in a modular approach and connected with renewable energy sources (wind, photovoltaics, and biomass) and variable power consumption entities (Figure 50.3). It was made of two 20 ft intermodal containers that were fitted with the system components: (i) storage unit and (ii) power unit. The storage unit contained the electrolyte reservoirs and associated balance of plant, while

---

[1] The project has received funding from the European Union's Horizon 2020 research and innovation program under grant agreement No 731239.

Figure 50.3 3D-model of 100 kWh/30 kW metal-free flow battery, including a storage and power unit (a). Photograph of the battery prototype set up at the Application Centre for Renewable Resources (b).

the power unit housed the cell stacks, the pump system for electrolyte transport, power electronics, and the battery management system (BMS). A Siemens S7 PLC was employed as fail-safe BMS controlling the actuators and sensors of the battery as well as providing two-way communication with the smart grid control system. The electrical system was based on a grid interface of three-phase 400 V (AC) at 90 A. A transformer was set to isolate the key system components from the electrical interface. Downstream the grid current was fed into a converter (rectifier-inverter) system that changes the alternating current (AC) of the grid to a direct current (DC) of lower voltage. This step was necessary as the cell stacks, which are the actual electrochemical power conversion unit of the battery, run on voltages lower than the grid voltage.

The battery system underwent comprehensive electrical performance tests, demonstrating its 100 kWh capacity and 30 kW power capability. Over the course of the experiments, the interaction of the battery system with the smart grid was analyzed. For several applications within the smart grid, the response time of the battery was an important parameter. With a duration of 600 ms for switching from charging to discharging all relevant functions of the smart grid could be performed. Finally, the full grid system was employed successfully to study prosumer business models, i.e. the application of a battery in the context of active consumers that also produce electricity and provide it to the grid, energy communities, or aggregators.

50.4 Raw Materials Outlook and Summary

Forecasts anticipate a sustained growth of the energy storage market of around 30% per year. The global market for stationary storage is expected to amount to 450 GWh in 2030, while the market for electro mobility will account for more than 3 000 GWh [53]. Concurrently, the demand for energy storage materials, in particular lithium, cobalt, and nickel minerals, will grow rapidly. Supply shortages for these key resources are expected, as production capacity must be multiplied. Supply bottlenecks will impact the markets as early as 2025 [54]. Besides raw material availability, the effects of metal mining dominate the ethical impact and carbon footprint of conventional batteries, such as lithium-ion batteries. In many parts of the world, they stand in opposition to the UN Sustainable Development Goals—whether decent work conditions, healthy living, or protection of the ecosystems. This is reflected in key parameters, such as Child Labor Proportion, Human Development Index, Corruption Perception Index, or SDG Index [55]. Apart from Li-ion batteries, the European Union also lists vanadium among the critical raw materials [56]. In the context of flow batteries, this translates to a demand of 5% of the world's production for 1 GWh installed storage capacity (2019) [57]. Hence, a new metal-free (organic) raw material base that may bring relief to this narrow battery metal market is needed. For example, bulk chemicals required for TEMPO are produced on the megaton scale and allow manufacturing in Europe. 1 GWh battery capacity requires only 0.001% of annual ammonia production and 0.07% of the acetone production, respectively [58, 59]. In the long run, organic raw materials may even be manufactured from renewable feedstock as the chemical industry transitions away from crude-oil-based petrochemical production toward biorefineries. Cereal straw and lignocellulose are identified as a suitable material basis for acetone, while cattle slurry may be used as a sustainable source of ammonia [55].

In summary, this chapter shows that it is feasible to take the basic concept of a novel, metal-free flow-battery chemistry from the laboratory to practical application. By carefully tuning the physicochemical properties of TEMPO derivatives, well-performing, robust, and readily available materials for energy storage can be provided. In the near future, the development will advance the presented prototypes toward an industrial product and make metal-free flow batteries available for widespread use as a sustainable energy storage system.

References

1 McNaught, A.D. and Wilkinson, A. (1997). *IUPAC. Compendium of Chemical Terminology (The "Gold Book")*, 2e. Oxford: Blackwell Scientific Publications.
2 Fremy, E. (1845). *Annales de Chimie Physique* 15: 408–488.
3 Piloty, O. and Schwerin, B.G. (1901). *Berichte der Deutschen Chemischen Gesellschaft* 34: 1870–1887.
4 De Nooy, A.E.J., Besemer, A.C., and van Bekkum, H. (1996). *Synthesis* 1996: 1153–1176.
5 Winsberg, J., Hagemann, T., Janoschka, T. et al. (2016). *Angewandte Chemie International Edition* 56 (3): 686–711.

6 Sánchez-Déz, E., Ventosa, E., Guarnieri, M. et al. (2021). *Journal of Power Sources* 481: 228804.
7 Volodarsky, L.B., Reznikov, V.A., and Ovcharenko, V.I. (1994). *Synthetic Chemistry of Stable Nitroxides*. Boca Raton, FL: CRC Press.
8 Call, L. (1977). *Pharmazie in Unserer Zeit* 6: 83–95.
9 Dagonneau, M., Kagan, E.S., Mikhailov, V.I. et al. (1984). *Synthesis* 1984 (11): 895–916.
10 Nakahara, K., Iwasa, S., Satoh, M. et al. (2002). *Chemical Physics Letters* 359 (5–6): 351–354.
11 Nishide, H., Iwasa, S., Pu, Y.-J. et al. (2004). *Electrochimica Acta* 50 (2–3): 827–831.
12 Satoh, M., Nakahara, K., Iriyama, J. et al. (2004). *IEICE Transactions on Electronics* E87C: 2076–2080.
13 Komaba, S., Tanaka, T., Ozeki, T. et al. (2010). *Journal of Power Sources* 195 (18): 6212–6217.
14 Nakahara, K., Iwasa, S., Iriyama, J. et al. (2006). *Electrochimica Acta* 52 (3): 921–927.
15 Lee, S.H., Kim, J.-K., Cheruvally, G. et al. (2008). *Journal of Power Sources* 184 (2): 503–507.
16 Oyaizu, K., Suga, T., Yoshimura, K., and Nishide, H. (2008). *Macromolecules* 41 (18): 6646–6652.
17 Kim, J.-K., Cheruvally, G., Choi, J.-W. et al. (2007). *Journal of the Electrochemical Society* 154 (9): A839.
18 Li, Z., Li, S., Liu, S. et al. (2011). *Electrochemical and Solid-State Letters* 14 (12): A171.
19 Janoschka, T., Martin, N., Martin, U. et al. (2015). *Nature* 527 (7576): 78–81.
20 Liu, T., Wei, X., Nie, Z. et al. (2015). *Advanced Energy Materials* 6 (3): 1501449.
21 Janoschka, T., Martin, N., Hager, M.D., and Schubert, U.S. (2016). *Angewandte Chemie International Edition* 55 (46): 14427–14430.
22 Liu, Y., Goulet, M.-A., Tong, L. et al. (2019). *Chem* 5 (7): 1861–1870.
23 Yu, X., Yu, W.A., and Manthiram, A. (2020). *ACS Applied Materials & Interfaces* 12 (43): 48654–48661.
24 Chang, Z., Henkensmeier, D., and Chen, R. (2017). *ChemSusChem* 10 (16): 3193–3197.
25 Winsberg, J., Janoschka, T., Morgenstern, S. et al. (2016). *Advanced Materials* 28 (11): 2238–2243.
26 Winsberg, J., Muench, S., Hagemann, T. et al. (2016). *Polymer Chemistry* 7 (9): 1711–1718.
27 Winsberg, J., Stolze, C., Muench, S. et al. (2016). *ACS Energy Letters* 1 (5): 976–980.
28 Zhou, W., Liu, W., Qin, M. et al. (2020). *RSC Advances* 10 (37): 21839–21844.
29 Hu, B., Tang, Y., Luo, J. et al. (2018). *Chemical Communications* 54 (50): 6871–6874.
30 Hagemann, T., Winsberg, J., Grube, M. et al. (2018). *Journal of Power Sources* 378: 546–554.
31 Luo, J., Hu, B., Debruler, C., and Liu, T.L. (2017). *Angewandte Chemie International Edition* 57 (1): 231–235.

32 Wang, H., Li, D., Xu, J. et al. (2021). *Journal of Power Sources* 492: 229659.
33 Nambafu, G.S. and Shao, M. (2020). *ECS Meeting Abstracts MA2020-01(3)*, p. 495.
34 Janoschka, T., Friebe, C., Hager, M.D. et al. (2017). *ChemistryOpen* 6 (2): 216–220.
35 Wylie, L., Blesch, T., Freeman, R. et al. (2020). *ACS Sustainable Chemistry & Engineering* 8 (49): 17988–17996.
36 Michaelis, L. and Hill, E.S. (1933). *Journal of General Physiology* 16 (6): 859–873.
37 Monk, P. (1999). *The Viologens*. Chichester, New York: Wiley.
38 Murugavel, K. (2014). *Polymer Chemistry* 5 (20): 5873–5884.
39 Gu, Y., Hong, W., Choi, W. et al. (2014). *Journal of the Electrochemical Society* 161 (12): H716–H721.
40 Watts, M. (2011). Paraquat. Pesticide Action Network Asia and the Pacific.
41 Wardman, P. (1989). *Journal of Physical and Chemical Reference Data* 18 (4): 1637–1755.
42 DeBruler, C., Hu, B., Moss, J. et al. (2017). *Chem* 3 (6): 961–978.
43 Bird, C.L. and Kuhn, A.T. (1981). *Chemical Society Reviews* 10 (1): 49–82.
44 Kosswattaarachchi, A.M. and Cook, T.R. (2018). *Journal of the Electrochemical Society* 165 (2): A194–A200.
45 Xing, X., Huo, Y., Wang, X. et al. (2017). *International Journal of Hydrogen Energy* 42 (27): 17488–17494.
46 Li, X., Li, J., Huang, C., and Zhang, W. (2019). *Electrochimica Acta* 301: 240–250.
47 Schwenke, A., Janoschka, T., Stolze, C. et al. (2016). *Journal of Power Sources* 335: 155–161.
48 Wei, X., Xu, W., Vijayakumar, M. et al. (2014). *Advanced Materials* 26 (45): 7649–7653.
49 Sukegawa, T., Masuko, I., Oyaizu, K., and Nishide, H. (2014). *Macromolecules* 47 (24): 8611–8617.
50 Small, L.J., Pratt, H.D., and Anderson, T.M. (2019). *Journal of the Electrochemical Society* 166 (12): A2536–A2542.
51 Tsehaye, M., Yang, X., Janoschka, T. et al. (2021). *Membranes* 11 (5): 367.
52 Navalpotro, P., Sierra, N., Trujillo, C. et al. (2018). *ACS Applied Materials & Interfaces* 10 (48): 41246–41256.
53 Heid, B., Kane, S., and Schaufuss, P. (2020). *Powering Upsustainable Energy - Building a More Sustainable Battery Industry*. McKinsey & Company.
54 Campagnol, N., Eddy, J., Hagenbruch, T. et al. (2018). *Metal Mining Constraints on the Electric Mobility Horizon*. McKinsey.
55 Brosowski, A. and Majer, S. (2021). *Whitepaper: Impact screening JenaBatteries*. Impacttelling.
56 Bobba, S., Carrara, S., Huisman, J. et al. (2020). *Critical Raw Materials for Strategic Technologies and Sectors in the EU*. Luxembourg: Publications Office of the European Union.
57 Polyak, D.E. (2020). *Mineral Commodity Summaries: VANADIUM*. U.S. Geological Survey.
58 Apodaca, L.E. (2020). *Mineral Commodity Summaries: NITROGEN (FIXED)-AMMONIA*. U.S. Geological Survey.
59 Acetone Market (2021). Future Market Insights.

51

Commercialization of All-Iron Redox Flow-Battery Systems
Julia Song

Co-Founder and CTO, ESS Tech Inc., Wilsonville, OR 97070, USA

51.1 Introduction

The target market for energy storage systems is projected to grow to a multi-billion-dollar industry in the next few years, as published by Guidehouse Insights (i.e. *Navigant Research*) [1] and Bloomberg NEF (https://about.bnef.com/) (see Figure 51.1). This demand is driven by increases in renewable energy penetration, area regulation requirements, high time of use (TOU) and demand prices, grid interruptions due to climate-driven severe weather events, and recent and anticipated legislation.

In 2011, Evans and Song founded ESS Tech Inc. (ESS) from their garage (https://essinc.com/about/) to develop and commercialize cost-effective grid-scale energy storage systems based on flow-battery technology to address the growing long-duration (4–12 hours) market. ESS initially focused on vanadium-based chemistry, developing a high-power cell and a compact stack design. Due to the high cost of the vanadium-based electrolytes, Evans and Song evaluated other flow-battery chemistries to address these cost challenges. Early academic literature on the iron-flow battery (IFB) indicated a potentially low-cost electrolyte approach using earth-abundant iron as the active material. Still, state-of-the-art power densities were low ($\sim 50\,mW\,cm^{-2}$), and cyclability was poor [2]. Given the early work ESS had done on a high-power density cell for vanadium-based chemistries, there was a clear opportunity to leverage this work to improve the iron-based chemistry.

With the support of the 2012 ARPA-e SBIR grant [3], ESS adapted its unique stack design to use iron salt electrolytes that cost less than 10% of the vanadium-carrying electrolytes. The resulting high-power cell design demonstrated a significant power density increase over reported IFB performance. In addition, ESS also addressed the IFB cycle life limitation by introducing the concept of a "Proton Pump" [4].

Flow Batteries: From Fundamentals to Applications, First Edition.
Edited by Christina Roth, Jens Noack, and Maria Skyllas-Kazacos.
© 2023 WILEY-VCH GmbH. Published 2023 by WILEY-VCH GmbH.

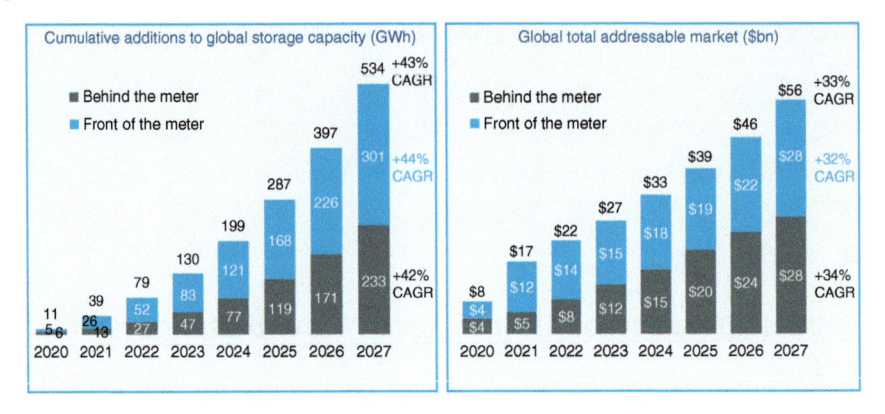

Figure 51.1 Strong demand growth is projected for energy storage systems. Source: Guidehouse Insights.

Since the ARPA-e grant, ESS has continued to drive both research and development and product commercialization to optimize IFB performance and cyclability and to reduce overall system cost. As of this writing, ESS has developed a suite of cells, stack, and system-level patents. ESS first launched its Energy Warehouse™ (EW™) product in 2015. The EW is a turn-key IFB energy storage system that includes power modules, electrolyte tank, power conversion, and all balance-of-plant components in a 40-foot ISO shipping container. Field deployments and demonstration projects of the Energy Warehouse product started in 2015.

In 2019, ESS launched its second-generation power module (S200), enabling a 50% increase in power density. The S200 power module was developed by ESS primarily for mass production and to allow for automated manufacturing. This advancement led to improvements in the EW platform and helped launch the Energy Center™ (EC™) platform in 2021. In conjunction with utility and EPC partners, the Energy Center platform was developed to fill a gap in the marketplace – utility-scale long-duration energy storage.

This chapter summarizes the key technical breakthroughs that have enabled ESS to commercialize its IFB-based energy storage products.

51.2 Background

A flow battery is a secondary battery that stores electrical energy in two soluble redox couples contained in external electrolyte tanks sized in accordance with the application requirements (see Figure 51.2). Due to the decoupling of the electrolyte (energy capacity) from the stack (power module), a flow battery offers numerous benefits over other traditional batteries [5]:

- The ability to change energy storage capacity without changing the whole system
- The ability to change system maximum power output with minimal system impact
- The reduction of part redundancy compared to other battery technologies

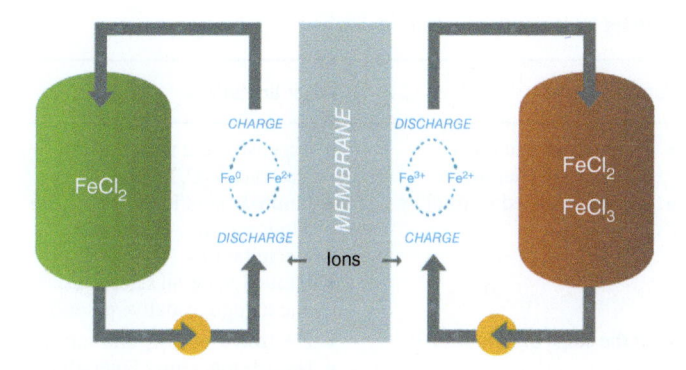

Figure 51.2 Diagram of an all-iron redox flow battery. Source: ESS Tech Inc.

In addition to the design benefits, there are also considerable cost and reliability benefits to this technology platform [6]:

- No precious metal is used in flow-battery electrodes
- The batteries can handle tens of thousands of deep charge and discharge (90%) cycles without performance degradation and capacity fade
- The electrolyte can be easily reused and recycled (high disposal value)

Some well-known examples of flow batteries are vanadium redox flow batteries (VRB), polysulfide bromide batteries, and uranium redox flow batteries. In addition to conventional flow batteries, a hybrid flow battery exists where only one electro-active component is dissolved in liquid and the other component is deposited as a solid layer. Some known examples of a hybrid flow battery are zinc–bromine and zinc–cerium batteries. A list of potential redox couples, with their maximum cell voltage, current development status, and major limitations, is presented in Table 51.1.

$$\text{Fe}^{2+} + 2e \underset{\text{discharging}}{\overset{\text{charging}}{\rightleftarrows}} \text{Fe}^0 \quad E^0 = -0.44\,\text{V} \tag{51.1}$$

$$2\text{Fe}^{2+} - 2e \underset{\text{discharging}}{\overset{\text{charging}}{\rightleftarrows}} 2\text{Fe}^{3+} \quad E^0 = +0.77\,\text{V} \tag{51.2}$$

Despite the benefits of flow-battery technologies and the size of the energy storage market, flow batteries currently occupy less than 1% of the grid-scale energy storage market [6]. The primary barriers are the cost for energy storage in terms of $/kWh, the cost for the power capacity in terms of $/kW, and battery performance in round-trip energy efficiency. To target the cost and efficiency issues of flow batteries, ESS initially studied several types of flow-battery chemistries, including VRB, but decided to pivot its development effort to the all-iron hybrid flow-battery (IFB) technology in 2012 because IFB had the potential to address all three above-listed barriers to commercialization.

As illustrated above, the initial discharged state of an IFB consists of dissolved ferrous ions (Fe^{2+}) in an aqueous solution. Upon charging, the externally stored electrolyte is pumped through the power module where reactions given by

Table 51.1 Flow-battery chemistries in comparison to IFB.

System	E_{cell} (V)	Current status	Major limitations
Vanadium–Vanadium	1.6	• Primarily studied redox flow-battery chemistry • Commercialized technology	• Expensive and variable raw material (V_2O_5) cost • Limited operation temperature window due to electrolyte precipitations • Limited material selections due to high oxidative potential.
Bromine–Polysulfide	1.5	• In the stage of field trial	• Electrolyte toxicity and safety • High-Temperature Operation • Material constraints due to high temperature and high potential • Safety issues
Iron–Chrome	1.0	• Developed by NASA • In the stage of field demonstration	• Expensive/mixed electrolytes • Electrolyte crossover issues • High requirements on membrane stability
H_2–Bromine	1.1	• In early research stage	• Electrolyte toxicity • Cost
Zn–Bromine	1.8	• In the stage of field demonstration • Demonstrated good performance	• Electrolyte toxicity • Battery life is limited due to Zn dendrite formation
Zn–Cerium	2.4	• Initially developed by Plurion, which went out of business	• Battery life is limited by Zn due to Zn dendrite formation • Limited material selections due to high oxidative potential

Eqs. (51.1) and (51.2) take place. Upon discharging, the reverse reactions take place. Because the negative reaction plates metallic iron (Fe^0) (Eq. (51.1)), IFB is considered a type of hybrid redox flow battery. The IFB technology was first characterized by Dr. Savinell in 1981 [2]. In that publication, the IFB battery used ferrous chloride salt in an aqueous solution with ammonium chloride as the supporting electrolyte. Several types of microporous plastic separators were used to divide the anode and the cathode. One of the primary functions of the separator is to allow cations and anions to travel through to preserve electroneutrality of the battery during charging and discharging. In the publication, the 100 cm² test cell was charged and discharged ~60 times with a maximum energy efficiency of 55% and a maximum power density of 50 mW cm^{-2}. It also reported that over cycles, the electrolytes became unstable, and battery performance degraded due to $Fe(OH)_3$ precipitation on the separator and electrode, which resulted in degraded IFB performance. To recover the performance, the battery was cleaned multiple times with acid rinse (to dissolve hydroxide).

In 2012 with the support of ARPA-e [3], the ESS team worked to overcome the reported issues with low performance, low-round-trip efficiency, and poor cyclability of the IFB technology. Key technology innovations are discussed below.

51.3 Key IFB Technology Breakthroughs and IFB System Commercialization

51.3.1 The Proton Pump

The most critical technical breakthrough that enabled ESS IFB system cyclability with no capacity fade is the concept and implementation of a Proton Pump patented by ESS in 2013 [4].

To understand the functions of a Proton Pump, we must start by understanding the side reactions that take place during IFB operations. During charge of an IFB, at the negative electrode, the ferrous iron plating reaction (Eq. (51.1)) competes with the hydrogen evolution reaction, Eq. (51.3) below, wherein protons in the electrolyte can accept electrons to form hydrogen gas, H_2. In addition, the plated Fe metal may corrode in an acidic electrolyte to produce ferrous ion, Fe^{2+}, and generate H_2, as shown in Eq. (51.4).

$$H^+ + e^- \leftrightarrow \frac{1}{2}H_2 \quad \text{(hydrogen evolution)} \tag{51.3}$$

$$Fe^0 + 2H^+ \leftrightarrow Fe^{2+} + H_2 \quad \text{(iron corrosion)} \tag{51.4}$$

Two issues arise from the above side reactions if left unmanaged. The first is electrolyte state-of-charge imbalance. For example, during charge, electrons for plating reaction are competed by hydrogen evolution reaction (Eq. (51.3)), or iron plating efficiency is also reduced by iron corrosion reaction (Eq. (51.4)). As a result, more Fe^{3+} is available in the positive electrolyte than plated metallic iron on the negative electrode during discharge. The second issue is electrolyte pH imbalance. Due to the above side reactions, the IFB negative electrolyte tends to stabilize at a pH range between 3 and 5 during charge. The formation of ferrous hydroxide ion ($FeOH^+$) and hydrogen evolution are limited. At the positive electrode, ferric ion, Fe^{3+}, has a much lower acid disassociation constant (pK_a) than that of ferrous ion, Fe^{2+}. Therefore, as ferrous ions are oxidized to ferric ions during charge, the positive electrolyte tends to stabilize at a pH less than 1. Over cycles, the loss of protons from Eqs. (51.3) and (51.4) can lead to electrolyte pH rise (protons transport relatively freely across the separator) to the point where Fe^{3+} starts to precipitate out of the positive electrolyte in the form of $Fe(OH)_3$. This precipitate is no longer electrochemically active, permanently reducing storage capacity. Additionally, the precipitate tends to foul the membrane separator and has the potential to clog the electrolyte flow paths impairing battery performance further.

For IFB to cycle with no capacity fade, ESS identified the need for an electrolyte health management system (EMS) to provide an efficient, simple, and cost-effective way to manage both the state of charge and the pH imbalance of the IFB electrolytes. The EMS will enhance the overall capacity, lifetime, and performance of redox flow-battery systems. Maintaining the positive electrolyte pH less than 1 in which the positive electrolyte remains stable, and maintaining the negative electrolyte pH over 3, in which the negative electrolyte remains stable with side reactions minimized, can reduce low IFB cycling performance and increase the efficiency of iron redox flow batteries.

An approach that addresses the electrolyte health issue caused by side reactions on the negative side comprises reducing the imbalanced ion in the positive electrolyte with the side reaction byproduct of the negative side, H_2. As an example, in an IFB system, the positive electrolyte comprising ferric ion may be reduced by the hydrogen gas according to Eq. (51.5):

$$Fe^{3+} + \tfrac{1}{2}H_2 \rightarrow Fe^{2+} + H^+ \tag{51.5}$$

In the IFB system example, by reacting the additional ferric ion from the positive electrolyte with the hydrogen gas generated from the negative side, the hydrogen gas can be converted back to protons and injected into the positive electrolyte, thereby maintaining a high pH in the negative electrolyte and low pH in the positive electrolyte, hence the name of "Proton Pump." Furthermore, by converting the additional ferric ion to ferrous ion, the state-of-charge of the positive electrolyte returns fully rebalanced with the state of charge of the negative electrode. The ESS IFB system is designed to be a sealed and closed-loop system so that the rate of "Proton Pump" (i.e. Eq. (51.5)) is completely self-limiting.

Although Eq. (51.5) is written for rebalancing electrolytes and balancing pH in an IFB system, the method of a proton pump may be generalized by Eq. (51.6) for many different flow-battery systems:

$$M^{x+} + \frac{(x-z)}{2}H_2 \rightarrow M^{z+} + (x-z)H^+ \tag{51.6}$$

In Eq. (51.6), M^{x+} represents the positive electrolyte M having ionic charge, x and M^{z+} represents the reduced electrolyte M having ionic charge, z.

With the implementation of the Proton Pump, ESS IFB flow-battery systems can:

- Cycle thousands of times without capacity or performance loss. As shown below in Figure 51.3 (ARPAe final report [7]), an ESS battery can cycle over 1000 cycles with no performance loss nor capacity fade. The same battery with same electrolyte demonstrated repeatable performance at the top of charge between cycle 1 and cycle 1000.
- Maintain electrolyte charge balance through chemical reduction of the imbalanced metal ions.
- Drive protons directly back into the positive electrolyte, thereby maintaining the pH of the electrolytes within stable ranges. One such example collected during an ESS field system operation is shown below in Figure 51.4, where blue line indicates electrolyte pH and red dots indicate system power.
- Minimize off-gassing of hydrogen from the system.

There are several key advantages of this simple rebalancing mechanism utilizing the hydrogen reduction reaction, which are given below:

- This chemical reduction reaction converts a 1 : 1 ratio of Fe^{3+} to Fe^{2+}, so no extra reactant species are generated that require secondary balancing cells.
- The reaction oxidizes evolved hydrogen back to protons in the electrolyte, thereby maintaining the electrolyte pH within a stable range.
- Since the amount of generated hydrogen is identical to the amount of imbalanced ferric ions, the reduction of ferric ions with recycled hydrogen completely balances the battery system.

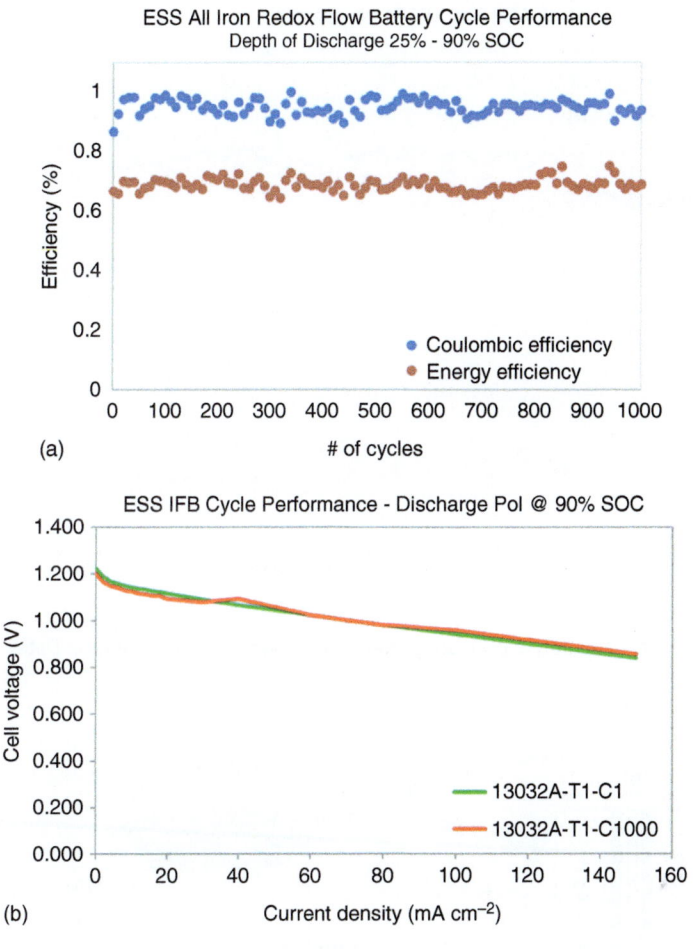

Figure 51.3 ESS IFB cycle performance (a) over 1000 cycles and top of charge polarization curves (b) at Cycle #1 and Cycle #1000. Source: Data from Ref. [7].

Going beyond just the concept, numerous "Proton Pump" reactor designs can be implemented to accomplish the desired outcome for an IFB system cycle operation. However, for system-level design trade-offs, these reactors' overall performance, cost, and reliability must be considered and optimized. For example, a trickle bed type of reactor design [4] is simple and reliable through a direct reaction interface. Still, the reactor rate is relatively low, and the overall implementation may result in a large footprint and high cost. A fuel-cell-like reactor design [8, 9] (Figure 51.5a) can also be used as this configuration demonstrates improved overall performance, and in addition, there is an opportunity to harvest additional discharge energy. However, like most fuel-cell designs, these reactors can be expensive due to the application of ion-exchange membrane and less reliable due to irreversible degradation mechanisms such as reverse current [10]. Other types of reactors that have been considered include a jelly-rolled type reactor design [11] (Figure 51.5b), or an in-tank passive reactor [12].

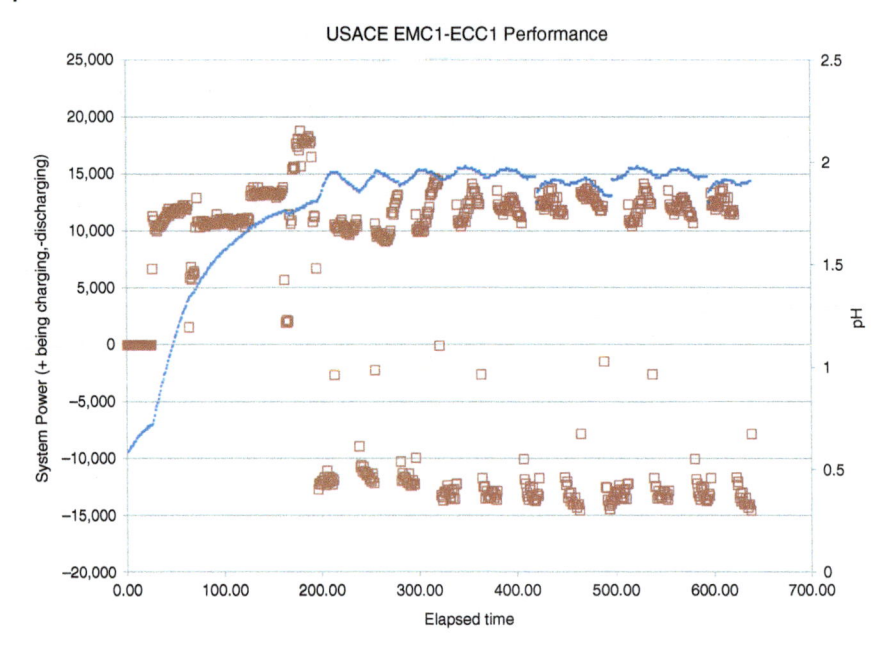

Figure 51.4 IFB system electrolyte pH management during system cycling. Source: Data from Ref. [7].

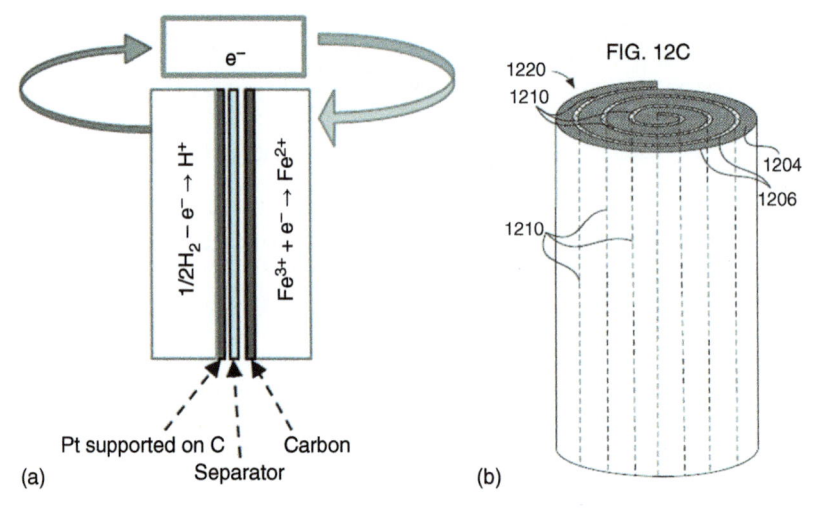

Figure 51.5 Examples of a fuel-cell type proton pump (a, US10,615,442) and a jelly rolled proton pump (b, US10,181,615).

51.3.2 IFB Power Module

Power modules are the heart of the IFB system where the electrochemical reactions take place. To realize the cost benefits of the IFB electrolytes, the IFB power module performance must be comparable to that of other flow-battery chemistries.

Repeat parts, such as bipolar plates, and separators must be designed and optimized to meet performance and minimize costs.

Under the ARPA-e grant, unique electrode designs were investigated to improve IFB performance. Forced convection [13] was used to pump positive electrolytes into a flow cell and across a redox plate. Forced convection ensures fresh, unreacted electrolytes to always be in contact with the electrode surface. Additionally, this configuration allows the entire electrode surface to be utilized while simultaneously removing any products formed.

The plating reaction at the negative electrode is the primary source of IFB performance loss and requires the most engineering effort to optimize. Iron plating kinetics is slow [2], so it requires high plating electrode surface area to reduce kinetic losses. But the benefit of high plating surface area diminishes as plated Iron reaches high plating density, i.e. *thicker plating*. In addition, because plated iron takes up physical space in the negative electrode compartment, adequate negative electrode space must be provided, so electrolyte flow is not impeded at high plating densities. This design effort will be a trade-off with ohmic losses due to low electrolyte resistivity. Finally, due to the negative side reactions (Eqs. (51.3) and (51.4)) that generate hydrogen, the IFB negative electrode needs to be designed with the capability to purge out hydrogen without entrapping bubbles adequately. ESS-patented electrode designs [13] significantly increase negative electrode surface area, without jeopardizing the required space for high iron plating densities or compromising ohmic resistance loss. Applying these unique electrode designs, ESS IFB cells have demonstrated significantly improved performance and round-trip efficiency (>70% at 200 mW cm^{-2}) [14], as shown below in Figure 51.6. ESS IFB electrode designs were selected as among the top 10 patents by C&EN 2020 Discovery Report – "Powering up" [15].

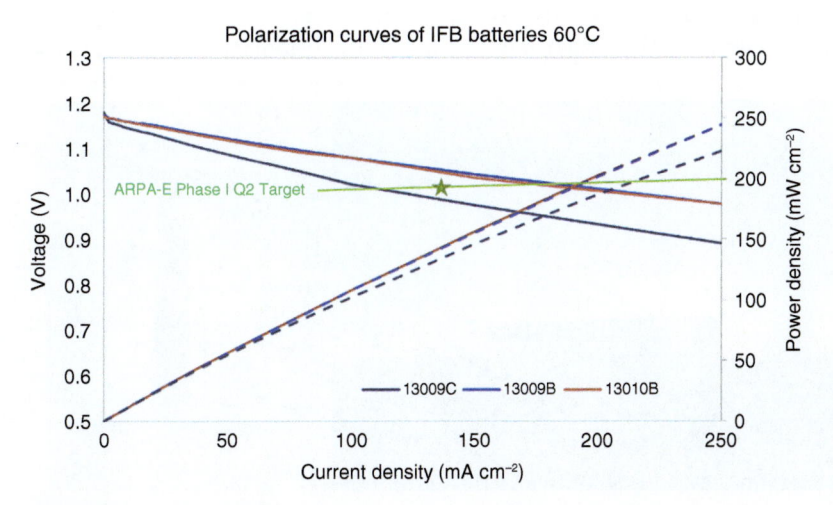

Figure 51.6 ESS IFB performance reported under ARPAe program. Source: Data from Ref. [14].

To scale up the subscale IFB performance into large IFB power modules, several key factors must be considered and optimized in the overall stack design to ensure meeting the following criteria:

- Uniform distribution of reactants within each cell
- Uniform temperature distribution within each cell
- Minimal resistive losses through-plane
- No reactant crossover or overboard leaks
- Minimum reactant pressure drops

One challenge for flow-battery power-module design is that all the cells are hydraulically connected through an electrolyte circulation path. This can be problematic as shunt current can flow through the electrolyte circulation path from one series-connected cell to another causing energy losses and imbalances in the individual charge states of the cells.

To prevent or reduce such shunt currents, properties of the electrolytes, such as electrical and ionic conductivities, must be characterized. Mathematical models can be developed to minimize any shunt currents and deal with any system-level or cell-level design trade-offs.

The two significant losses that require analysis and optimization are pumping losses and shunt current losses. The pumping losses arise from pumping the positive and negative electrolytes into and out of the cells. The shunting currents are due to the electrolyte being conductive and small shorts developing due to the electrolyte touching all the cells. There is an optimal design to minimize these two losses, and it can be defined as $\min(\sum \text{Shunting Losses} + \sum \text{Pumping losses})$ [16].

Since 2013, ESS has released two generations of full-scale IFB power module designs to production. From 2013 to 2018, Gen1 IFB power modules (S100), with an active area of ~660 cm^2, were used in all ESS' Gen I 50 kW/400 kWh EW systems for early field demonstration projects (two installation examples are shown in Figure 51.7). These systems were installed and tested by C&I customers (grid-connected and microgrids), military, and utility customers for a variety of use

Figure 51.7 ESS Gen1 50 kW/400 kWh deployments in customer site A (left), where ambient temperature goes below freezing in winter, and customer site B (right), where ambient temperature is over 110F in the summer. Source: ESS Tech Inc.

cases. The S100 power modules and systems enabled ESS to validate IFB technology readiness in real-world operations and demonstrate its capability to connect to a grid or microgrid and respond in real time to a range of grid controllers to satisfy various grid operation user cases. It also exposes IFB systems to extreme ambient environmental conditions, such as above 110F daily desert heat or below-freezing temperatures in the winter, to validate IFB-wide operation conditions without the need for any air conditioning system. These field deployments also exposed some weaknesses of S100 power modules. For example, S100 power modules used O-rings for sealing purposes, which turned out to be problematic and over time caused overboard electrolyte leakage/seepage issues. In addition, S100 power modules were built by hand with machined components, so their bill of materials and labor costs were high.

In 2019, ESS successfully launched its commercial power module (S200) design providing nominally 12 kW power for up to 12 hours of energy storage. These S200 battery module represents a 5x power improvement over the earlier S100 module design. All major components of S200 battery modules are injection or compression molded, significantly reducing cost of these repeat parts. As a result, the cost of the full S200 IFB battery module in $/kW is less than 50% of S100 modules. Furthermore, the S200 battery module is fully glued up with no O-rings, eliminating all possible sources of electrolyte leaks. Finally, S200 power modules were designed with specific mechanical features to enable robotic pick-n-place automation and minimize human operations.

Figure 51.8 below shows the S100 and S200 power modules on display in the ESS lobby at their headquarters in Wilsonville, OR. Based on the S200 IFB power modules, ESS launched its Gen II EW system with a 90 kW peak power interface and 400 kWh stored energy as well as its EC product using 333 kW Power Train

Figure 51.8 ESS S200 power module and S100 power module on display in its Headquarter Lobby. Source: ESS Tech Inc.

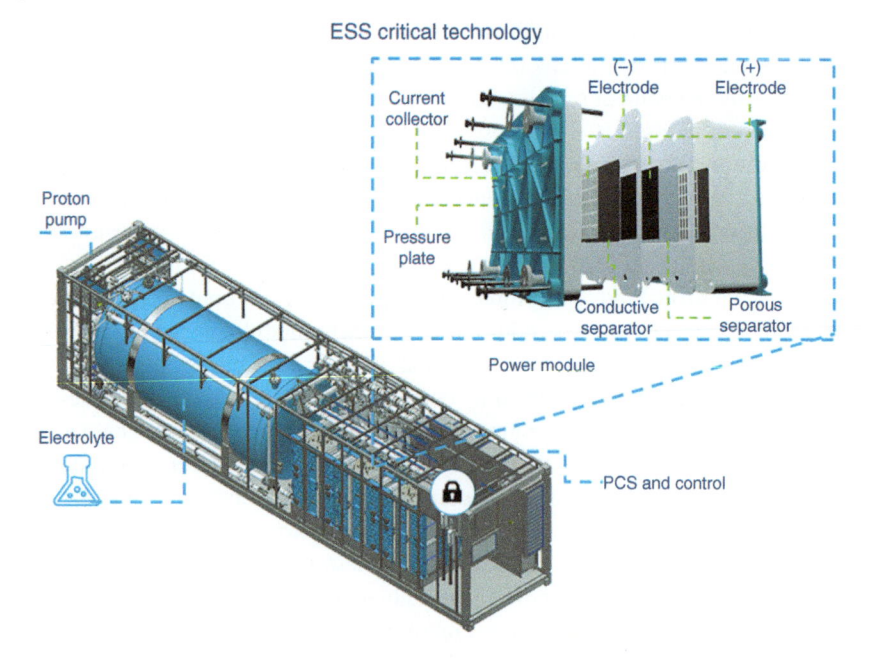

Figure 51.9 Illustration of subsystems of an all-iron redox flow battery. Source: ESS Tech Inc.

skid packed with S200 IFB power modules (https://www.globenewswire.com/news-release/2021/02/10/2173287/0/en/ESS-Inc-Announces-the-Energy-Center-a-Flexible-Scalable-and-Environmentally-Sustainable-Long-Duration-Battery-Storage-System.html).

51.3.3 System Optimization and LCOS

In addition to power modules, electrolytes, and proton pumps, a complete IFB system also includes the Balance of Plant (BoP) subsystem, sensors, thermal management, the Power Conditioning System (PCS), and the control system, which communicates internally to all subsystems and externally to customer's site controller. Figure 51.9 below illustrates the internals of an ESS EW system with its major subsystems labeled. Other system-level benefits, advancements, and optimizations of IFB technology will be discussed in this section. The EC product is essentially a customer-designed, scaled-up version of the EW, so all features discussed in this section are based on the EW.

When evaluating an IFB system against other energy storage technologies, it is important to go beyond just power module and electrolyte cost alone and take a comprehensive view by using the Levelized Cost of Storage (LCOS) equation below:

$$\text{LCOS} = \frac{\sum \text{CapEx} + \sum \text{Installation} + \sum \text{Disposal} + \sum \text{O\&M}}{\sum \text{Annual Usable KWh}}$$

CapEx	Cost of power modules, electrolytes, BoP, power electronics, etc.
Installation	Shipping, Cost of Pad, Safety, labor, connection to grid, etc.
Disposal	Removal of spent materials or batteries, shipping, recycling, or disposal
O&M	Operation cost of monitoring, upkeep, labor, travel, supplies, etc.
Usable kWh	Energy transacted through the battery that encapsulates round-trip efficiency, usable state of charge

As shown, the LCOS of a given energy storage system is a result of the overall system capital expense, cost associated with system installation and site permitting, regular O&M cost, end-of-life disposal cost, as well as overall annual usable energy from the system, which includes state-of-charge and round-trip efficiency.

The overall **CapEx** cost of a flow-battery system varies based on system power requirement (i.e. power module cost), system capacity and electrolyte composition (i.e. electrolyte cost), BoP, and PCS components. As one could imagine, the larger the capacity of an energy storage system, the more dominant the electrolyte cost is in the cost of the entire system. The smaller the capacity of an energy storage system, the more dominant the power module cost is in the cost of the entire system. The Capex advantages of an IFB system are broken down in Chapter 5.4, and the main advantage of the IFB system over other battery technologies is its low-cost iron salt electrolytes. Therefore, moving toward a longer duration (4–12 hours) storage makes the IFB systems more attractive due to their low electrolyte cost.

As storage assets become paired with renewable generation, the ability to reliably shift renewable energy to peak periods enables the hybrid asset to perform as both a low-cost energy source and a firm peaking asset. Conventional four-hour Lithium-ion systems can perform this role on some days when the renewable asset generates enough to charge the battery. However, this is not 100% of the time. A long-run-time battery can serve the peak load for multiple days in a row when the irradiance was low, or the wind didn't blow. Resource planners are likely to value this capability highly because it provides superior resiliency over the lithium-ion solution.

The **Installation** cost includes costs related to shipping, logistics, site preparation, and site permitting. One benefit of the IFB system that is not commonly aware is that its electrolyte can be shipped completely in dry form and hydrated locally. In 2016, ESS deployed one of its early IFB systems to the US Army Corp of Engineers to handle long-duration storage in forward operating bases (FOB). The USACE chose the IFB battery specifically because it can be shipped in a dry state, with water added on location at the destination (https://essinc.com/us-military-to-test-mission-critical-ability-of-dry-deployment-flow-battery/). This reduces the weight of the transported battery by 60% compared to other flow or traditional batteries. In addition, because iron salt is readily available globally, this also opens the possibility to ship IFB systems around the world dry, and source, load, and hydrate electrolyte materials locally.

Due to its water-based electrolyte, IFB systems pose a lower fire risk than Lithium-ion systems (https://vrbenergy.com/wp-content/uploads/2019/11/Energy-Storage-

Table 51.2 Safety standards for IFB products.

Safety	Standards	Highlights
Electrolyte safety	EPA	• IFB electrolyte disposal process consists of neutralization, filtration (to remove iron salt precipitates), and disposal
Electrical safety	UL 9540 [17]	• Pass electrical shock prevention • Pass short circuit testing
	IEC 62932 [18]	• Pass ground fault prevention
Hydrogen safety	UL 9540	• Pressure safety manifold includes a burst disc, a vacuum break, pressure monitoring, and pressure venting setup • Inside the tank, H_2 concentration is kept above UFL. In addition, there is no ignition source, and there is water vapor present
	IEC 62932	• Outside of the tank, H2 plumbing is validated during FAT. During operation, forced ventilation ensures inside the container is maintained below LFL
Secondary containment	UL 9540	• EW container floor is sealed • Drain ports are provided to customers
	IEC 62932	• Leak sensors installed for leak detection

Safety-Monitor-October-2019.pdf; https://www.nfpa.org/-/media/Files/News-and-Research/Fire-statistics-and-reports/Hazardous-materials/RFFireHazardAssessmentLithiumIonBattery.ashx) and can be easily sited and permitted for installation with space constraints and onsite fire safety requirements. Many utility purchasers need to site storage projects in areas where Lithium-ion may be considered difficult to permit. This may include areas subject to wildfire or distribution-level locations in dense cities or close to residential property. As a result, the installation costs of an IFB system are significantly lower than others.

Table 51.2 below summarizes the key safety standards used in IFB system certification and site permitting.

In 2019 ESS also participated in a study comparing the environmental impacts associated with the production of three different flow-battery technologies: vanadium redox flow batteries (VRFB), zinc–bromine flow batteries (ZBFB), and all-IFB [19]. The ESS S100 IFB power module was used for this study. A cradle-to-gate life-cycle analysis (LCA) was performed to systematically study and provide quantitative measurements for the environmental costs associated with each life-cycle stage/process (i.e. material extraction, component manufacturing, etc.). Specifically, eight impact factors were studied: global warming potential (GWP), ozone depletion potential (ODP), fine particulate matter (PM), acidification potential (AP), freshwater eutrophication potential (EP), Freshwater ecotoxicity potential (ETP), abiotic resource depletion potential (ADP), and cumulative energy demand (CED).

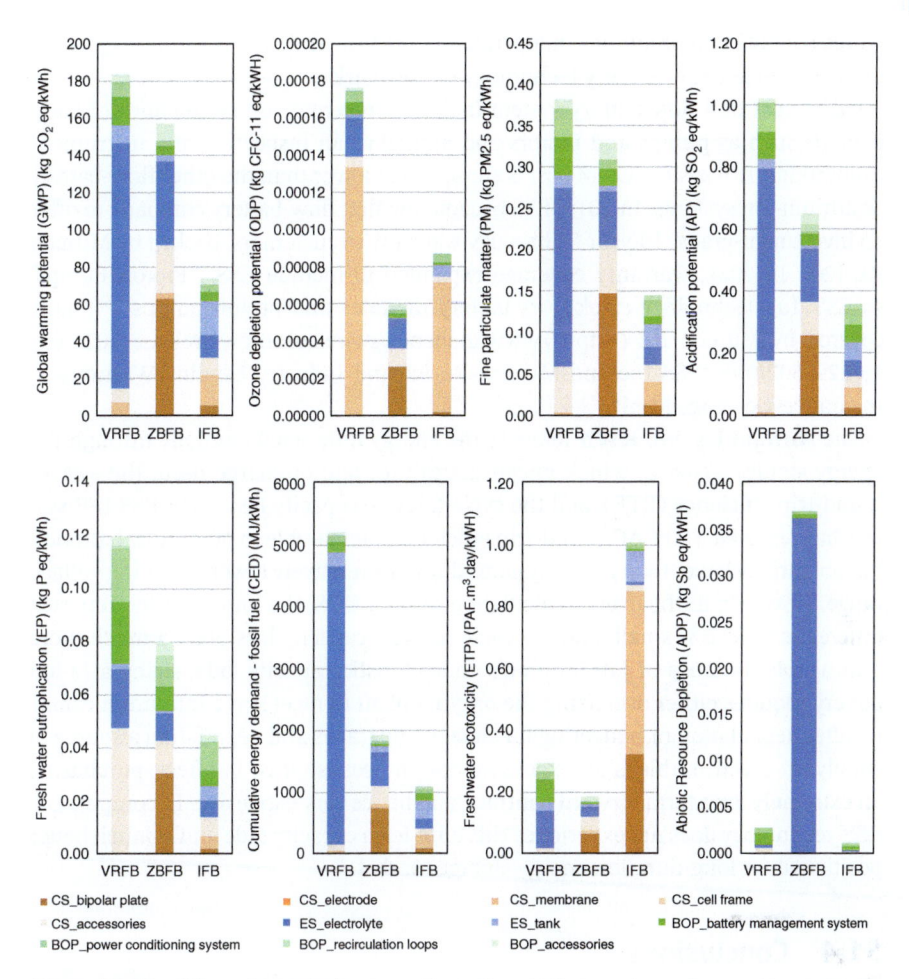

Figure 51.10 Life-cycle analysis among three flow-battery systems. Source: He et al. [19]/with permission from Elsevier.

As shown above, in Figure 51.10, the article concluded that the **IFB system exhibits the lowest impact scores** in six of the eight impact categories, except ozone depletion potential and freshwater ecotoxicity potential. The ZBFB system has the lowest impact scores for ozone depletion potential and freshwater ecotoxicity potential but the highest for abiotic resource depletion potential. The VRFB system exhibits the highest impact scores for global warming potential, ozone depletion potential, fine particulate matter, acidification potential, freshwater eutrophication potential, and cumulative energy demand. For the IFB, the impact on the global warming potential is largely due to the production of the electrolyte tank and the power module cell frame of S100 power modules. The S200 power module has been redesigned to accommodate more sustainable materials resulting in much better scores overall. This study has independently confirmed that IFB technology is the most sustainable flow battery technology on the market.

O&M cost of a flow-battery system includes preventative maintenance of components such as pumps and motors and annual clean inspection and instrument calibrations. IFB system O&M cost is similar to or lower than the other flow-battery or lithium-ion systems. In 2019, ESS became the first flow-battery company to offer an investment-grade, 10-year technology warranty to customers, backed by Munich RE [20]. In 2021, warranty coverage expanded to include ESS' "Proton Pump," so ESS' full technology stack – its battery modules and proton pumps – is now covered by Munich RE (https://www.globenewswire.com/news-release/2021/09/16/2298502/0/en/ESS-Inc-and-Munich-Re-Expand-Industry-Leading-Warranty-Insurance-Coverage.html).

The **Annual Usable KWh** refers to the energy transacted annually through the energy storage systems, which encapsulates the state of charge used, the system round-trip efficiency (RTE), and the cycle-to-cycle capacity loss. THE ESS IFB system has reported a 70% AC round-trip efficiency comparable to other flow batteries, but potentially lower than Li-ion systems due to its relatively lower operating voltage range. However, in the overall calculation of the LCOS, the impact of the reported difference in RTE is small, and the cycle-to-cycle capacity loss plays a much more critical role. Because of Lithium-ion's characteristic capacity fade, sizing a Li-ion battery requires either oversizing the original plant to meet some minimum capacity after degradation or budgeting for capacity upgrades and negotiating a source of supply. In practice, this is an expensive option because it is, in effect, purchasing an extremely long-term forward contract, and hence very expensive. In comparison, ESS' technology does not experience this long-term capacity fade and is much better positioned for long-duration energy storage market.

51.4 Conclusions

Utilizing earth-abundant iron, salt, and water, ESS has spearheaded the commercialization of IFB systems since 2012 [21]. The IFB system delivers safe, reliable, long-duration energy storage, with a 25-year operational life, and near-zero capacity degradation backed by one of the largest international insurers, Munich Re.

ESS' IFB system is at technology readiness level (TRL) 8 – *Deployment of Technology in Final Form*. Once public information on successful commercial deployments is released, IFB energy storage systems will reach the final TRL 9 – *Normal Commercial Deployment*.

On October 11th 2021, ESS successfully began trading on the New York Stock Exchange with the ticker symbol "GWH" after closing its combination with the company's SPAC (Special Purpose Acquisition Company) merger (https://www.renewableenergyworld.com/storage/ess-becomes-first-long-duration-energy-storage-company-to-list-on-nyse/#gref), becoming the first U.S. long-duration energy storage system company to be listed on NYSE. The ESS IFB system is well-positioned to reach full commercialization, allowing the company to meet the needs of a rapidly growing long-duration energy storage market.

References

1. Navigant Research (2017). http://www.navigantresearch.com/tag/energy-storage-systems (accessed December 2017).
2. Hruska, L.W. and Savinell, R.F. (1981). Investigation of factors affecting performance of the iron-redox battery. *Journal of the Electrochemical Society* 128: 1.
3. ARPAe, https://arpa-e.energy.gov/technologies/projects/iron-flow-battery. Project term: 01 October 2012–30 August 2017.
4. Song, (Yang) J. and Evans, C.E. (2014). Method and system for rebalancing electrolytes in a redox flow battery system. US 9,509,011, 11 December 2014 and issued 29 November 2016.
5. Yang, Z., Zhang, J., Kintner-Meyer, M.C.W., and Lu, X. (2011). Electrochemical energy storage for green grid. *Chemical Reviews*.
6. Perry, M. (2012). *Renaissance in Flow-cell Technologies*. ARPA-e Flow Battery Workshop, Washington DC.
7. ESS ARPAe Final Report, 2017.
8. Song, (Yang) J. and Evans, C.E. (2014). Method and system to maintain electrolytes stability for all iron redox flow batteries. US10,615,442, 11 December 2014 and issued 29 November 2016.
9. Song, (Yang) J. and Evans, C.E. (2014). Method and system to maintain electrolytes stability for all iron redox flow batteries. US10,680,268, 11 December 2014 and issued 29 November 2016.
10. Reiser, C., Bregoli, L., Patterson, T. et al. (2005). A reverse-current decay mechanism for fuel cells. *Electrochemical and Solid-State Letters* 8 (6): A273–A276.
11. Song, (Yang) J. and Evans, C.E. (2019). Method and system for rebalancing electrolytes in a redox flow battery system. US10,181,615, 6 October 2016 and issued 15 January 2019.
12. Selverston, S., Savinell, R.F., and Wainright, J.S. (2016). In-tank hydrogen-ferric ion recombination. *Journal of Power Sources* 324: 674–678.
13. Evans, C.E. and Song, (Yang) J. (2017). Redox and plating electrode systems for an all iron hybrid flow battery. US9,614,244, 6 March 2014 and issued 4 April 2017.
14. ESS (2012). ARPAe Program Report Phase I Q2.
15. p24, C&EN Discovery Report' Powering Up', Q2 2020.
16. Evans, C.E. and Song, (Yang) J. (2014). Internally manifolded flow cell for an all-iron hybrid flow battery, US9,685,651, 6 March 2014 and issued 20 June 2017.
17. UL9540. https://standardscatalog.ul.com/ProductDetail.aspx?productId=UL9540.
18. IEC62932. https://webstore.iec.ch/publication/28334.
19. He, H., Tian, S., Tarroja, B. et al. (2020). Flow battery production: materials selection and environmental impact. *Journal of Cleaner Production* 121740. https://doi.org/10.1016/j.jclepro.2020.121740.
20. Peter Röder (2019). Battery performance now insurable – Munich Re coverage paves the way for renewable energy. Munich RE – Not if But How.

https://www.munichre.com/en/company/media-relations/media-information-and-corporate-news/media-information/2019/2019-03-07-battery-performance-now-insurable-innovative-munich-re-coverage-paves-the-way-for-renewable-energy.html.

21 ESS Inc. The ESS Energy Warehouse™. ESS, Inc – Catalyzing a Cleaner Future. [Online] https://essinc.com/energy-storage-products/.

52

Application of Hydrogen–Bromine Flow Batteries: Technical Paper

Wiebrand Kout and Yohanes A. Hugo

Elestor BV, P.O. Box 882, 6800AW, Arnhem, The Netherlands

52.1 Introduction

The hydrogen–halogen flow battery is a promising storage system, primarily due to its fast and reversible kinetics and potentially low system cost. R&D efforts are mostly focused on the H_2/Br_2 system due to its high-power density, high round-trip efficiency, and high-solubility aqueous electrolyte [1].

Low costs can be achieved for both the energy and the power domain. The abundant availability of bromine and hydrogen means that the cost for the active materials can be below €10/kWh with upscaled electrolyte production. Up to 17 kW m^{-2} power density has been proven in cells, which makes the H_2/Br_2 cell one of the most power-dense flow batteries.

Disadvantages compared to other flow batteries are the relatively complex system design and the vulnerability of the H_2 catalyst. Safety aspects are important in the system design because Br_2 is poisonous and H_2 is explosive.

The energy domain, i.e. the storage of electrolytes and H_2, is discussed in the next paragraph. The power domain, including membranes, electrodes, cell stacks, and balance of system, is addressed in paragraph 3. Paragraph 4 deals with the practical application of H_2/Br_2 flow batteries, permittability, and provides a future outlook.

52.2 Energy Domain

52.2.1 Electrolyte

An energy density between 50 and 55 Wh l^{-1} is possible with an electrolyte composition of 1.5 M Br_2–3.6 M HBr at 100% SoC. The theoretical upper limit exceeds 300 Wh l^{-1}, however, achieving this requires a HBr start concentration of >20 M, which is probably impractical. The energy density could rise to

Flow Batteries: From Fundamentals to Applications, First Edition.
Edited by Christina Roth, Jens Noack, and Maria Skyllas-Kazacos.
© 2023 WILEY-VCH GmbH. Published 2023 by WILEY-VCH GmbH.

80–150 Wh l^{-1}, which is possible with 8.8–12.6 M HBr start concentration and well-developed systems and stacks.

Highly concentrated electrolytes not only reduce the required tank volume but also the cell voltage. Application of the simple Nernst equation is not accurate for predicting the cell performance because complex formation in highly concentrated electrolytes strongly influences the cell voltage [2]. Complexation also strongly influences the performance in membraneless H_2/Br_2 cells [3].

The Br_2 vapor pressure rises with concentration, which increases risk. Bromine-complexing agents (BCA's) reduce the vapor pressure but do so at the expense of cell performance and the added cost for the BCA. 1-Ethyl-1-pyridinium bromide can be used as BCA and will increase safety in case of electrolyte spills. However, it was found that the BCA cations interact with the sulfonate groups of the perfluoro sulfonic acid (PFSA) membrane, resulting in reduced membrane conductivity [4].

Practical experience in building several prototype systems has shown that the electrolyte tank is an important system component. The electrolyte tank can be more costly than electrolyte itself, e.g. when glass-lined steel tanks are used. Plastic tanks that store the electrolyte at near atmospheric pressure are preferred, however, care should be taken that no H_2–Br_2 gas mixture escapes during operation. Caution must be exercised to ensure that the mechanical properties of the polymer tank are not affected by Br_2 over time. When screening materials for construction, it should be considered that the gas above the liquid in the tank is often more corrosive and damaging to the material than the liquid itself.

52.2.2 Hydrogen

The H_2 flowing out of the cells during charging contains impurities. Electroosmotic drag causes accumulation of so-called *cross-over liquid* in the H_2 electrode during charging. The cross-over liquid contains 2–3 M HBr. Cross-over Br_2 is reduced by the catalyst layer and is usually not detected in the cross-over liquid, only if the state of health of the H_2 catalyst decreases. A system developer has the choice to either purify the H_2 before storing it in a standard H_2 tank or to coat this H_2 tank with a suitable epoxy or another coating to make it sufficiently corrosion resistant. A catalytic recombiner can be used to react any remaining Br_2 to HBr [5].

To reduce the cost of the storage system, it is desirable to remove the H_2 tank. Two possible options exist to eliminate the need for a separate H_2 tank, which are as follows:

- A H_2/Br_2 system can be connected to a H_2 pipeline if gas purity is ensured. This eliminates the need for H_2 storage, which again reduces cost.
- Because of the large difference in density, a single tank can be used for both the hydrogen and the electrolyte. A pilot system with this single tank is developed for the first time within the H2020 MELODY project.

52.3 Power Domain

52.3.1 Membrane Electrode Assembly

52.3.1.1 H_2 Electrode

A precious metal catalyst, often Pt or Pt–X alloys, is needed for catalyzing the hydrogen evolution and oxidation reactions. The catalyst precious metal loading can be less than 0.05 mg cm^{-2} to sustain current densities of about 0.5 A cm^{-2} during a short period of testing [6]. However, a catalyst loading above 0.1 mg cm^{-2} Pt or Pt–X catalyst loading is commonly used for redundancy. The main failure mechanism is a gradual Pt catalyst degradation–dissolution that results in insufficient Pt catalyst loading to maintain the desired current density. But as long as the catalyst loading is above this minimum threshold, the Pt degradation–dissolution rate is not influencing cell performance [7]. The Pt content in the cross-over liquid can be measured to predict the cell lifetime.

Catalyst dissolution mainly occurs when insufficient H_2 is present around the catalyst particles. The cells are most vulnerable at the end of the charge cycle, when the amount of cross-over liquid peaks, and also during standby or maintenance. A reduction of the hydrophobicity of the H_2 electrode over time is possible. Unscheduled stops and system malfunctions can have dramatic knock-on effects on the state of health of the cell stacks. If no hydrogen is present because of an error or safety shutdown, catalyst degradation can occur within minutes, which permanently damages the cell stacks. If a stack needs to be stored or transported after it has been tested with bromine electrolyte, it is advised to thoroughly rinse both the electrolyte and hydrogen compartments with deionized water until the pH is neutral, and to leave both compartments filled with clean water until the stack is again installed in the system.

Several countermeasures exist to reduce catalyst degradation and dissolution. These include:

- Applying a small current when the system is not used, also called "trickle charging," ensures that the catalyst is always surrounded by H_2 gas.
- Alloying Pt with another metal can reduce the dissolution rate.
- A nanoscale polydopamine coating can stabilize the surface of the Pt catalyst and act as an efficient barrier against Br_2 chemisorption and corrosion [8].

52.3.1.2 Membranes

The first field-tested H_2/Br_2 systems used nanoporous proton conductive membranes consisting of nanosized silica powder and polyvinylidene fluoride (PVDF) binder. These nanoporous membranes achieve conductivity by the electrolyte, which fills the pores of typically 1.5 nm diameter. These membranes can be low-cost, however, the maximum pressure difference across the membrane is determined by this pore size [9].

Later systems used dense PFSA membranes. Reinforced long-chain PFSA membranes with an ion exchange capacity (IEC) around 0.8 meq g^{-1} and a thickness of

75 µm or less are preferred [10]. The electrospinning process can be used to produce suitable membranes and it improves the blend membrane morphology due to fast vitrification that precludes the polymer–polymer phase-separation that occurs with solution casting [11]. Dual-fiber PFSA/polyphenylsulfon (PPSU) electrospun membranes showed promising performance, which indicates that they might be an alternative to conventional PFSA membranes [12]. The electrospinning process was also used to make membranes with sulfonated polyether ether ketone (PEEK) instead of PFSA.

52.3.1.3 Liquid-Side Electrode

The liquid-side electrode has fast kinetics but relatively slow mass transport. Carbon cloth or carbon paper can be used. Providing sufficient active surface area for the bromine reaction, e.g. by applying carbon nanotubes to the electrode, and reducing the mass-transport distance from the flow field to the membrane improves performance and allows a lower electrolyte flowrate [13]. Using a thermally activated carbon cloth or paper and applying a microporous layer with high surface area carbon are other ways to increase the reaction area.

52.3.2 Stack

An important stack design consideration is preventing flooding with cross-over liquid of the H_2 catalyst. Therefore flow-through designs are not advised for the H_2 side if the active area exceeds 35 cm^2. With a flow-through design, the removal of cross-over liquid from the center of the cell becomes too difficult, which results in rapid catalyst dissolution. A serpentine flowfield is preferred for the H_2 side.

During discharge, the cell performance can be improved if the H_2 flowrate is increased with a recirculation pump or a venturi. In situ measurement of localized current densities showed that a stoichiometric flow factor of at least 1.5 for the hydrogen during discharge prevents a large current density gradient, with low current spots near the H_2 outlet. During charge, no H_2 oversupply is needed [14].

Electrochemical H_2 compression inside the cells is possible and is often more efficient than using an external adiabatic compressor. The backflow of H_2 through the membrane into the electrolyte should be considered, as this will decrease the coulombic efficiency with decreasing membrane thickness for a given pressure difference. However thicker membranes will reduce the voltaic efficiency. Finding the optimum differential pressure requires a thorough analysis of this trade-off.

Interdigitated flowfields work well for the liquid side. Intercalation of Br_2 into the cell materials, such as Viton or ethylene propylene diene monomer (EPDM) gaskets and graphite composite cell plates, can lead to swelling and formation of cracks. The rate of intercalation increases with the free Br_2 concentration and temperature in a nonlinear way. Ex situ tests are useful to find the threshold concentration above which this intercalation occurs. It is important to design the liquid flowfield in such a way that stagnant electrolyte is avoided. A stationary plug of electrolyte in a cell can lead to locally high Br_2 concentration, which in turn leads to intercalation and swelling of cell parts.

Up to 1.7 W cm^{-2} power density has been proven in cells, see Figure 52.1. This performance was measured by Elestor at 60 °C and electrolyte composition of 2.3 M Br$_2$ and 2 M HBr.

52.3.3 Balance of System

The combination of elevated H$_2$ pressure, Br$_2$ corrosion resistance, H$_2$ explosion safety, and low cost makes designing an economically viable H$_2$/Br$_2$ system challenging. ATEX components, which are required if the H$_2$ pressure exceeds 0.5 bar, can significantly increase balance of system cost.

Plastic materials often have good corrosion resistance compared to metals, but will increase the fire risk. Installation of an automatic CO$_2$ extinguishing system is advised. PVDF is a suitable material for the electrolyte flow loop, for items such as leak trays, lower-cost polymers such as high-density polyethylene (HDPE) and PP can be used.

High air ventilation rates improve H$_2$ safety but reduce both Br$_2$ safety and system efficiency. Designing a safety strategy that takes these kinds of trade-offs into account is essential. Under no condition should it be possible for Br$_2$ fumes to escape the system, this includes not only during operating the system but also during maintenance work. The system design should enable easy and thorough flushing of stacks and the electrolyte flow loop in case stacks or components need to be replaced.

Sourcing the right bidirectional inverter is a challenge for all flow-battery system developers. The cost of these bidirectional inverters reduces with a higher input voltage. A DC input voltage of 400–1000 V is typical for the most cost-effective MW-class inverters. This means that at least 500 cells should be connected electrically in series, which necessitates proper management of shunt currents.

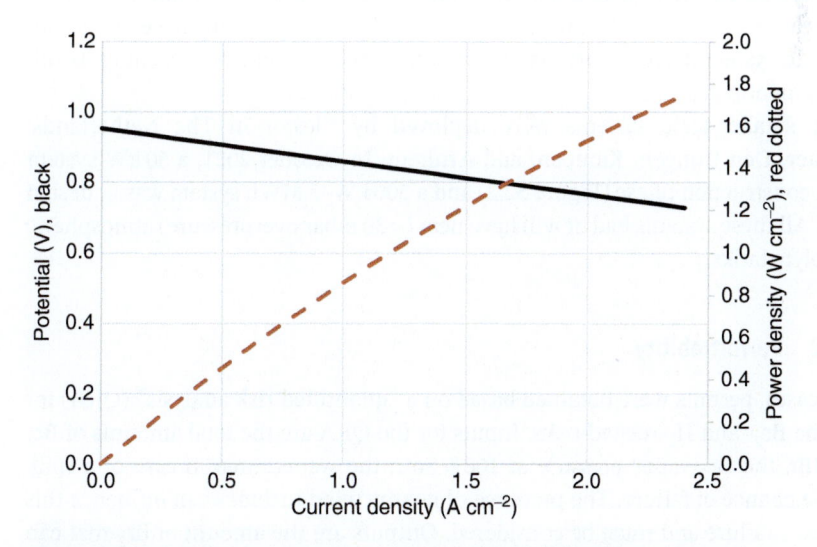

Figure 52.1 Polarization and power-density curve of an Elestor H$_2$/Br$_2$ cell.

Figure 52.2 Pilot system in Arnhem, The Netherlands.

52.4 Application

52.4.1 Field Tests

The largest H_2/Br_2 system built to date was the *Flowbox* system with 150 kW and 900 kWh. It was operated near Marseille in a joint project by Enstorage (IL) and Areva (FR). It was a pressurized and pressure-balanced system, at 100% SoC, both the H_2 and the electrolyte tanks reached 17 bar. This approach enabled the use of low-cost porous PVDF membranes but made the system more expensive to build. Use of ATEX components, certification of specialty pressurized components (including the stacks), and a glass-lined steel-tank-storing electrolyte at 17 bar increased the cost of the system. Slow intercalation of Br_2 into the stack's Viton gaskets led to stack failures. Economic problems at both developers, not related to system itself, unfortunately, prevented the system from reaching its full development potential.

Four smaller-scale systems were deployed by Elestor in The Netherlands: Deventer, Den Dungen, Kampen, and Arnhem. In October 2021, a 50 kW system was in construction phase (Figure 52.2) and a 500 kW–3 MWh system was in design phase. All these systems had or will have near (<30 mbar overpressure) atmospheric electrolyte tanks.

52.4.2 Permittability

In all cases, permits were obtained based on a "quantified risk analysis" (QRA) for both the Br_2- and H_2-related risks. Inputs for the QRA are the total amounts of Br_2 and HBr, the Br_2 vapor pressure at 100% SoC, the wetted area in case of a spill, and the chance of failure. The presence of nearby wind turbines can influence this chance of failure and must be considered. Outputs are the amount of Br_2 that can be released into the air and the safe distance. This safe distance was calculated by a specialist engineering firm using software for atmospheric dispersion modeling. No "vulnerable objects" (e.g. houses) are allowed within a safe distance. This means that permits for installing H_2–Br_2 flow batteries cannot be obtained for every location. The H_2–Br_2 technology is, therefore, best suited for larger-scale systems at more remote locations.

52.4.3 Outlook

Future upscaled systems can have a small footprint (Figure 52.3). The electrolyte tank footprint can be >1.5 MWh m^{-2} if an electrolyte storage tank can be built with a height of 20 m and the electrolyte energy density is 80 Wh l^{-1}. If the system can be connected to a H_2 pipeline the H_2 storage does not add to the footprint at all. The small footprint and low-active material cost could enable future bidirectional powerplants based on H_2/Br_2 flow-battery technology. Upscaled systems of 300 MW to 2 GW with 100- to 150-hour duration might turn offshore wind power into baseload power. Coupling to a MW class water electrolyser is possible, using shared H_2 infrastructure. In case high-quality heat is available, H_2/Br_2 cell stacks can be used for thermochemical H_2 production. Over 40 years ago the sulfur bromine cycle was proven for thermochemical–electrochemical H_2 production in a pilot plant in Italy. The electrochemical step was identical to the charging of the H_2/Br_2 flow battery [15]. It is also possible to make a dual-mode system, capable of switching between flow battery and water electrolyser operation.

52.5 Conclusion

The H_2/Br_2 flow battery has low-cost active materials and power-dense stacks. The active materials do have safety aspects that must be addressed thoroughly, without losing the cost advantage. This is the main economic challenge of technology.

Figure 52.3 Impression of a 320 MW cell stack skid.

The main technological challenge is to ensure the lifetime of the H_2 catalyst. Apart from finding better catalysts, there is improvement potential in membranes, electrodes, cell stacks, and systems.

The H_2/Br_2 flow-battery technology is most likely to be used in large-scale, long-duration systems operated by professional users. Deep integration with future H_2 infrastructure and electrolysers is possible and could strengthen the business case.

References

1 Cho, K.T., Tucker, M.C., and Weber, A.Z. (2016). *Energy Technology* 4: 655–678.
2 Wlodarczyk, J.K., Küttinger, M., Friedrich, A.K., and Schumacher, J.O. (2021). *Journal of Power Sources* 508: 230202.
3 Ronen, R., Atlas, I., and Suss, M.E. (2018). *Journal of The Electrochemical Society* 165: 3820–3827.
4 Küttinger, M., Brunetaud, R., Włodarczyk, J.K. et al. (2021). *Journal of Power Sources* 495: 229820.
5 Peled, E., Blum, A., Aharon, A., et al. (2012). Energy storage and generation systems, 2012, US patent no. US2012/0299384.
6 Zalitis, C.M., Sharman, J., Wright, E., and Kucernak, A.R. (2015). *Electrochimica Acta* 176: 763–776.
7 Hugo, Y.A., Kout, W., Sikkema, F. et al. (2020). *Journal of Energy Storage* 27: 101068.
8 Saadi, K., Nanikashvili, P., Tatus-Portnoy, Z. et al. (2019). *Journal of Power Sources* 422: 84–91.
9 Peled, E., Blum, A., and Goor, M. (2009). *Encyclopedia of Electrochemical Power Sources* 3: 182–191.
10 Hugo, Y.A., Kout, W., Sikkema, F. et al. (2018). *Journal of Membrane Science* 566: 406–414.
11 Abbasi, S., Forner-Cuenca, A., Kout, W. et al. (2021). *Journal of Membrane Science* 628: 119258.
12 Yarlagadda, V., Dowd, R.P. Jr., Park, J.W. et al. (2015). *Journal of The Electrochemical Society* 162: 919–926.
13 Lin, G., Chong, P.Y., Yarlagadda, V. et al. (2016). *Journal of The Electrochemical Society* 163: 5049–5056.
14 Martinez Cantu, B.B., Fischer, P., Zitoun, D. et al. (2021). *Energies* 14: 4945.
15 van Velzen, D., Langenkamp, H., Schütz, G. et al. (1980). *International Journal of Hydrogen Energy* 5: 131–139.

53

Some Notes on Zinc/Bromine Flow Batteries
Bjorn Hage

BH Consulting Ltd., 4A Richardsson Rd, Coogee, WA 6166, Australia

53.1 Modern Large Scale ZB Flow Battery

Figure 53.1 is a picture of a 500-kWh discharge capacity system installed at the CSIRO Energy Centre, Newcastle, NSW, Australia.

The system consists of ten 50 kWh modules individually connected to the building power system. The modular design provides for a reliable battery where malfunction of one battery will only reduce the max power capability. A faulty module can be readily replaced or repaired without disruption to the rest of the battery operation.

Another advantage of the modular layout is that the battery control system can engage only the number of modules necessary at any one time. This improves battery efficiency as the modules that are running will do so at a high power level where the efficiency is optimised.

53.2 Energy Storage Cost

The most crucial parameter for successful renewable energy storage is the energy storage cost (ESC) [$/kWh/cycle] – the cost to store and return one unit of electric energy. Once true full-scale storage cost is at par with renewable energy generation costs, say US$ 0.05/kWh for photovoltaic energy, we have truly commercial night-time solar energy, i.e. at US$ 0.10/kWh.

This goal requires a low-cost and long-life battery. One concept to achieve such a truly competitive ESC is a large-scale flow battery with a service life of 20 years or more. As active cell substrates, for any battery, will likely not last that long, particularly under continuous deep cycle real-life use, it is anticipated that the flow battery cell stacks must be replaced or rebuilt during the battery life. Table 53.1 is an outline of how truly game-changing ESC can be achieved.

Any battery will require very pure and high-quality electrolyte. The zinc–bromine flow battery (ZBFB) electrolyte is already, at low production volumes, commercially available at around US$ 10/kg. The high-energy density means that only 81 is

Flow Batteries: From Fundamentals to Applications, First Edition.
Edited by Christina Roth, Jens Noack, and Maria Skyllas-Kazacos.
© 2023 WILEY-VCH GmbH. Published 2023 by WILEY-VCH GmbH.

Figure 53.1 500 kWh ZB flow battery consisting of ten grid-connected 50 kWh battery modules, ZBB Technologies Ltd. 2010.

Table 53.1 20 year+ 'battery life' giving competitive stored renewable energy.

Years	Cycles	US$/kWh			
		Capex			
System life: 20 years Stack life: 5 years	365 per year	Stacks	Electrolyte	The Rest	ESC
5	1825	50	50	50	0.08
10	3650	100	50	50	0.05
15	5475	150	50	50	0.05
20	7300	200	50	50	0.04

required per kWh of stored energy, at 2.25 M (moles per litre zinc bromide salt concentration). Further work may allow much higher concentrations potentially giving even lower costs. Electrolyte cost of US$ 50/kWh is not an unreasonable target, Table 53.1.

Producing good quality ZBFB flow-battery stacks at US$ 50/kWh capacity is possible but will require major polymer technology development and large production volumes. The ZBFB technology has the advantage of using no expensive materials. It can therefore be, somewhat crudely estimated, that a 5 kWh capacity stack that will weigh 25 kg can be produced at a cost of US$ 250 = US$ 10/kg or US$ 50/kWh (Table 53.1).

'The Rest' in Table 53.1 represents pumps, tanks, plumbing, control, containment, etc. to make a complete battery. The estimation is simply based on

US$ 50 000 per MWh battery discharge capability. Note that 1 MWh would, in this context, still represent only a small part of a multi-MWh capacity battery.

53.3 The Gould/ERC ZBF Battery

During the 1970s Gould Corporation in the US was developing a system, in parallel and separate from the more well-known EXXON system. The work involved basic studies and engineering work in a program cost-shared with the US Department of Energy (DOE).

Energy Research Corporation (ERC), Connecticut, US, acquired the technology in the early 1980s and continued the work, with ongoing support from the US government through DOE/Sandia National Labs.

Murdoch University (MU), Perth, Western Australia, where zinc/bromine work had been going on for some years already, initiated by Professor AJ Parker, founding professor, Chemistry, joined ERC in a shared R&D collaboration.

53.4 The Gould/ERC ZBF Battery Stack

The bipolar stack utilised a high surface area activated carbon felt flow-through positive electrode substrate to support the bromine/bromide reactions. The felt is bonded to the bipolar plates, that separate each cell, using a conductive solvent-based glue bond. The design originally used graphite sheet bipolar plates which were brittle and difficult to bond to the moulded polymer flow frames. As a replacement ERC and MU developed a compression moulded, carbon black filled 'carbon plastic' bipolar plate with a thickness of ~1 mm.

The 872 cm^2 stack design utilise a single injection moulded uPVC (un-plasticised PVC) flow frame to hold the bipolar plates and to form the flow channels for both cell halves. uPVC has been found to be, generally, more stable to bromine, compared with PP (polypropylene). The flow frame has the catholyte (positive electrolyte) flow channels on one side and the anolyte channels on the reverse side, see Figure 53.2.

The original design used a system where the catholyte was entering at the sides of the vertically located flow frames and exiting from two stack manifolds. The idea is to purge any complex phase, by allowing it to sink down and exit through the bottom manifold. The top exit manifold was to allow any gas bubbles to escape. It was, however, found that this design did not significantly reduce the amount of complex phases trapped inside the stack. It possibly also resulted in some gas entrapment, at the lower part of the cells. The side inlet concept was therefore abandoned for a conventional bottom-to-top flow design, i.e. the mid-bottom inlet manifold was not used.

The stack design is a compression assembly with seals formed where protruding sealing ridges, on the flow frames, seal against the 'gasket area' of the membrane. This gasket, black appearance in Figure 53.3, is formed by dipping and then drying the micro-porous silica-filled polyethylene membranes (Daramic®, 0.9 mm) in a cold-set silicone.

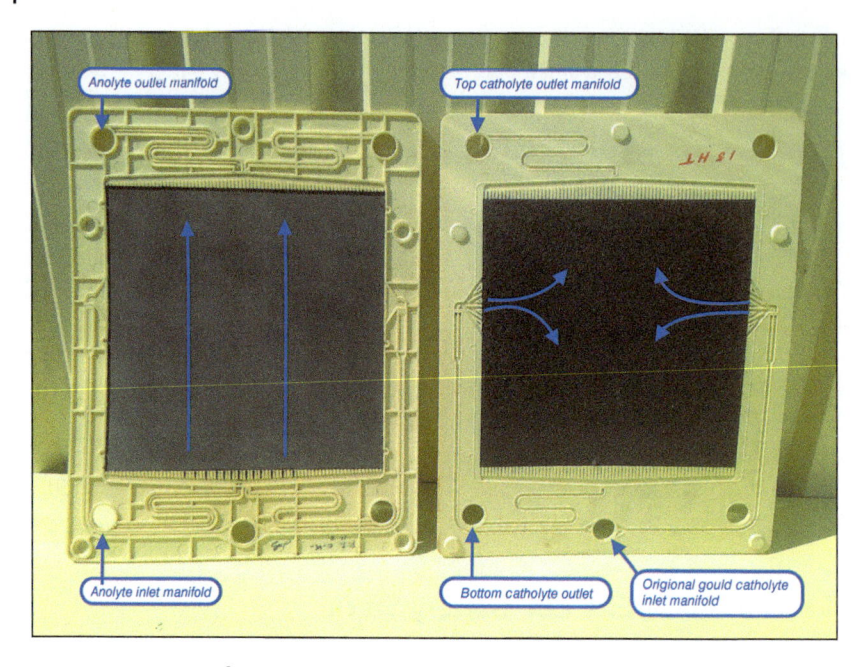

Figure 53.2 872 cm² flow frame and bipolar plate/carbon felt assembly.

Figure 53.3 Membrane/seal assembly.

53.5 The Gould/ERC ZBF Battery Stack Electrolyte System

The gould/ERC bromine complexing agent (BCA) is, just like the EXXON BCA, soluble in the electrolyte until it complexes with bromine, at which point it precipitates as a polybromide 'oil', a separate and dense phase that sinks to the bottom of the electrolyte tank.

One difference between the two technologies is that the Gould/ERC BCA is complexed with bromine before being used in battery cycling. This way the bromine complex (polybromide oil) is fully formed, as a separate phase in the catholyte tank, already at 0% State-of-Charge (SoC).

While the positive aqueous electrolyte bromine concentration is in the range of $7–15\,g\,l^{-1}$ over a cycle, the complex phase bromine concentration ranges from 1200 to $1600\,g\,l^{-1}$. The idea is that the complex phase is kept in the tank, away from the membranes, and unable to reach the zinc deposited in the cell stack.

53.6 The Gould/ERC ZBF Battery Stack Electrolyte Circulation System

The bromine complex buildup in the carbon felt and membranes reduce cell efficiency by increasing the bromine inventory available for self-discharge. Reducing this buildup, i.e. preventing any complex phase retention in the cell stack would improve system efficiency. This was a major objective of the early ERC work.

During charge and discharge, it is also essential to achieve mixing between the complex phase (polybromide oil) and the aqueous part of the electrolyte. This is to allow bromine to be complexed, during battery charge, and released to the electrolyte during discharge.

Initially, the mixing was achieved using a third pump, the 'polybromide' pump in Figure 53.4. This pump was later replaced with a tank design where the catholyte returned from the stack enters through the complex phase tank store. The new tank was designed to efficiently separate, by gravity, the heavier complex phase and return only the aqueous catholyte to the battery stack.

The new system was said to offer inherent safety advantages. *'as almost all of the bromine in the battery is stored in the polybromide, and since the complex is not circulated in any flow lines the risk of a spill of this hazardous material is greatly decreased'* (ERC report to Sandia; SAND90-7016 May 1990)

The elimination of complex phase loading of the carbon felts and the microporous membranes was however never very successful. It turned out that, even with the new tank, a considerable percentage of non-complexed BCA was present in the aqueous electrolyte at 0% SoC.

The presence of aqueous, non-complexed, BCA caused the formation of bromine complex (polybromide) inside the cells, where it could accumulate in the carbon felt and membranes during charge. It was found that, after extended periods of cycling,

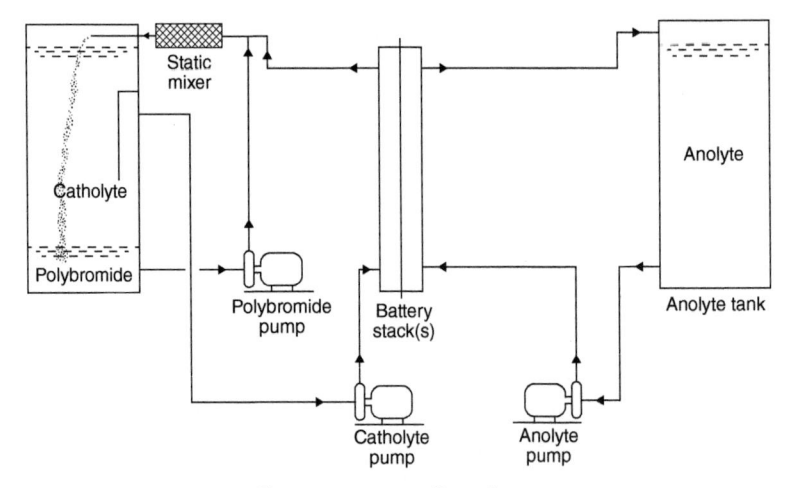

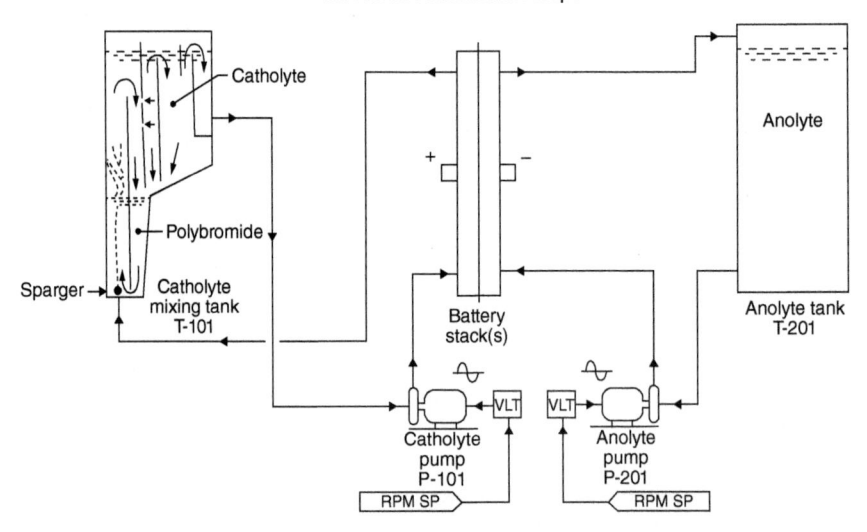

Figure 53.4 Electrolyte circulation systems (ERC report to Sandia; SAND90-7016 May 1990).

as much as one-third of the normal discharge capacity could be discharged without running the catholyte circulation pump. This indicates that $1/3$ of the bromine produced during charge was complexed and captured inside the stack.

53.7 Shunt Currents – Are They for Real?

Shunt currents are ionic currents that occur in batteries where several cells share the same electrolyte. In a typical flow battery, the currents will run, through the flow channels, in the direction from a cell with a higher potential towards cells with lower voltage, see Figure 53.5.

Figure 53.5 Shunt currents.

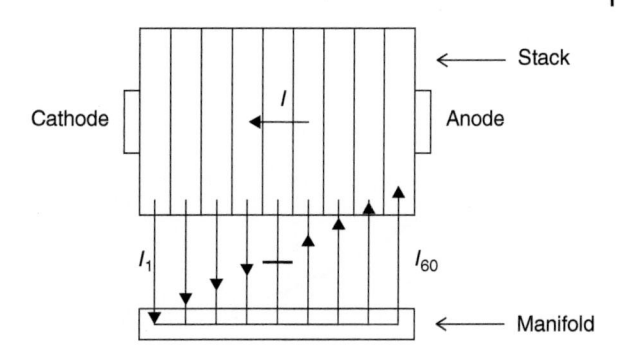

A large battery energy storage system where all cells share the electrolyte has many advantages. If everything is equal between cells, such as electrolyte concentrations and temperature, the cells should, in theory, also perform equally. This is important where several cells are connected, in series. A cell that is running empty, before other cells, will be reverse charged by the other cells. This may cause cell damage and will spread to other cells if cycling continues without cell balancing cycles.

A simplified calculation shows that the cell current drain, from shunt currents, will be maximum at the centre cells and that this drain is proportional to the total number of cells squared, see Table 53.2.

To estimate actual shunt current losses, a test was performed on a 60-cell, 2500 cm² stack (ZBB Energy Corp ~2002). The battery was fully charged, 4.5 hours at 50 A (225 Ah), and then quickly rinsed out and cut apart. The zinc deposited on each bipolar plate was weighed, see Figure 53.6.

At 225 Ah each electrode should hold 274 gr of Zn less an amount lost from shunt currents and self-discharge during the charge cycle. We can see that many cells hold

Table 53.2 Shunt current calculation.

I_n – Shunt channel current for cell #n [A]	$I_n = \dfrac{V}{R}\left(\dfrac{1}{2} - \dfrac{n}{N}\right)$
V – Total stack voltage [V]	
N – Total number of cells [#]	
R – Shunt channel resistance [Ω] i.e. the manifold will sit at half the total stack voltage and there will be zero current running between the mid cell and the stack manifolds	$I_{cell} = \displaystyle\int_0^n I_n = \dfrac{V}{R}\left(\dfrac{n}{2} - \dfrac{n^2}{2N}\right)$
I_{cell} – Shunt current cell drain as a function of cell number	$I_{cell} = \dfrac{VN}{8R}$
i.e. The cell current drain will be highest at the mid-stack cell [$n = N/2$]	$\left(n = \dfrac{N}{2}\right)$
E_0 – Cell voltage [V]	$I_{cell} = \dfrac{E_0 N^2}{8R}$
i.e. The maximum current drain is proportional to the number of cells squared	$\left(n = \dfrac{N}{2}\right)$

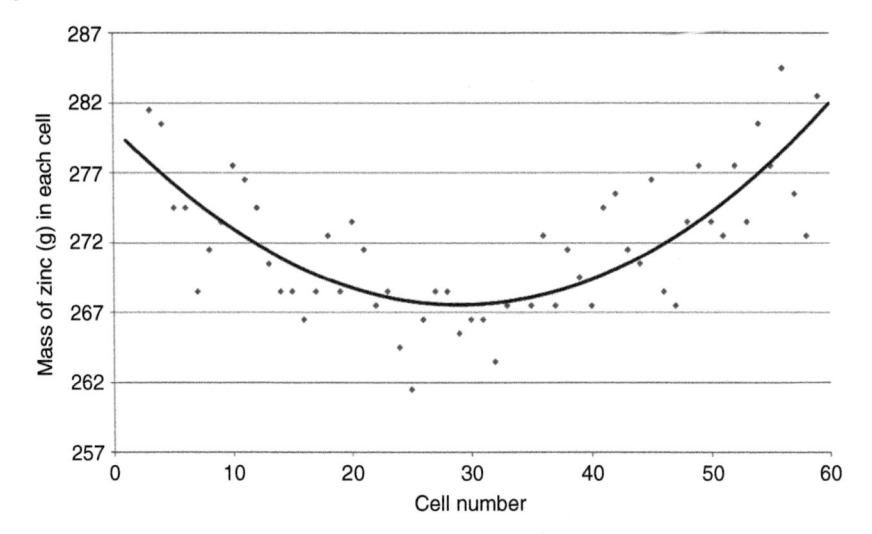

Figure 53.6 Amount of zinc as a function of cell number, 100% SoC.

more than 274 gr of Zn. This is likely Zn left from previous cycling, i.e. the stack held some zinc from previous cycles, not necessarily visible on the total stack voltage since some cells may be reverse charged.

It is interesting to note that the metallic zinc coating in this fully charged $ZnBr_2$ battery (90 mAh cm^{-2}) represents only a thin coating, with a theoretical thickness of 0.15 mm. In reality, the zinc coating measures up to twice this thickness, indicating a typical Zn plating porosity of up to 50%.

With a shunt current resistance (R in Table 53.2) of approximately 750 Ω/cell, for this stack design, the theoretical loss, at the centre cell, from shunt currents, should be approximately 1 A or about 2% of the stack charge current. This should, in theory, mean that we should find about 5.5 gr of Zinc less in the centre of the stack compared with the end cells.

The scraping off and weighing of the zinc from each cell was difficult and not very accurate (perhaps within +/−5 gr). Some cells, including the terminal electrodes (cell #1 and #60) were so damaged during the stack postmortem that weighing of the Zn deposit could not be performed.

Even if the data are not so accurate, it is visible that centre cells typically contain less Zn compared with end cells, say 5 gr +/−5 gr less ...

Conclusion must be that shunt currents do exist ! - and must be considered in flow-battery stack design. The practical coulombic efficiency (Ah out/Ah in) can be considerably reduced by shunt currents as the useful discharge must be terminated when the centre cells are empty as continued discharge will reverse these cells. At this point, end of discharge, a rapid rise in electrolyte temperature can typically be observed. This is from reversed cells that plate zinc straight into the still bromine-rich catholyte half cells. Zinc and bromine react with significant heat being generated.

53.8 Can Battery Efficiency be Over 100% ?!

Of course, it cannot. But it is true that an electric grid-connected battery can return significant energy to the grid without having increased the grid load very much during charging.

A battery energy storage system can return more 'real energy' than the 'real energy' consumed during charge. This 'battery benefit' is due to the AC/DC conversion during charge when the 'apparent power' component can be 'phase compensated' and used to charge the battery.

This may be significant and valuable since the utility can often only charge its customers for the 'real power' component delivered whilst it is the 'apparent power' load that will limit the capacity of network distribution equipment, such as wires and transformers. A distribution grid with a small phase angle, i.e. a power factor (pf) close to 1, can carry more real power before an expensive upgrade is necessary.

An analogy that has proved to be useful in explaining this is the beer glass, Figure 53.7.

The reactive power – the foam on top of the beer – is of no use to the consumer and reduces the network capacity to carry real power. Grid engineers strive to reduce the reactive power just like bar staff are challenged to keep the amount of foam to a minimum.

A test, to prove that a battery storage cycle can, in effect, turn reactive power into real power; was performed on a 400 kWh ZBF battery string located and used as substation support at a suburban substation in Melbourne, VIC, Australia.

Figure 53.7 The beer glass is the utility grid equipment, it has a finite capacity that can be used for beer (real power) or for foam (reactive power).

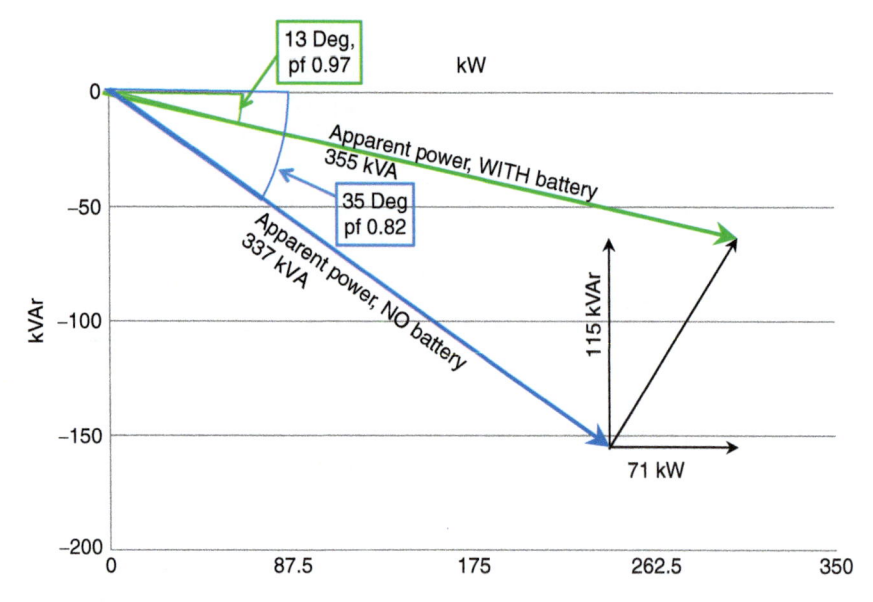

Figure 53.8 Power correction during charge of a 200 kWh battery.

One 400 V DC 'string' battery was charged, from 12:00 noon (on 16 July 2003) until approximately 4 : 30 p.m., at a constant DC current of 150 A. During this charged period the imaginary power (kVA) injection, created by the AC to DC battery charger, corrects the grid power factor from about 0.8 to close to 1. Figure 53.8 shows the typical phase angles at the substation with and without the battery being charged. Green arrow and green curve, Figure 53.8 is the 'test day', when the battery was cycled. Blue is for the 'reference day' – same week-day, one week earlier with no battery cycling.

Note that during the battery charge period the substation load (apparent power) was only increased by an average of about 18 kVA due to battery charging.

From 12:00 noon until 4:30 p.m. on 14 July 03, the battery was discharged at 150 A DC, see Figure 53.9. For a period in the middle of the discharge period, it was necessary to reduce the discharge rate to 75 A DC. This was due to the grid voltage climbing towards unacceptable levels.

This is an often overlooked aspect of end-of-grid battery discharge at end-of-grid locations where voltage reduction cannot be achieved through 'tap changing' of transformers. Battery discharge may not be possible if it pushes the grid voltage above acceptable levels. In Australia, this is typically at ~430 V (nominal voltage 415 V).

Discharge was continued until the battery was empty at about 8:30 p.m. During the battery discharge period, the substation load was effectively reduced by 50–100 kVA. The $ZnBr_2$ flow battery had returned 204 kWh of DC energy to the grid.

The test demonstrates an interesting use of a battery energy storage system. Over 200 kWh of real power was returned to the grid over the 4.5 hours discharge period and the substation grid load was reduced by 50–100 kVA during this period.

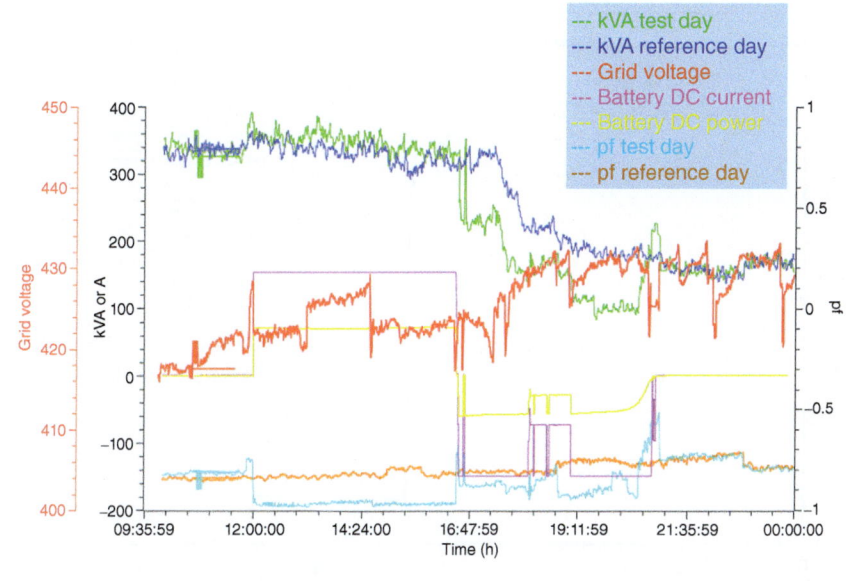

Figure 53.9 Returning 204 kWh DC from a charge cycle with minimal load increase to the substation transformer (pf – power factor, pf 1 = phase angle 0 Deg).

This energy has been stored without significantly increasing the apparent load of the grid during battery charging.

If nothing else, the above example demonstrates that battery efficiency is not the one and only parameter to take into account.

53.9 Zinc–Bromine Battery Efficiency

When comparing the efficiency of two electricity storage systems it is essential to be very careful regarding how the efficiency is defined. Ideally, two batteries must be similar in size and subject to the same cycle regime at the same temperature, etc. It must be clearly outlined how the efficiency values are defined. See Figure 53.10 for details on a typical deep discharge cycle for a ZBF battery.

The battery achieved a coulombic efficiency (CE) of ~88%. This is simply the useful (>30 V) discharge Amp-hours divided by the charge Amp-hours. Similarly, a voltaic efficiency (VE) of 87% was calculated as the average discharge voltage divided by average charge voltage, see Figure 53.10, 50 A discharge, cycle #64.

Note how much lower the discharge voltage is when the same battery is later discharged at twice the discharge current, 100 A (cycle #92). This is a typical feature of the ZBFB where the electrolyte resistivity is comparatively high. Note also that the coulombic efficiency is largely the same, the discharge duration is halved due to the doubling of the current.

It is interesting to attempt to understand where the losses come from and Figure 53.11 is an estimate of this. Note that as much as 40% of the voltaic losses

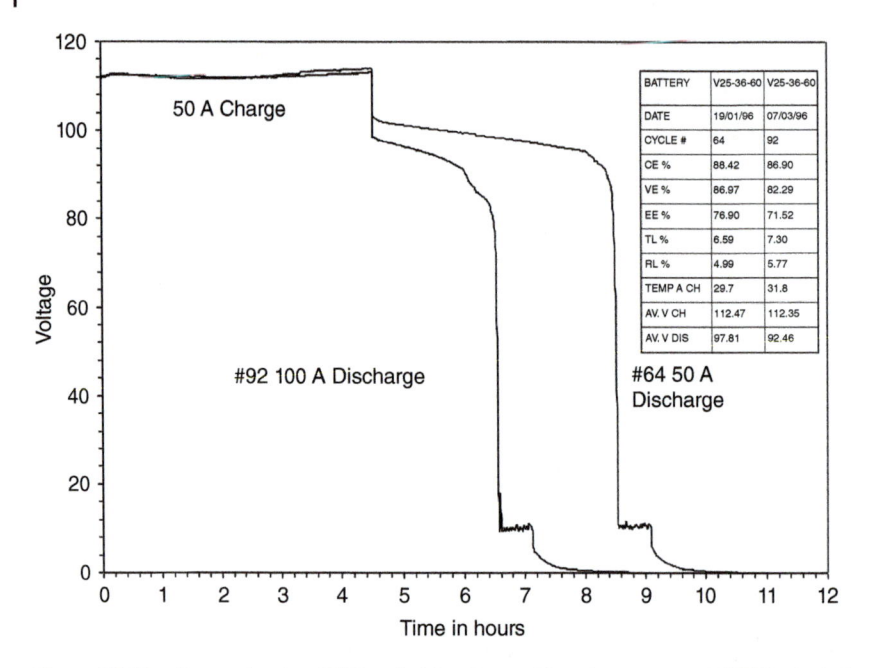

Figure 53.10 Test cycle plot, 2500 cm², 60 cells, auxiliary (pump power) NOT included, TEMP After charge is electrolyte temp. – ZBB Energy Corp. ~1996.

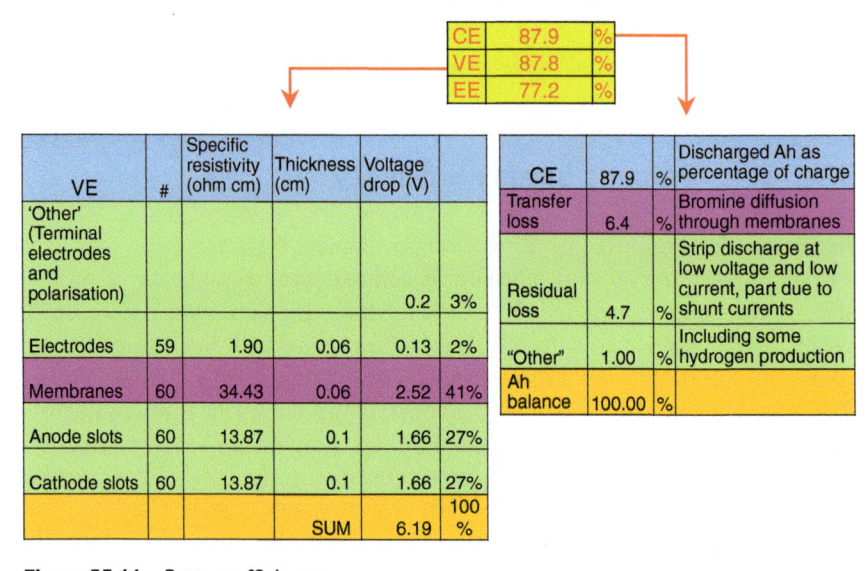

BATTERY	V25-36-60	V25-36-60
DATE	19/01/96	07/03/96
CYCLE #	64	92
CE %	88.42	86.90
VE %	86.97	82.29
EE %	76.90	71.52
TL %	6.59	7.30
RL %	4.99	5.77
TEMP A CH	29.7	31.8
AV. V CH	112.47	112.35
AV. V DIS	97.81	92.46

CE	87.9	%
VE	87.8	%
EE	77.2	%

VE	#	Specific resistivity (ohm cm)	Thickness (cm)	Voltage drop (V)	
'Other' (Terminal electrodes and polarisation)				0.2	3%
Electrodes	59	1.90	0.06	0.13	2%
Membranes	60	34.43	0.06	2.52	41%
Anode slots	60	13.87	0.1	1.66	27%
Cathode slots	60	13.87	0.1	1.66	27%
				SUM 6.19	100%

CE	87.9	%	Discharged Ah as percentage of charge
Transfer loss	6.4	%	Bromine diffusion through membranes
Residual loss	4.7	%	Strip discharge at low voltage and low current, part due to shunt currents
"Other"	1.00	%	Including some hydrogen production
Ah balance	100.00	%	

Figure 53.11 Battery efficiency.

are from the ionic resistance of the microporous membranes. This resistance is very much determined by the resistivity of the electrolyte, the porosity, and wettability of the membranes (how well the membrane can soak up and be filled with the electrolyte)

The membranes are also responsible for the major part of the coulombic efficiency loss, i.e. from bromine diffusing through the membranes and reacting with plated zinc (self-discharge). It clearly makes sense to work towards a system with more conductive electrolyte and more efficient membranes.

In terms of electrolyte conductivity, it is also interesting to note that the bromine storage agent, in above case MEP (methyl ethyl pyrrolidinium bromide) contributes significantly to the resistivity of the electrolyte system, see Figure 53.12. Note that the graph represents electrolytes at 0% SoC when the MEP is fully in aqueous solution. Dotted lines represent the measured resistivity of a typical low-cost microporous silica/polymer membrane. Note how the electrolyte resistivity can be lowered considerably by the addition of supporting agents. Ammonium and potassium can be used with similar effects.

An advantage of microporous membranes is their low-cost and proven long life. Limited work has been carried out with more advanced, so-called ion-exchange membranes and it is indeed possible to achieve considerable improvement in coulombic efficiency using an ion-exchange membrane. This author has carried out limited such experiments on non-flow $ZnBr_2$ battery cells, see Figure 53.13.

The two cells are identical except for the membrane, current density is low, $2\,mA\,cm^{-2}$. Non-flow $ZnBr_2$ batteries (NFZB) fall outside the scope of this publication but it is worth noting that NFZB's can generally achieve considerably

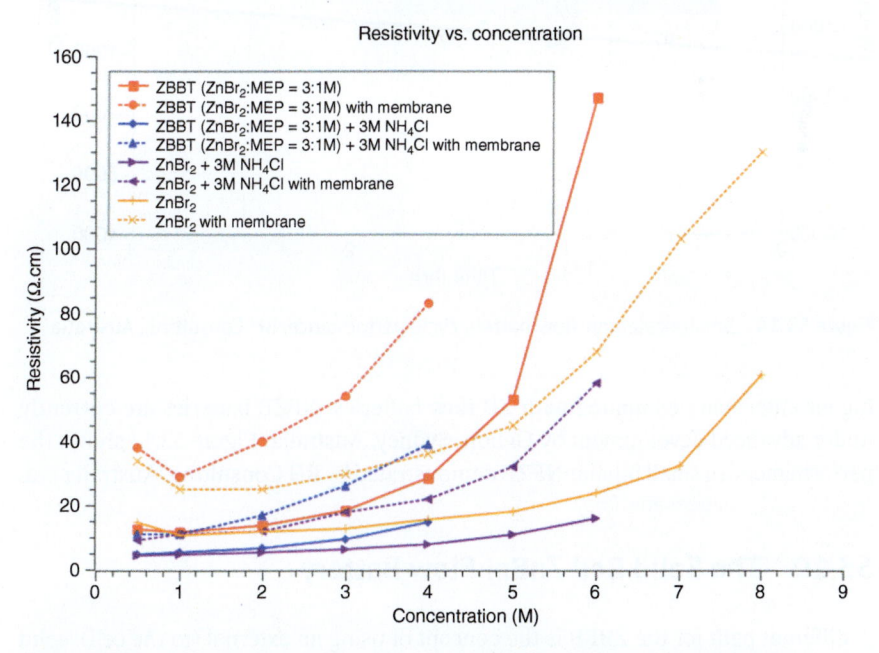

Figure 53.12 Electrolyte resistivity (M – moles per litre of $ZnBr_2$ in the solution).

Ah_{out}/Ah_{in} = CE

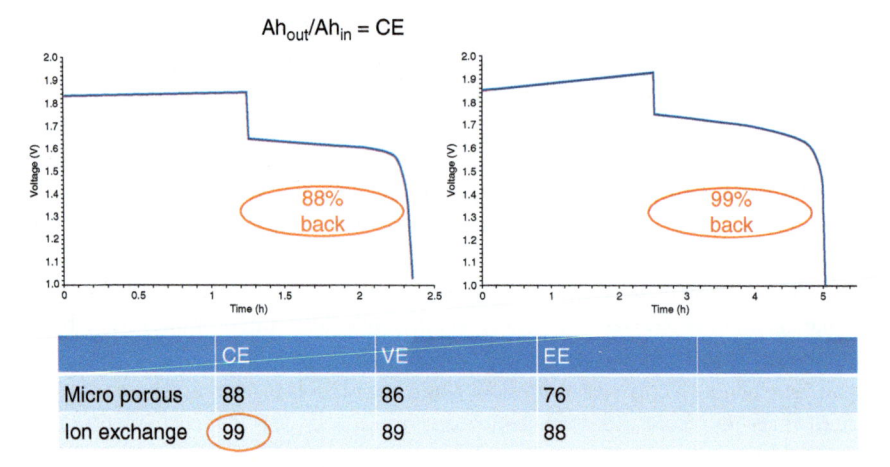

	CE	VE	EE
Micro porous	88	86	76
Ion exchange	99	89	88

Figure 53.13 Cell test with microporous vs ion-exchange type membrane, non-flow ZB battery.

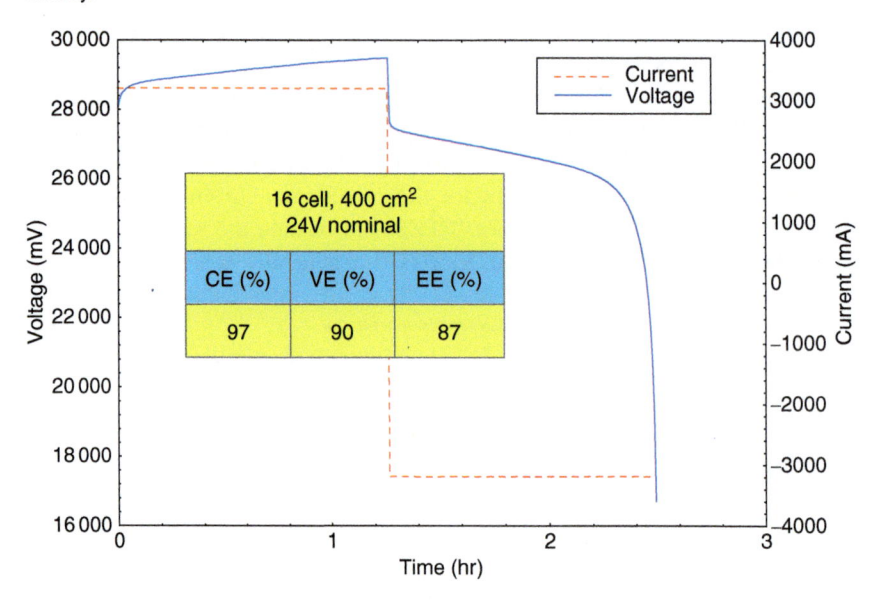

16 cell, 400 cm² 24V nominal

CE (%)	VE (%)	EE (%)
97	90	87

Figure 53.14 Small-scale non-flow battery cycle performance, BH Consulting, Australia Ltd.

higher efficiencies compared with ZB flow batteries. NFZB batteries are currently under advanced development by Gelion, Sydney, Australia. Figure 53.14 shows the performance of a small bipolar NFZB demonstrated by BH Consulting, Australia Ltd.

53.10 The Solid Bed $ZnBr_2$ Flow Battery

A different path for the ZBFB is the concept of using an external (to the cell) 'solid bed' for the bromine complexing, see Figure 53.15.

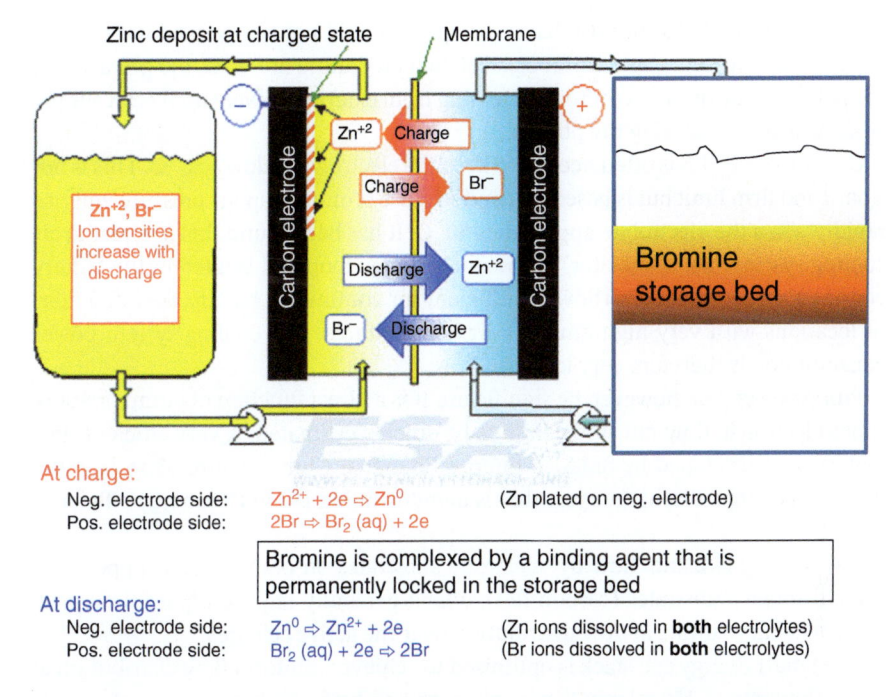

Figure 53.15 Principle of solid bed ZnBr$_2$ flow battery, BH Consulting, Australia Ltd.

Using this concept, only the aqueous catholyte will be circulated through the cells and a significant reduction in electrolyte resistivity will give improved voltaic efficiency. It has been shown that continuous current densities of 20 mA cm^{-2} and above can be sustained also with low-cost positive cell substrates. The electrolyte flow system complexity is reduced, since no mixing/recirculation of complexed bromine is necessary, and the battery will require less pump power. Safety is enhanced with no possibility of any complexed bromine escaping from tank or from electrolyte circulation system leaks. With no complexing agent, in the electrolyte, there is also the possibility of using a higher concentration of ZnBr$_2$ salt, without corresponding increase in voltaic losses. see Figure 53.12.

53.11 Flow Battery Auxiliary Power

Flow-battery cycle data are often given as DC to DC without Auxiliary power. This may be excused while we are dealing with experimental systems and small-scale demonstrations, i.e. any auxiliary power necessary for pumps and fans may be proportionally high due to the small system size.

Heating may be required, but generally only if the battery is exposed to very low ambient temperatures and at times when the battery cannot be operated, i.e. every battery will be able to keep itself warm, with appropriate insulation, simply from its own losses.

A 'conventional' ZBFB, MEP-based complex, should not be operated with frozen electrolytes, this sets a minimum electrolyte temperature at ∼8 °C, the point when the polybromide oil starts to freeze. Heating from external power supply can simply be done just by operating the pumps.

Cooling of a ZBFB is often needed to keep the electrolyte below ∼50 °C. This is not a hard and firm limit but is based on caution since bromine vapour pressure will rise rapidly when the electrolyte approaches 60 °C. It has been found that a 'direct to air cooling system', i.e. a 'radiator' in the anolyte flow loop and located in the battery containment ventilation airflow is sufficient for continuous battery operation also in locations with very high ambient air temperatures. The cooling system power consumption is therefore very low.

Pump power can, however, be significant. It is a direct function of pump pressure times electrolyte flow rate and can thus be quickly estimated. A very efficient flow battery was developed by SMART Energy, Shanghai China. Figure 53.16 is a plot of a short charge–discharge cycle that is demonstrating DC to DC energy efficiency of 80%.

This battery is using 'flow by' electrode substrates on both sides, as opposed to 'flow-through' substrates (carbon felt), which generally require higher flow pressures to achieve uniform flow distribution over the active cell area.

The Smart Energy cell stack is optimised to achieve a uniform flow distribution at a low flow rate of 150 ml min^{-1} per cell, each cell half, at a pump pressure of only 40 kPa. At a combined, pump and pump motor efficiency of 50% the pump power is still a significant 12 W and the pump energy expended during the one hour cycle represents ∼2 % of the discharge energy. The conclusion is that flow batteries must be carefully designed for low flow rates and low stack hydraulic pressure drop.

Note also that the pump energy, 12 Wh, is more than the 'Zinc Strip' (residual loss) energy, 5 Wh, (or 2.8% of the charge amp-hours), electrically discharged at a low current after the 20 A main discharge. The 2.5% 'Transfer Loss' represents charge Amp-hours not recovered, i.e. self-discharge due to bromine diffusion inside the cell and possibly a minute amount of hydrogen evolution.

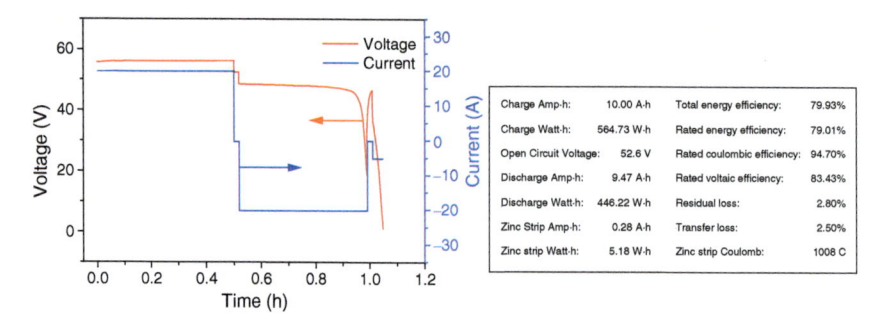

Charge Amp·h:	10.00 A·h	Total energy efficiency:	79.93%
Charge Watt·h:	564.73 W·h	Rated energy efficiency:	79.01%
Open Circuit Voltage:	52.6 V	Rated coulombic efficiency:	94.70%
Discharge Amp·h:	9.47 A·h	Rated voltaic efficiency:	83.43%
Discharge Watt·h:	446.22 W·h	Residual loss:	2.80%
Zinc Strip Amp·h:	0.28 A·h	Transfer loss:	2.50%
Zinc strip Watt·h:	5.18 W·h	Zinc strip Coulomb:	1008 C

Figure 53.16 Cycle plot and efficiency analysis, Smart Energy 30 cell, 1000 cm^2 battery.

53.12 Summary

Flow batteries are recognised as a leading contender for large-scale stationary renewable energy storage and the zinc–bromine flow battery may well be the front-runner amongst a host of battery technologies.

The most crucial parameter for successful renewable energy storage is the ESC [US$/kWh/cycle] – the cost to store and return one unit of electric energy. The ultimate target is US$ 0.05 kWh per cycle and has not yet been reached, by any technology, flow battery, non-flow battery, or hydrogen (production-storage generation).

The zinc–bromine flow battery certainly has potential to be produced at low-cost. The energy density is relatively high (requiring less electrolyte) and all materials, including the electrolyte can be produced at low-cost.

Long service life with more than 2000 continuous deep cycles has been demonstrated, at industrial scale and with minimum performance decline. Typical real-life return DC/DC efficiency, at four hours deep discharge, can be expected to be around 77%.

The non-flow zinc–bromine flow battery, as demonstrated by BH Consulting, Australia Ltd. (this author) and others, has now reached 'near-commercial' status. The technology offers simplicity and potentially even lower cost than its flowing sibling. Long and stable cycle life has been demonstrated and scale-up efforts are underway. The technology is about as safe and nonflammable as any battery can be.

54

Mobile Applications of the ZBB
Gerd Tomazic

Battery Consultant, A-8680 Mürzzuschlag, Hofkirchergasse 4, Austria

54.1 Introduction

The sudden and dramatic increase of the oil price by the OPEC states in 1974 created a worldwide shock. As a consequence, an intensive search for alternative energy sources and for possibilities to store energy was initiated. Especially in the USA where the economic and social life was strongly dependent on full mobility everywhere and at any time, a great amount of money was spent to develop batteries to power electric vehicles. Available systems, for instance lead–acid batteries, Ni–Zn-batteries, and Ni–Cd batteries, were found to have severe disadvantages, especially regarding energy density, achievable range, and toxic electrolyte components.

Due to its high cell voltage of 1.82 V, the zinc–bromine redox system was seen as an attractive battery candidate. In 1983, the Studiengesellschaft für Energiespeicher und Antriebssysteme (SEA), a member of the Austrian Industries Technologies Division, was buying a license from the Exxon Research and Engineering Group and took over the knowhow for the ZBB technology. In the same year, a project team was formed by EXXON consisting of five members located in the US, Japan, Australia, and Austria. The project goal was to make the ZBB system ready for mass production. A valuable condition for this cooperation was that during the cooperation period, all results won during the project work had to be demonstrated periodically during ZBB-symposia, and the results of the work could be used mutually by every partner. By this measure team, internal secrets were avoided.

Every team member of the newly formed EXXON team could freely decide how to manage the project work mainly consisting of the fulfillment of two tasks:

- Study and development of electrochemistry, material development, and battery testing.
- Design of components, fabrication of batteries, integration of batteries in EVs, and testing of EVs in public traffic.

Flow Batteries: From Fundamentals to Applications, First Edition.
Edited by Christina Roth, Jens Noack, and Maria Skyllas-Kazacos.
© 2023 WILEY-VCH GmbH. Published 2023 by WILEY-VCH GmbH.

Due to SEA's industrial background and the lack of a well-established company internal electrochemical task force, the second task initially was the main focus for SEA. At start, the electrochemical activities were performed by consultants of the TU-Graz and Vienna. Battery testing in the lab and electrochemical investigations step by step were taken over after the volume of the testing work did not allow to outsource it anymore.

54.2 Scheme of the Zink–Bromine Battery

Figure 54.1 shows the scheme of a zinc–bromine battery system with a bipolar stack.

If thinking of the application for EVs, the following requirements have to be considered when comparing to the standard storage application:

- EV-designers want to offer maximum power for the drive train. For this reason, the battery system voltage should be as high as possible. For fabrication reasons, stacks with more than about 80 cells are not recommended (warpage of electrodes and tightness of stack). This results in the requirement for several stacks that have to be combined into stack towers where all stacks are electrically connected in series. High voltage means high shunt currents (SC). High SC requires careful observation during battery operation.
- Battery weight is of great importance for the EV application. Every single component has to be investigated regarding its potential to save weight. Too strong reduction of wall thickness is restricted due to bromine penetration to the outside.
- Batteries have to be integrated into the EV platform. This can require complex shapes for the design of the reservoir and has to consider that the full exchange of electrolyte volume is guaranteed during loop circulation.
- The heat exchanger within the zinc reservoir has to be very efficient to ascertain that, together with the EV-heat exchanger, all heat can be dissipated under all operation conditions.

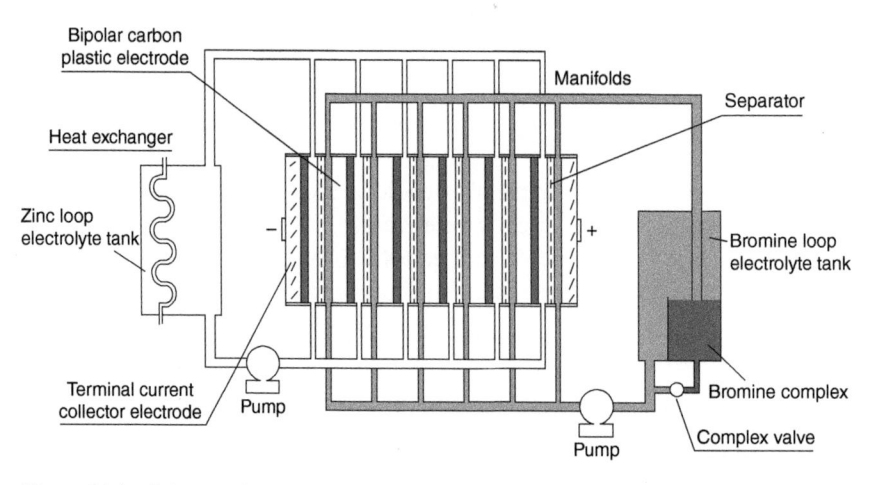

Figure 54.1 Scheme of zinc–bromine battery.

- Real-life EV application does not allow to strip after every cycle. This means that EV application for a ZBB requires battery operation modes that are a combination of random cycling with stripping controlled by the open-circuit voltage and the state-of-charge (OCV–SOC) relation. Due to the detrimental effect of dendrite formation within a stack, random cycling is the most critical operation mode for the battery system and by that the most important lifetime-limiting factor.
- Mobile application in public traffic requires additional measures regarding safety. This is true in the same manner as it is demanded for fossil-fueled cars. Electrolyte leakage and measures in case of a fire have to be considered specially.

54.3 EV-Application Issues

The evaluation of the performance of an EV usually is based on the following properties:

Battery type, energy density (Wh/kg = Range), power density (W/kg = Power), and Ragone diagram = relation between energy density and power density.

Table 54.1 shows the nonexhaustive list of issues have to be included in the program of R&D work. The SEA team consisted in 1983 of 7 members and at the end

Table 54.1 Themes to be handled during R&D work of batteries for electric vehicles.

Electrochemical properties	**Geometrical properties**	**Safety**
Capacity	Volume energy density	Crash behavior
Voltage	Volume power density	Voltage protection
Number of cells	Design density	Electrolyte leakage
Voltaic, Coulombic Efficiency	Upscaling effects	Mechanical stress
Self-discharge	Downscaling effects	Electrolyte Temp.
Stand losses		Leakage handling
Corrosion processes	**Operating properties**	Fire handling
Gas-build-up	Charging technology	
Operation temperature	Charging profile	**Cost**
Deep temperature behavior	Stripping after every cycle	Battery
Maximum temperature	Random cycling	Renewal
	Overcharge, deep discharge	Periphery
Lifetime	Trickle charge	Deposition
Calendar life	SOC measurement	Recycling
Cycle Life	OCV management	Commercialization
Overcharge sensitivity	Thermal management	Maintenance
Deep-discharge sensitivity		
Influence of standing time		

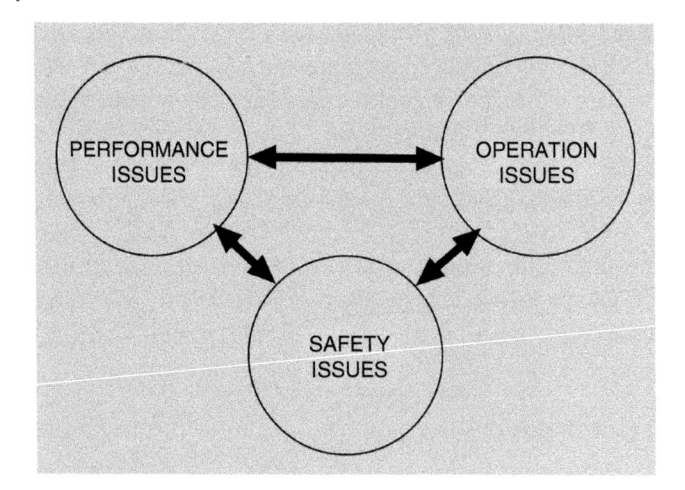

Figure 54.2 Battery operation relevant issues.

of the project of about 60 members. It is obvious that not all issues could be handled sufficiently under these circumstances.

Figure 54.2 shows that performance, battery operation, and safety are interdependent on each other.

54.3.1 Performance

The main influence on battery performance is given by the electrochemical performance of the stack and electrolyte components. Nernst equation, Fick's law, conductivity of electrode material and electrolyte, separator diffusion, voltaic and Coulombic efficiency, voltage, single-cell voltage, open-circuit voltage, as well as in-depth measurements, for instance, voltammetry, and impedance spectroscopy are the means to study electrode kinetics.

Handling of this "scientific" issues is not foreseen within the scope of the paper.

In case of the ZBB there is a further strong influence given by cell geometry.

Figure 54.3 shows a cross section of the ZBB cell with the mechanical parameters influencing cell performance. The difference to a real flow battery (the ZBB is a hybrid-flow battery with flow by cells) is the change of the zinc-half-cell gap (0.3–0.5 mm) during charge and discharge.

This changes the flow rate within the half-cell. The smaller the gap the less electrolyte is provided. If too much zinc is deposited, the flow rate can be too low to provide sufficient electrolytes. Dendrites can be the consequence.

The sketch above shows only a small portion of one cell. In a stack, there are many cells, effects can be accumulative, preferred electrolyte pathways through all cells can be formed, and specific areas within the cells can be "closed" by zinc. The worst case, a short through the stack from one collector to the other collector, can occur.

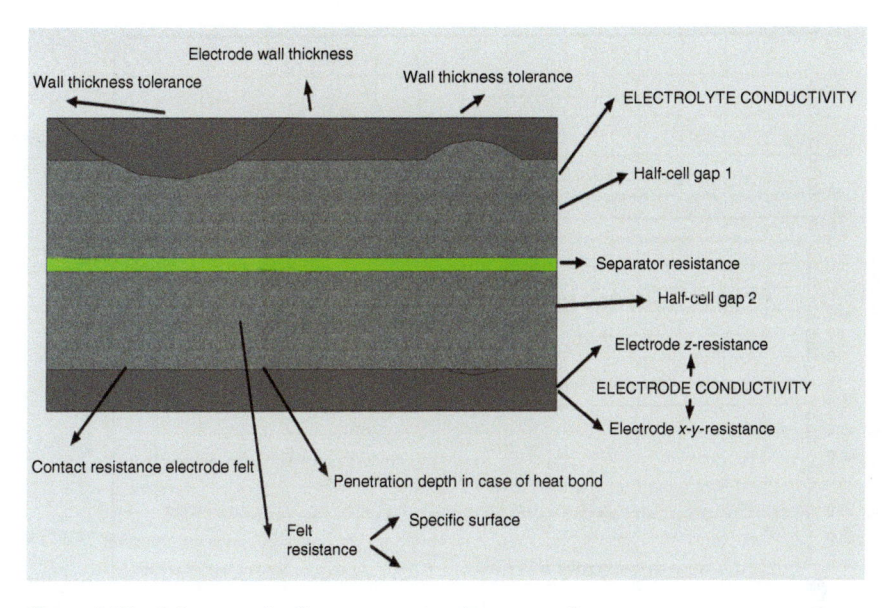

Figure 54.3 Influence of cell components on battery performance.

54.3.2 Operation

Among the many issues belonging to this group, the charge/discharge procedure is of **utmost importance**.

54.3.2.1 Stripping After Every Cycle (SAEC)

After stripping no zinc–metal is present in the stacks. This means that every problem with precipitated zinc created during the cycles before is eliminated. Stripping by that can be seen as a "healing"-cycle. It has to be emphasized that reports in the literature about a very high number of lifetime cycles predominantly are based on stripping after every cycle (SAEC).

Figure 54.4 shows charge/discharge curves for cycle numbers 28 381 and 641. Cycling mode was SAEC. The stable performance is impressive.

54.3.2.2 Random Cycling (RC)

The battery is charged and discharged according to the requirements for usage. This is the standard procedure for the EV application. Despite it is necessary to perform from time to time a very thorough stripping cycle of extra-long duration. The more imbalanced charge cycles have accumulated, the more is the amount of imbalance in zinc precipitation, and the greater is the probability of reversed cells, which is a result of partial overdischarge of individual cells. The problem can be mitigated by the addition of a surplus of bromine to the electrolyte, which contributes to the erasion of precipitated zinc within the "wrong" half cell. The electrochemical stripping, in this case, is supported by the chemical corrosion of zinc by bromine.

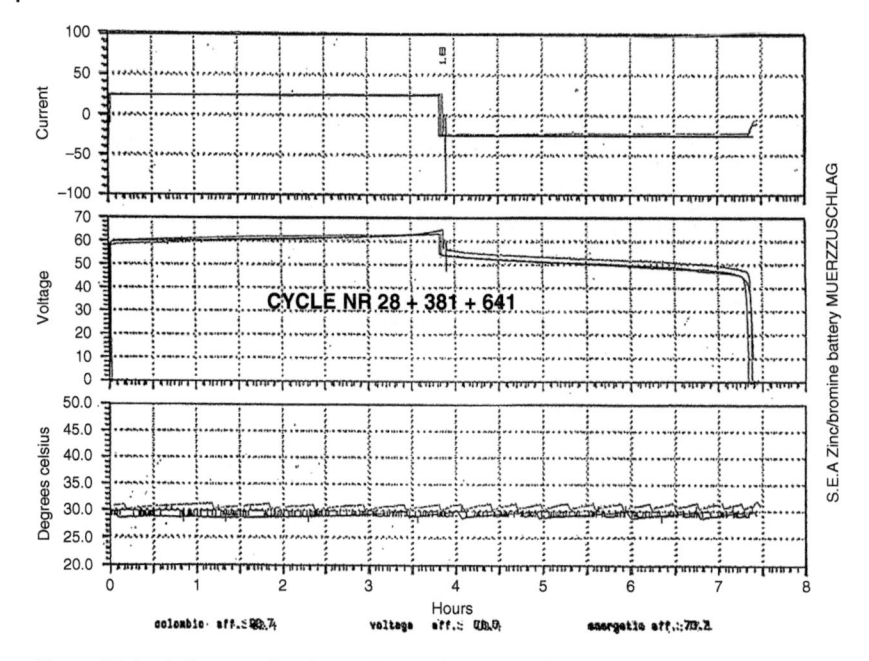

Figure 54.4 Influence of cycle-number on battery performance.

It is obvious that the geometric quality of the stack cells defines the number of RC that can be made until stripping is required.

Experience of SEA has shown that up to 200 RC procedures are possible without SAEC.

54.3.2.3 Standby

When stopping the pumps about 80% of the bromine is stored in the bromine reservoir and 20% is remaining in the stacks. This bromine allows immediate current draw but is also a self-discharge potential. To avoid too great voltaic losses, the stack bromine has to be "washed-out" from the stack before pumps are stopped. SEA developed a procedure to reduce the bromine content to less than 5% by a combination of pump stop and settling periods. Through this procedure the increase of the electrolyte temperature within the stack is small great loss in capacity is avoided.

54.3.2.4 Shunt Current Interruption

Whenever the battery is showing voltage, shunt currents (SC) occur within the stacks. SC is responsible for zinc maldistribution in the stack cells. SEA developed a procedure to empty the manifolds during standby periods resulting in significant lower voltage drop at the stack during standby periods. Figure 54.5 shows the voltage drop for a stack with and without SC interruption.

54.3.2.5 State-of-Charge (SOC) Measurement

Figure 54.6 a relation between the open-circuit voltage (OCV) and the state of charge (SOC), which is defined by the Nernst equation.

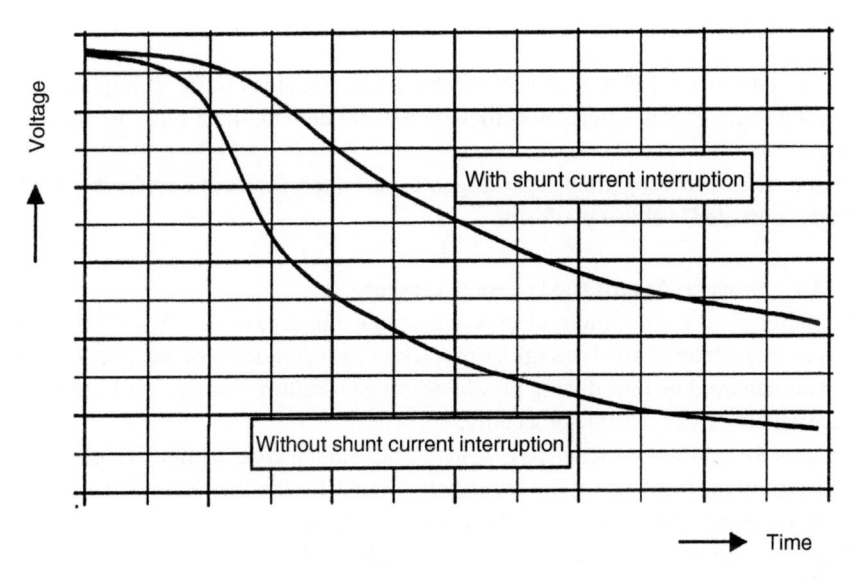

Figure 54.5 Effect of shunt current interruption.

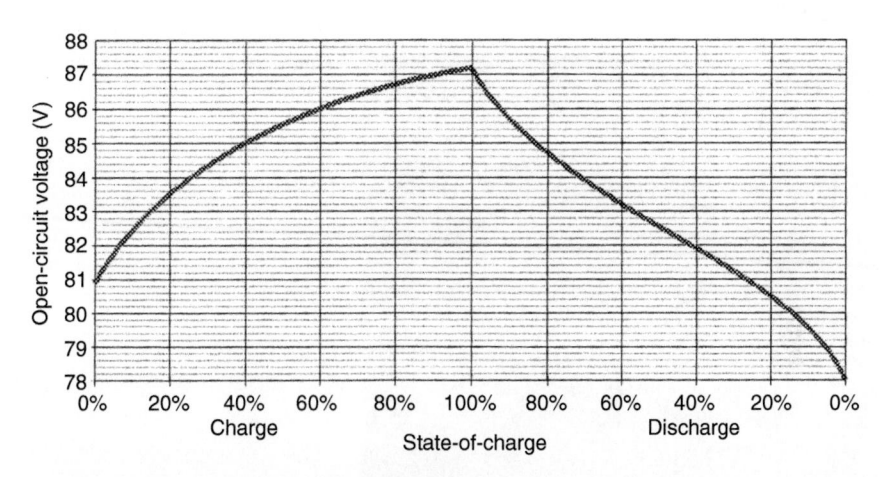

Figure 54.6 State of charge measurement.

Through this, the measurement of the SOC is possible by measuring the OCV. This is true for a specific electrolyte composition. The relation has to be calibrated from time to time using the fully stripped battery as the base for 0% SOC.

54.3.2.6 Thermal Management

As a thumb rule, it is said that corrosion doubles per a 10° C increase in the electrolyte temperature. The standard operation temperature for the ZBB is between about 15 and 40°C. Operation of the battery at temperatures up to 60° is possible but due to the usage of PE components, detrimental life-limiting effects can occur.

The heat exchanger positioned within the zinc reservoir can be designed to allow heat dissipation for all EV applications.

54.3.3 Safety

As it is well-known, safety under all thinkable situations is of utmost importance for mobile application. Efforts to achieve safety are demanding a huge amount of money.

SEA restricted investigations regarding safety to battery issues and preferably used already "admitted" vehicle platforms.

54.3.3.1 Measures Against and in Case of Leakage

The main task at start of the project was to create a battery design that allowed to verify a system **free of any leakage** from stacks, reservoirs, and loop periphery. This goal was achieved by introducing plastic-welding procedures for any working step, where welding was a feasible procedure.

Figures 54.7–54.9 show front-face-welded stacks and a cross section at right. The bent shape of the front face at the cross section indicates the shrinkage that guarantees strong compression after attachment of the endplates.

Figures 54.10 and 54.11 show a welded double-stack tower (2 + 2) without and with external manifolds. Stack towers can be connected electrically in series as well as in parallel.

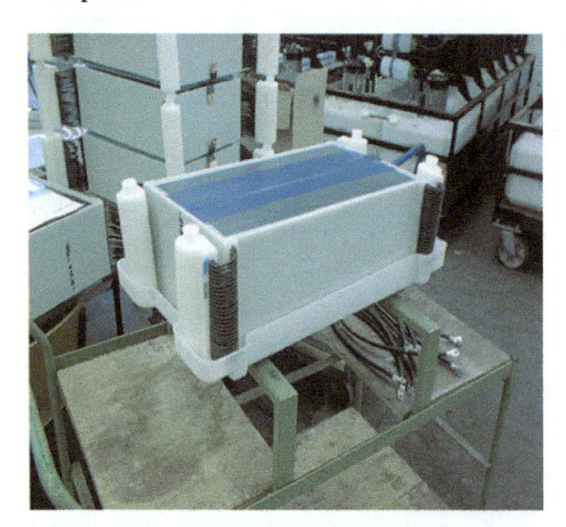

Figure 54.7 Front-face welded stack with manifolds.

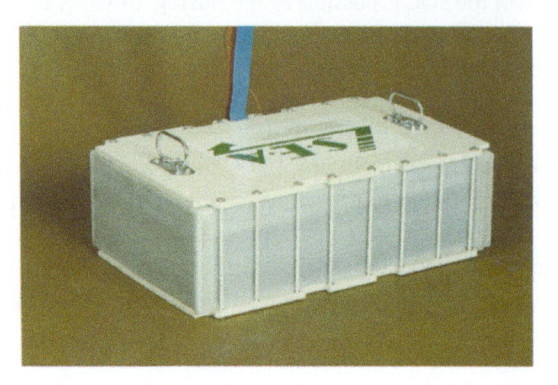

Figure 54.8 Front-face welded stack with end-blocks.

Figure 54.9 Cross-section through front-face welded stack.

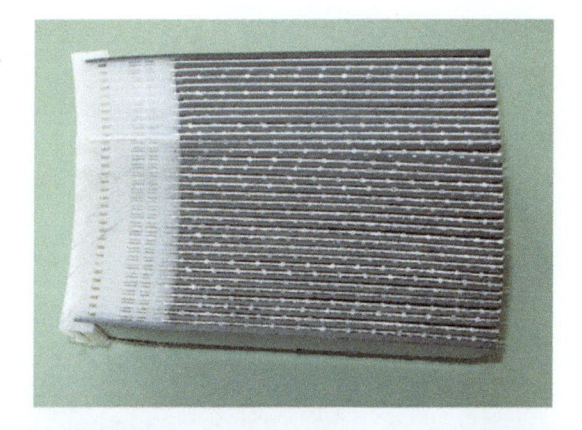

Figure 54.10 Fully assembled double-stack towers.

Figure 54.11 Stack towers with attached black heat-exchangers.

Figures 54.12 and 54.13 show seamless rotational sintered reservoirs with machined openings for pump housing, loop- and heat-exchanger connections. Sintered-low-density materials require a wall thickness of about 6 mm to avoid bromine diffusion.

Figure 54.14 shows the tubular heat exchanger within the zinc-reservoir and in Figure 54.15 the reservoir internal spider tubing, a measure to reduce shunt currents significantly. Very efficient heat exchangers are provided with more than 2000 m tubing length.

Figure 54.12 Low-profile reservoirs with internal heat exchanger.

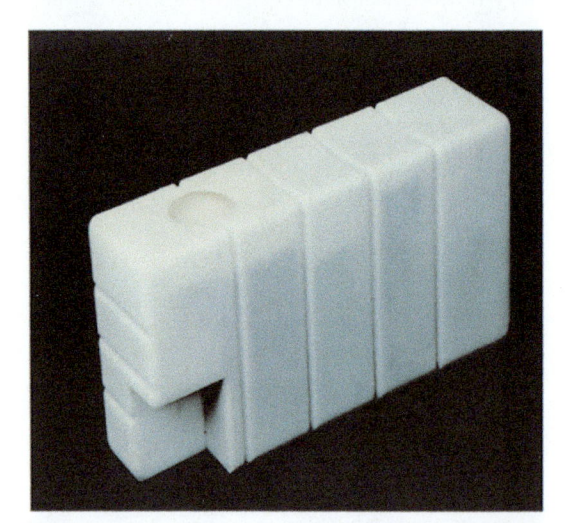

Figure 54.13 High profile zinc-reservoir.

Figure 54.14 Zinc-reservoir with internal heat exchanger.

Figure 54.15 ZBB-reservoirs with internal spider tubing.

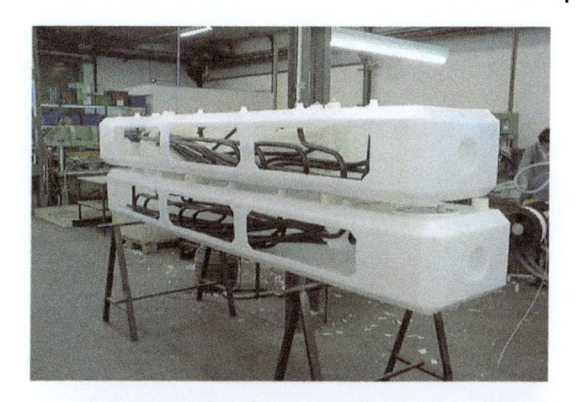

Figure 54.16 Rubber-type manifold-to manifold-connector.

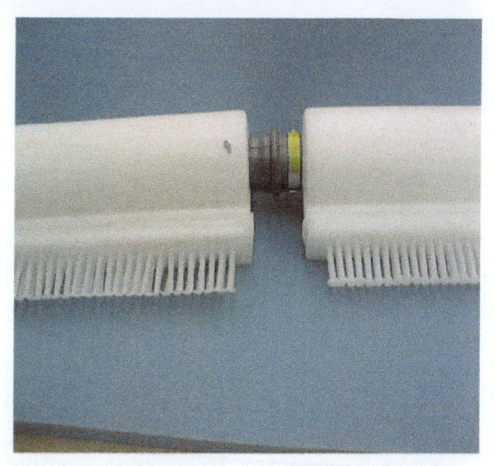

Figure 54.16 shows a remaining possible source of leakage where welding is not possible because of the requirement to create a detachable connection. At left stack, manifolds are connected by pushing elastomer tubes over the tubular ends to be connected. The same is true for the connections between stacks and manifold channels. In both cases, the connections are found to be extremely reliable and there is no reason to change the solution all over the years. A great advantage is the flexibility of the connection system.

At right stack, a more advanced version combines the elastomer tubes into a single unit.

The solutions shown in Figures 54.16–54.18 allow to fulfill the performance requirement to produce and assemble leakage-free batteries independent of the battery type and battery size.

54.3.3.2 Crash Test

Freedom of leakage does not mean that the battery is suited and certified to be used in public traffic. Admission to public traffic demands a lot of further test procedures where the crash test is certainly the most important.

Figure 54.21 shows crash tests performed by using a rig that allowed to release a battery from a height of 11.7 m (achieving 52.9 km h^{-1}). The battery used was a

Figure 54.17 Rubber-type u-tube manifold to stack connector.

Figure 54.18 Rubber-type multiple connection bars.

Figure 54.19 Low profile battery in Hotzenblitz-EV.

Figure 54.20 Hotzenblitz reservoir under pressure test.

Figure 54.21 Rig to perform Hotzenblitz crash test.

15 kWh/144 V low-profile zinc–bromine battery, which was tested in the Hotzenblitz car (Figure 54.19). Figures 54.19 and 54.20 show the battery assembled in a container and a reservoir under pressure test. In a first step, the battery was tested by crashing into a glass foam container experiencing about 60–70 g damping force. To adjust this deceleration, a steel leg with a plate of defined area was assembled to the container. In a second step, the same battery was falling without damping to bottom experiencing more than 1000 g. After the damped fall, the battery could be charged/discharged without leakages; in the undamped case, the reservoir broke at welding seams and could not be used anymore.

54.3.3.3 Vibration Test
During mobile application, the battery is under steady influence of vibrations and shocks. To investigate a possible influence on the stability of materials and of welding seams, a ZBB was tested by two methods, which are as follows:

- **Sinus pulses**: Frequency 22 Hz, acceleration 3 g, duration two hour
- **Shock**: Half-sinus, vertical axis, 6 ms, acceleration 23 g 1000 shocks

No optical changes were experienced. No influence on battery performance was found.

Cross section through the short after cleaning from the carbonized part

Figure 54.22 Cross section through stack after battery short.

54.3.3.4 Electrical Short

The consequences of an artificially produced battery short for a ZBB were tested by investigating the electrolyte temperature and optical changes of battery components. A battery with a 32-cell stack was fully charged, and 40 cycles using the SAEC mode were performed. After stopping the pumps and without washout of the stack, a hard short was verified through contacting the poles. A current of more than 400 A could be measured and the electrolyte temperature increased to 120 °C.

The safety-relevant observation is shown in the following photo (Figure 54.22):

After disassembly of the endplates and after cleaning of the carbonized area on top of the stack a crater shaped hole was identified. The very interesting experience: Due to the formation of a molten cylinder area around the crater, no spill of electrolyte occurred.

Considering the fact that stack electrolyte is only a small portion of the total electrolyte, this non-spill behavior indicated that a short is not an extremely dangerous event.

54.3.3.5 Hydrogen Recombination

Every battery having aqueous electrolyte produces hydrogen during charge. The amount of hydrogen is influenced by the charge procedure and/or by ions, which reduce hydrogen overvoltage. To avoid a dangerous hydrogen accumulation, for instance, in garages, a hydrogen recombinator based on platinized carbon cloth was developed. The recombinator transfers hydrogen into water.

Further safety-relevant investigations are as follows:

Fire and explosion hazards
Absorption of spilled electrolyte
Evaporation and atmospheric dispersion of electrolyte components
Electrolyte disposal
Recycling of electrolyte

54.4 Testing of Electric Vehicles

Battery developers are used to test batteries in the laboratory by fully automated equipment all over the year without interruption. For this purpose, test stands are equipped with all technical means necessary to charge, discharge, operate the batteries in all operation modes, and to control and monitor every event around the clock.

Testing of EVs is more challenging because it is necessary to have all necessary equipment on board. A self-consistent battery controller is responsible for battery operation as well as for the handling of safety-relevant situations.

54.4.1 Controller

Figure 54.23 shows the final result of the controller development at ELIN using the SPS-Eldatic 1000. In a second step, the controller was miniaturized. All operational and safety-relevant requirements (650 variables) were considered.

54.4.2 Basic Design at Project Start

- Vertical stack(s) between two reservoirs.
- Fully welded stacks with external channel tubing.
- Box that partially covers tubing allows usage as heat exchanger. Number of stacks between the reservoirs or number of cells per stack is chosen according to specified voltage (Figures 54.24–54.26).
- Active electrode area 1200 cm^2.
- Capacity per 8 cell stack = 1.25 kWh, per 32 cell stack 5 kWh.
- Disadvantage of vertical cells: Bromine accumulates at bottom of cell and prevents uniform mixture of complex and aqueous phase.

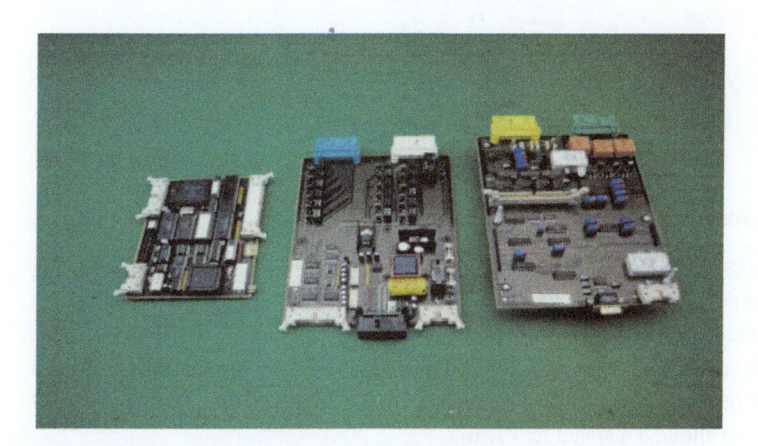

Figure 54.23 Battery controller developed by ELIN.

Figure 54.24 Early vertical stack design study at SEA.

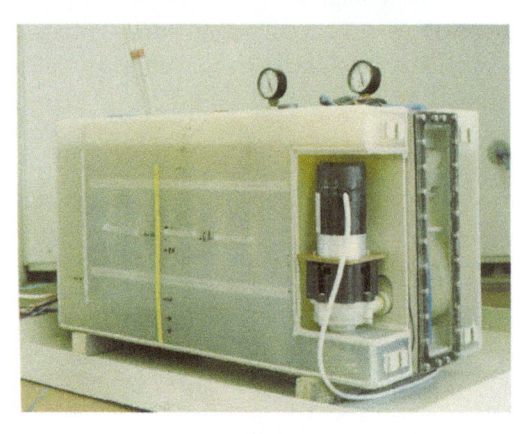

Figure 54.25 First SEA-battery realized according to Figure 54.24.

54.4.3 Colenta Test Vehicle

On May 5,1985, the first zinc–bromine battery was installed worldwide into the test vehicle shown in Figure 54.72. The 4 kWh/48 V battery shown in Figures 54.28 and 54.29 showed good performance regarding its battery functions, early problems were experienced purely from the electrotechnical side. The vehicle was also used to install and test a 2× (3 kWh/36 V) ZBB allowing a range of up to 80 km. A 2 kW electrical motor allowed a speed of 30 km h^{-1}.

Despite the restricted speed, the vehicle was used for a long time for everyday requirements of the company. The total driving distance was about 8000 km.

Figure 54.26 First double vertical stack battery realized at SEA.

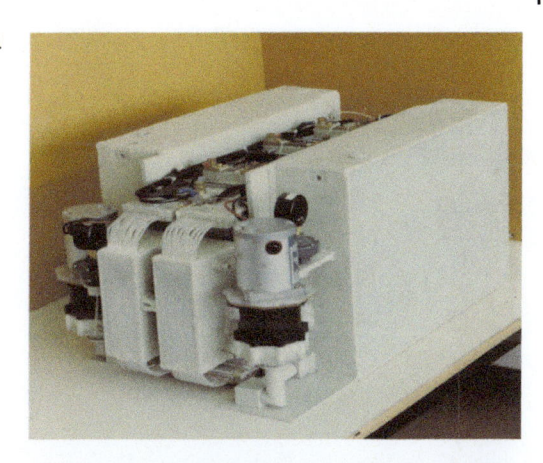

Figure 54.27 Front view of Colenta test vehicle.

Figure 54.28 Back view of Colenta test vehicle.

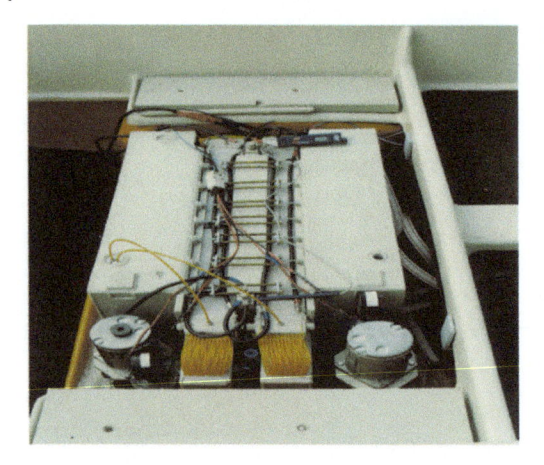

Figure 54.29 Battery in Colenta test vehicle.

Figure 54.30 Golf City Stromer-test vehicle.

54.4.4 City Stromer

The City Stromer was an EV-prototype version of the VW Golf (Figure 54.30).
The following batteries were installed into the City Stromer:

Figure 54.31: 20 kWh/108 V ZBB provided by EXXON.
Figure 54.32: 16 kWh/110 V ZBB with L-shaped reservoirs.
Figure 54.33: 16 kWh/110 V ZBB with tubular reservoirs and horizontal pumps.

All batteries for the City Stromer showed poor Coulombic performance and only a few test runs could be made. The EXXON battery showed several leaks and had to be assembled very soon.

The two SEA batteries showed a bad Coulomb performance with elevated electrolyte temperature as a consequence. The reason for this low performance was assumed to be the result of low-quality separators.

The negative experience with the batteries for the City Stromer initiated a refurbishment of the battery design.

Replacement of the Daramic separator by the Japanese ASAHI separator: Testing of the ASAHI separators showed up to 93% Coulomb efficiency due to hydrophilic

Figure 54.31 Golf equipped with 20 kWh EXXON battery.

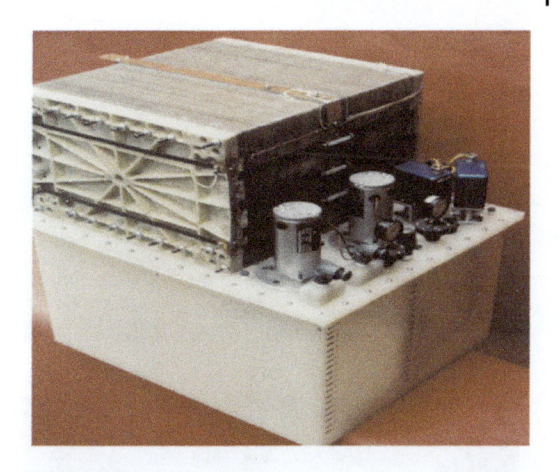

Figure 54.32 16 kWh-battery with L-shaped reservoirs.

Figure 54.33 16 kWh battery with tubular reservoirs.

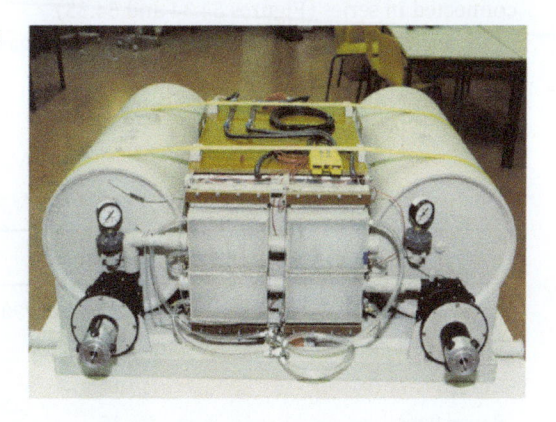

surface properties. This resulted in a significant lower heat loss during battery operation.

Second advantage: Complex phase is not soaked by the hydrophilic separator; easy washout from cell is possible.

Third advantage: Easy washout allows usage of horizontal stacks. No 4-way valve is necessary.

The decision was made to not use vertical stacks anymore.

Figure 54.34 Mini-EL-City equipped with ZBB and lead acid batt.

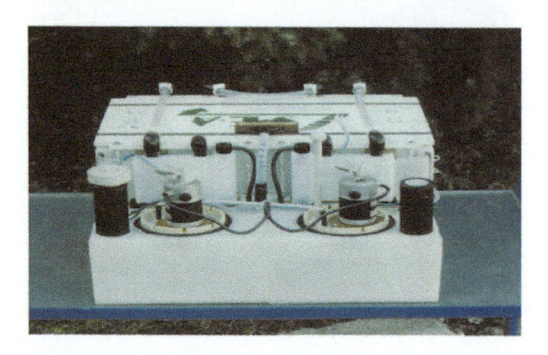

Figure 54.35 5 kWh/49 V ZBB for Mini El City.

54.4.5 Mini-El City

As a standard, the mini-el is equipped with a lead–acid battery, 3 pieces 12 V/90 Ah connected in series (Figures 54.34 and 54.35).

To install a 5 kWh/49 V ZBB, the following measures were necessary:

– Installation of a 48 V controller
– Installation of a cooling loop

Comparison of the MINI-EL city performance

	Lead–acid	Zinc–bromine
Battery	3 pcs 12 V/90 Ah	1 pcs 5 kWh/48 V
Weight (kg)	94	87
Charge Time (h)	14	4.5
Range (km)	35	120
Speed average (km h^{-1})	33	33
Speed top (km h^{-1})	40	55
Consumption (batt) kWh	3.57	3.75
Consumption (grid) kWh	7.14	6.66

The performance difference was convincing. Considering the significant lower charging time of the ZBB means the practical range is a multiple of that of the lead–acid range.

54.4.6 Colenta Mini Cab Bus

Comparison of Colenta bus performance

	Lead–acid	Zinc–bromine
Battery	8 pcs 12 V/110 Ah	1 pcs 20 kWh/96 V
Weight (kg)	308	330
Charge Time (h)	12	5
Range (km)	1.run 67 2.run 53	1.run 172 2.run 187
Speed average (km h^{-1})	38	42
Speed top (km h^{-1})	83	82
Consumption (batt) kWh	10.8	10.9
Consumption (grid) kWh	19.6	19.5

The Colenta bus was delivered with the standard series-connected lead–acid gel batteries (8 pcs 12 V/110 Ah//96 V) (Figures 54.36 and 54.37).

Total battery weight 308 kg.

Charge time 12 hours, about 12 kWh energy from the grid.

Test runs with the lead–acid battery were done in the vicinity of Mürzzuschlag.

A 20 kWh/96 V ZBB was installed into the Colenta bus. To enable battery operation, a thermal management system (identical to those installed in standard cars) had to be assembled.

The lead–acid battery showed a little bit better power behavior at begin of discharge, which is typical for lead–acid. After some km, the ZBB showed better power performance. Voltage drop of LA to 40 V required stop of test to avoid detrimental

Figure 54.36 Colenta mini cab bus.

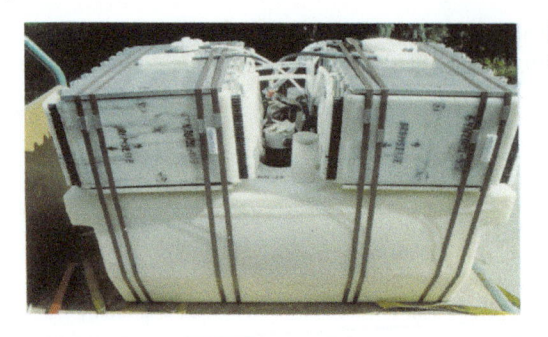

Figure 54.37 ZBB for Colenta mini cab bus.

effects. As it was experienced in other tests, the range of LA decreases dramatically cycle per cycle if too low voltage is allowed.

54.4.7 Hotzenblitz Commuter Car

The 15 kWh/144 V battery was the first low-profile ZBB foreseen to be assembled into a car platform underneath the seats of the driver. Special measures had to be made to install thermal management and to allow to apply the shunt current interruption principle during pump-off periods. The car was equipped with an AC asynchronous 16 kW motor and achieved a range of 150 km h^{-1} and a top speed of about 120 km h^{-1} (Figures 54.38–54.40).

As described in Section 54.3.3.2, this battery type was crash tested and showed stable performance after the test at 60–70 g average.

54.4.8 ELIN-VW Bus

The ELIN-VW bus was equipped with two series-connected ZBB of 22.5 kWh/216 V each.

Maximum charging voltage: 288 V. Maximum discharge current 200 A. Total weight 740 kg.

Advanced free programmable ELIN-Eldatic controller. Charger: maximum 40 kW.

Regenerative breaking was allowed to support the standard car breaking procedure. This feature was experienced to be very comfortable and allowed to reduce the

Figure 54.38 Hotzenblitz EV.

Figure 54.39 Hotzenblitz EV cross-section with low profile ZBB.

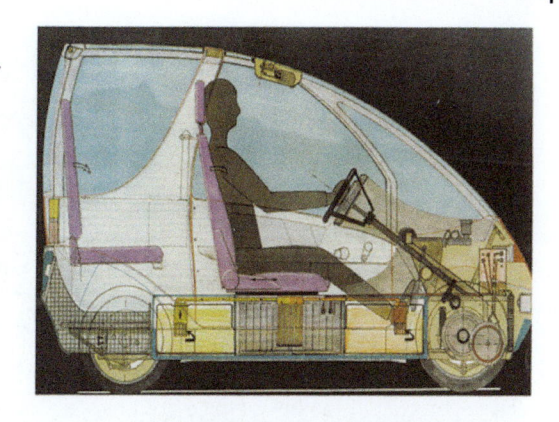

Figure 54.40 Hotzenblitz low-profile ZBB.

speed almost to zero. Due to the enhanced car weight, the brakes of standard vehicles used as EVs, in general, have to be seen as a weak point of the system.

The bus was tested at SEA for about a month and then was operated for another month by the Austrian Post Services to deliver parcels to customers. After this activity, the bus was tested in Vienna City by SEA for a further month without experiencing a major difference in performance. The total distance driven during this test period was about 3000 km, and the average consumption was about 600 Wh km^{-1} (Figures 54.41–54.43).

54.4.9 Fiat Panda

Operation and testing of the Fiat Panda was a high-level challenge because all activities were done in cooperation with the Austrian Verbundgesellschaft (Austrian Electricity Board [AEB]). SEA was providing the ZBB and defined the requirements for the battery operation. AEB provided the drive train and the motor controller (Figure 54.44).

At begin of the project, SEA did a lot of test runs and realized improvements, AEB verified the requirements defined by SEA. Then, only AEB tested the Panda.

Figure 54.41 ELIN-VW-bus with 2 × 22.5 kWh/216 V ZBB.

Figure 54.42 ZBB assembled in ELIN-VW-bus.

For battery developers, the standard case is testing the batteries in the laboratory. The second stage is testing the battery in a mobile application without the support of the lab equipment. The third most challenging step is to transfer the battery to a partner without having direct access to the battery anymore.

During the first test phase, a 15 kWh/72 V ZBB was assembled into the Panda. Finally, this battery was replaced by a 20 kWh/96 V ZBB (Figure 54.45). To save space in the car's inner, the sheet of the trunk partially was cut out to allow to lower down the battery to the lowest possible level (Figure 54.46). Parallel to the battery-relevant adaptations, a great lot of work was done by AEB to optimize the drive train, the controller, the onboard charger, the DC/DC converter for the pumps,

54.4 Testing of Electric Vehicles

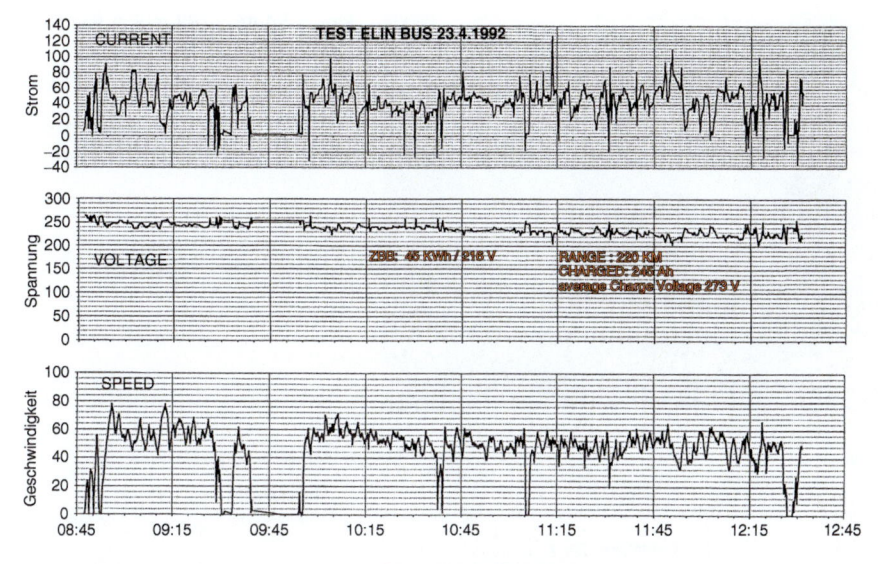

Figure 54.43 Performance data for ZBB in ELIN-VW-bus.

Figure 54.44 Fiat Panda during solar race.

Figure 54.45 15 kWh/72 V ZBB assembled in Fiat Panda.

Figure 54.46 ZBB in Fiat Panda, view from underneath.

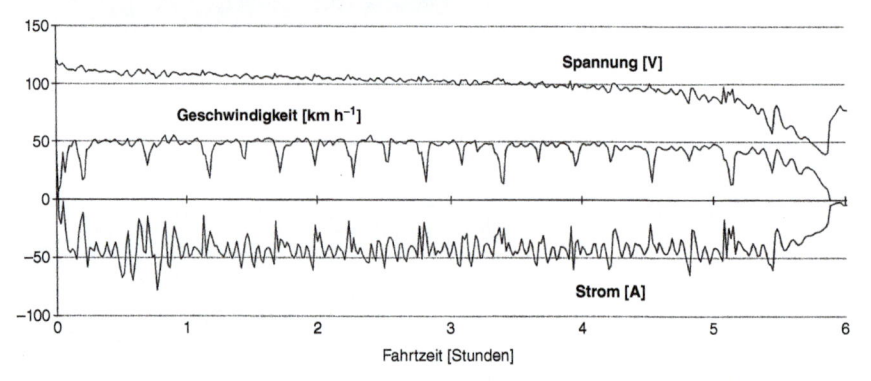

Figure 54.47 Voltage, speed, and current during Panda test-run.

and the management of battery operation. Testing in real traffic was done regularly, and due to the high range (standard was 150–200 km per charge), about 25 000 km was achieved in a short period.

The diagram in figure 54.47 shows the data for a record test runs where a range of 262 km at an average speed of 50 km h^{-1} was achieved with the Panda of SEA. Thinking of a charge time of about three hours means that the range is not a real restriction for the usage of an EV.

Aside from the standard on-road tests, AEB together with SEA participated in several special events, the most important are as follows:

- Austro Solar crossing Grossglockner High Alpine Road (2571 m) 1990
- Austro Solar 1992,
- Austro Solar 1994

54.4.9.1　12 Electric Hours of Namur 1991

In contrast to most test runs with EVs, which are performed in low traffic areas to prove a large range, the electric hours of Namur were planned and performed

Ergebnis: 12 Stunden von Namur
27. / 28. September 1991

N.	Fahrzeug	Kat.	Chauffeur	Batterie	Charge	Gewicht	Runden	Fahrtzeit	Stehzeit	km-12 h	V Durch.	Bemerkungen
17	MTC VIDYUT	A	Meulenberghs	Pb	E	160	57	11:52	0:07	242	20.4	Moped mit Austauschbatterie
14	Hercules-TU Berlin	A	Rochlitz	NiCd	E	105	25	6:21	5:38	104	16.32	Fahrrad mit Austauschbatterie
15	NEC MELEX	A	Guidot	Pb	N	370	37	10:27	1:32	157	14.98	Golf Fahrzeuge
9	NEC MELEX	A	Deckers	Pb	N	370	36	10:38	1:21	153	14.41	
8	NEC MELEX	A	Remacle	Pb	N	370	25	7:24	4:35	105	14.21	
22	VEHICULE DIESEL	B	Landrain	DIESEL	--	1800	60	11:57	0:02	256	21.41	Konventionelle Fahrzeuge
21	VEHICULE A ESSENCE	B	Colignon	ESSENCE	--	845	59	11:44	0:15	254	21.61	zum Vergleich
19	SCHOLL Optima	B	Scholl	NiCd-Pb	E	670	63	11:46	0:13	267	22.66	Austauschbatterie
4	MTC SURVA	B	Knobloch	Pb	E	1330	61	11:35	0:24	256	22.13	austauschbarer Batterieanhänger
16	ASNE VW Transporter	B	Jacobs	NaS	N	2200	61	11:59	0:00	258	21.54	Pritschenwagen, 2 Stk B11 56 kWh
6	SEA Fiat panda	B	Schauer	ZNBr	N	1060	59	11:42	0:17	253	21.66	ZnBr 96 V / 15 kWh
2	ELCAT	B	Ryynänen	Pb	N	1250	58	12:00	0:00	246	20.47	Blei-Starter 72 V 18 kWh
7	VOLTA	B	Trehello	NiCd	N	1120	57	10:53	1:06	244	22.38	
20	Renault Master	B	Chaidron	NiCd	N	3070	54	10:19	1:40	230	22.30	
1	VOLTA	B	Fouchier	Pb	N	1270	43	9:07	2:52	184	20.21	
10	MTC SURVA	B	Van der Linden	Pb	N	1170	41	8:43	3:16	175	20.08	wie 04 jedoch ohne Austausch
3	Fridez Pinguin 4	B	Nagel	Pb	N	770	40	8:45	3:14	172	19.62	Pb - Gel - Batterie
12	Fridez Tavria	B	Guisse	Pb	N	1185	37	8:08	3:51	159	19.53	Rußland-Ungarn-Schweiz
5	VUB PGE TM	B	Minnaert	Pb	N	1470	35	7:27	4:32	150	20.09	Universität Brüssel
11	VUB PGE 3P	B	Van Muylem	Pb	N	1170	33	6:52	5:07	140	20.40	
18	PASQUINI BOXEL	B	Van Mierlo	Pb	N	1070	4	0:54	11:05	17	19.09	
23	ALTROBUS	C	Cecchetti	Pb	N	13000	43	--	--	186	--	

Figure 54.48 Results for "12 Electric hours of Namur 1991."

to find out the range of an EV that is driving along with a predefined round course within the urban area of Namur. The rules to follow were very simple (Figure 54.48):

> -find out the achievable range when driving during two predefined 6-hour periods on two following days. Charging can be performed at any time, at any place with any selected charger. It is even allowed to exchange the batteries, either by a real exchange or by exchanging a trailer, which is equipped with new fully charged batteries.

After the two six-hour periods, the total range is measured and published.

As a very special feature, two cars with combustion engines were allowed to participate under otherwise identical conditions.

The result was extremely astonishing and remarkable:

- in urban traffic, the maximum range is practically independent of the vehicle if sufficient energy is provided. The average speed is in the range of 21 km h^{-1}.
- this statement includes the vehicles with combustion engines!!

Remark: The Panda (Pos 6) was 17 minutes charged on the first day, no charge on the second day.

Experiences with 20 kWh Zinc Bromine Battery Schauer Gerd 26th ISATA, Sept.1993, Aachen 93 EL 019

Figure 54.49 Space frame of a Daewoo-EV.

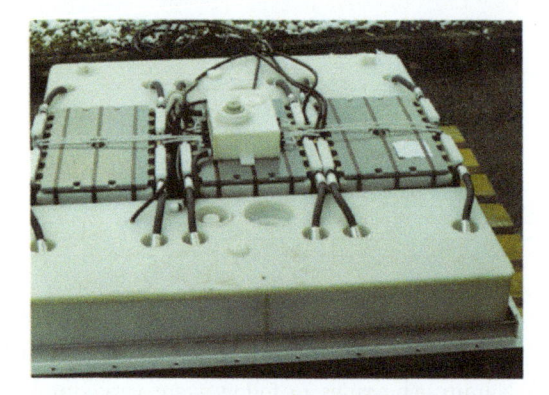

Figure 54.50 Low profile ZBB for Daewoo.

Figure 54.51 ZBB assembled in Daewoo space frame.

54.4.10 DAEWOO Electric Vehicle

The space frame of a DAEWOO EV design study was delivered from Boston to SEA Mürzzuschlag to be equipped with a ZBB, and in addition, the electric motor which had been delivered from Great Britain had to be installed.

The A 33 kWh/288 V low-profile battery with integrated shunt current integration system was installed. As can be seen in the figures 54.49, 54.50 and 54.51, the battery did not restrict the passenger cabin so that all standard installations could be done without additional measures.

Testing was not performed by SEA; the vehicle was delivered immediately back to the USA. No written data about the performance of the vehicle are available.

Figure 54.52 35 kWh/380 V-ZBB in Geo Prism sedan California.

Figure 54.53 Geo PRISM sedan ready for race in Phoenix.

Figure 54.54 Geo PRISM sedan: view from the backside.

54.4.11 Geo PRISM Sedan: 35 kWh Battery Delivery

A 35 kWh 380 V ZBB was delivered to the UC-Davis University/CA. The integration work for the battery was carried out by Dr. David Swan, Institute of Transportations Studies, University of California at Davis (Figures 54.52–54.54).

The UC-Davis University team participated in several races with great success:

- Third in the Phoenix race with 180 km in two hours. Outperformed by far all EVs equipped with lead–acid batteries.
- First in the clean air rally 1994, achieving 690 of 700 points outperforming even natural gas-operated cars.
- Fourth in the race between New York and Philadelphia 1995, achieving a maximum range of 270 km per charge.[1]

---

1 Design and Vehicle Integration of an advanced Zinc Bromine Battery David H. Swan and J.T. Guerin University of California, Davis.

54.5 Summary

Chapter 54 describes the R&D activities of the Austrian Company SEA during the period from 1983 to 2003 to develop the zinc–bromine battery system, especially regarding the EV application. It is shown that the EV application of the battery demands a lot of additional measures compared to the stationary energy-storage. This is true for the battery operation but especially true regarding the guarantee of safety in traffic.

Several alternative battery developers also tried to enter the new EV-market but after a short time, they had to accept that the lead–acid battery system was the only one that was available and was accepted despite showing not the best performance. The NiCd battery was banned for toxicity reasons, and the high-temperature batteries showed too high thermal losses.

For the zinc–bromine battery, this had the consequence that the lead–acid traction battery was the measure for competitiveness. The high-energy density, design flexibility, and fast charge ability finally proved the superiority of the ZBB. The "lead–acid range" of 60–70 km could be extended to a "ZBB range" of 250 km+.

The chapter describes the test results for several vehicles beginning from a simple three wheeler platform up to a Geo PRIZM sedan with a 35 kWh/380 V ZBB. Wherever it was possible, the tests were made using the original lead–acid batteries at first and comparing the data after installation of a ZBB. The results were convincing regarding ZBB performance.

The 12 hours of Namur, which included participants using combustion engine cars, resulted in the remarkable knowledge that the "possible" range in urban areas was "**practically independent**" of the car type. The average speed was found to be in the range between 22 and 25 km h^{-1}. If this result of the year 1991 is valid up to now then this would mean that the automotive industry produces the "wrong" vehicles for most places of the world. An application adapted fleet of small commuter cars with a 5 kW drive train based on a flat-rate system together with train and bus would dramatically decrease present traffic (and global warming) problems.

As a final remark, it has to be mentioned that Li battery systems, which did not play a role during the EV-activities of SEA, have been developed to such superior systems that (perhaps with the exception of the NiMeH battery) all other battery candidates of the project time do not play a major role. All existing potential ZBB developers are focusing their development goals on the fabrication of large energy-storing units.

Index

a

absolute vacuum scale (AVS)   101
acetonitrile   46, 829, 855, 856, 984, 1117
acid-base titration   309
acid-doping activation   136
AC power systems   4
AC resistance   310
activation barrier   230, 1116
activation overpotential   62, 385, 386, 457
all-iron flow battery (IFB)   792, 794, 795
   economics   803–507
   electrolyte factors   796–800
   hybrid flow battery performance   807–809
   membrane factors   800–803
all-vanadium flow battery (VFB), (G1)   675–677
all-vanadium redox cell patent   509–510
all-vanadium redox flow battery (VFB)   463
alternating current cyclic voltammetry (ACCV)   252, 254, 255
aminoxyl anion   1117
aminoxyl radicals   1010
amphoteric ion exchange membranes (AIEM)   131, 139, 594–595
anion exchange membranes (AEMs)   76, 137, 139, 309, 320, 326, 593–594, 643, 801
Annual Usable KWh   1142
antibonding states   294
antimalarial drug   985
anti-Stokes scattering   284

aqueous organic flow batteries (AOFBs)   897
   advantages   898–900
   aza-/azo-aromatics   910
   cost   905–906
   crossover   901
   lifetime   901–903
   membrane   906–907
   metal coordination complexes   911–912
   molecular engineering   904–905
   nitroxide radicals   910
   performance   915–918
   pH imbalance   907
   properties
      fade rate   913
      general decomposition mechanisms   913
      operation pH   913–914
      redox active molecules   915
      redox potential   912
      solubility   912–913
   redox aromatic carbonyl compounds   908–909
   solubility   900
   toxicity   907–908
   viscosity   900–901
area specific resistance   598
Argand diagram   246
aromatic backbone-based membranes   134–135
arrays of stacks   89

Flow Batteries: From Fundamentals to Applications, First Edition.
Edited by Christina Roth, Jens Noack, and Maria Skyllas-Kazacos.
© 2023 WILEY-VCH GmbH. Published 2023 by WILEY-VCH GmbH.

Index

atomic absorption spectroscopy (AAS) 521
atomic orbitals (AOs) 342
availability 1070
average oxidation state (AOS) 104
Avrami equation 113

b

battery control and management system
battery management system (BMS) 648
battery stacks 517
battery storage technologies 20
Beer's law 598
behind-the-meter (BTM) 84
benzathiadiazole (BTZ) 984
benzophenone (BP) 984
bimodal porosity distribution 265
bipolar plates 75, 471, 567
bipolar redox organic molecules 987
bipolar stack 713
birth to death cost cycle 71
Born–Oppenheimer approximation 338, 339
bottom-up modelling approach 472
boundary conditions
Bragg's law 216, 287
Bromine Complexing Agent (BCA) 1146, 1157
bromine reservoir 736
Butler–Volmer equation 62, 113, 230, 231, 357

c

capacity decay rate 92
capacity retention 324, 326
capital costs 472–473
capital expenditure (CAPEX) 71, 92
carbon-based electrodes
  activation treatment 752
  catalyst 753–754
  GF/CF/CP 749–752
carbon-based electrolytes 747
carbon black nanomaterials 609
carbon corrosion 268, 579, 580
carbon electrode
  electrolyte flow in 274–275
  injection of electrolyte 268–273
  material 269
carbon felt (CF) 749
carbon felt electrodes 88
carbon fibers 264
carbon functionalisation 290
carbon/graphite felts 564
carbon materials 281
  functionalisation of 288–290
  imaging techniques 297
  infrared spectroscopy 296–297
  Raman spectroscopy 283–286
  structure of 283
  surface area determination and porosity 298–299
  surface chemistry of 282, 288
  temperature-programmed desorption (TPD) 291–292
  TG/TGA 292–293
  thermal method 290–291
  TPR/TPO 292
  X-ray photoelectron spectroscopy 293–296
  X-ray powder diffraction (XRD) 286–288
carbon paper (CP) 565–566, 749
carbon particle size 282
carbon plastic electrodes 713
casting method 596
catalyst corrosion 698
catalyst dissolution 1147
cation exchange membranes (CEMs) 76, 132, 320, 326, 590–593, 643
  aromatic backbone-based membranes 134–135
  non-perfluorinated membranes 134
  perfluorinated membranes 132–134
  polyimide-based membranes 135–136
CellCube 1087
  innovation 1095
  installation 1092
cell geometry 721
cell overvoltages 55

cell potential applied 274, 275
cell-scale modelling
   battery performance models 383–384
   lifetime models
     capacity fade through membrane transport 389
     theoretical and practical capacity 388–389
   Ohmic resistance 385
   overpotentials and mass transport problems 385–386
   system efficiency modelling
     Coulombic efficiency 387–388
     pumping losses 387
     round-trip efficiency 388
cell-scale models 383
cell stacks 73, 75
cell voltages 21, 55, 60, 198, 199, 202, 469
   components of 202–204
centralised generation 4, 5
centrifugal pumps 76
cerium 819
   electrode materials 821–823
   positive electrode reaction 820–823
   redox reactions 820–821
charge/discharge cycling 264, 319–320, 580, 641–642
charge/discharge processes 355
charge transfer reaction 231
charging/discharging cycles 627
chemical hazards, VFBs 178–180
chemically regenerative fuel cells 36
chemical shift 216
chemical stability 129, 130, 599
chemical vapor deposition (CVD) 282, 298
Chinese national standards 171, 173
chronoamperometry (CA) 568
City Stromer 1188
CNC machining 88
coarse-grained molecular dynamics (CGMD) 361
cobalt 931
Colenta bus 1191

commercial membranes 1122
component level, RFB 470–471
compressed air energy storage (CAES) 19
compressed air to spin 18
computational flow dynamics (CFD) 87
computational methods 996
concentration imbalance 642–644
constant current density 264
constant phase element (CPE) 247
continuous electrolyser 158
continuum scale 381
conventional battery mode 102
conventional diffusion cell 128
copper-chloride species 858
copper chloro-complexes 855
copper deposition 837
copper hexacyanoferrate (CuHCF) 111
copper-lead 203
copper perchlorate-acetonitrile system 856
copper perchlorate salts 855
corrosion resistance 218
Cottrellian flux 242
Coulomb counting 636–637
Coulombic efficiency (CE) 55, 135, 544, 598–599, 607, 1163
cradle to cradle 662
cradle to gate 662
Crank–Nicolson method 236
critical governing theory 360
cross-over liquid 1146
cryogenic cycle 19
Cu/Cu Redox Flow Batteries (CuRFBs) 855, 856
   cell components 858
     electrodes 864–866
     electrolytes 858–864
     separators 866–868
   chemistry 856–858
   economy and sustainability 870–871
   stack configuration 868–870
current collectors 75
current efficiency (CE) 320–321

cyclic voltammetry (CV)   622, 838, 843
  diffusion domain approximation   237–240
  and electrochemical impedance spectroscopy (EIS)   221
  LSV at planar electrodes   233–235
  measurement   231–233
  Randles–Ševčík relations   233–235
  real-space simulation approach   240–242
  simulating strategies for   235–239
cyclic voltammogram   283, 542, 793

d

data-driven models
  algorithms   486
  computed data sets   490–491
  data   489
  experimental data sets   491
  finger prints   492
  quality and limitations   489
  representation   489
  solubility   496–499
  training data   490
D band   284
DC power systems   4
Debye–Hückel limiting law   59
decoupling HER and OER   104
degree of sulfonation (DS)   135
de-Levie transmission   248
density functional theory (DFT)   336, 339–343, 357
desorption temperature   291
1,4-dialkoxybenzene derivatives   980
dialysis membranes   1122
diaphragm-like films   30
dielectric continuum   358
differential electrochemical mass spectrometry (DEMS)   274
diffractogram   287
diffusion domain approximation   237–240
Directive 2006/66/EC   171
discharge capacity   324, 326
discontinuous process   158
discrete element method (DEM)   366
dissipative particles dynamics (DPD)
  methodology of   364–365
  RFB method   366
distributed generation   4, 5
distributed ohmic resistance   248
distribution of relaxation times (DRT) analysis   249–252
divinyl benzene   595
Donnan exclusion effect   137
Donnan exclusion principle   320
dual-flow circuit flow battery   101–105
dynamic stack thermal model   450–452

e

economic modelling methods   472
  capital cost   472–473
  total cost of storage   473–474
economic system definition   468
electrical double layer (EDL)   361, 362
electrical energy   54, 1027
electrical energy storage (EES)   54
electrical hazards, VFBs   180–187
electricity   3–4
  generation and system control   7
  storage   8–10
  time shifting   10
electricity markets   765
electricity storage   9
electric shock, risk of   183
electric vehicle (EV)   1185
  DAEWOO   1198–1199
  geo PRISM sedan   1199
electroactive materials   21
electrocatalytic properties   283
electrochemical analysis   231
electrochemical cell   103, 337, 513
electrochemical conversion step   55
electrochemical energy efficiency   91, 608
electrochemical energy storage (EES)   54, 69, 355, 674
electrochemical flow batteries   202
electrochemical impedance spectroscopy (EIS)   245, 310, 327–328, 568, 617

cyclic voltammetry (CV) and 221
distribution of relaxation times (DRT)
    analysis 249–252
  interpreting 246–247
  Kramers–Kronig relations 251–252
  normalization method 248–249
  Paasch model 248
  principles and advantages of 245–246
electrochemical level, RFB 469–470
electrochemically reversible 234
electrochemical methods 229
  advanced electroanalytical techniques 252–255
  cyclic voltammetry (CV) 231–245
  definitions 229–230
  electrochemical impedance spectroscopy (EIS) 245–252
electrochemical rate constant 574
electrochemical rebalancing 104
electrochemistry 54
electrode-development
  battery types 736
  collector 724
  handling of shunt current 730–734
  loop periphery 728–730
  reservoir 735–736
  separator 723
  stack welding 724–728
electrode-electrolyte interface 548
electrode kinetics 231–232
  in conventional three-electrode cells 570–571
  effects of electrochemical pretreatment 571–575
  electrochemical and thermal pretreatment 575–576
electrode materials 568
  morphology of 263–266
electrode overpotentials 55
electrode potentials 58, 60
electrode reactions 61, 66, 837
electrode warpage 721
electrolysers 156
electrolyte agitation 194
electrolyte balancing system 775
electrolyte circulation 838
electrolyte conductivity 545
electrolyte degradation 620
electrolyte flow 89
electrolyte flow rate 89, 645
electrolyte flowrate control 457–458
electrolyte health management system (EMS) 1131
electrolyte imbalance 641–642, 647
electrolyte imbibition 270
electrolyte management system (EMS) 773
electrolyte rebalancing 104
electrolytes recycle 74
electrolyte stability 552
electromagnetic compatibility (EMC) 180
electromotive force 90
electron-beam irradiation approach 140
electron density 341
electronic structure methods 339
electron stoichiometry 72
electron transfer coefficient 230
electron transfer process 356–360
electrospinning 566
electrostatic Coulomb forces 337
electrostatic interaction 344
electrotechnical field 160
ELIN-VW-Bus 1192
energy bandgap 99, 100
energy capacity 103
Energy Center™ (EC™) 1128
energy consumption rate 71
energy density 37
energy domain
  electrolyte 1145–1146
  hydrogen 1146
energy efficiency (EE) 55, 76, 135, 323–324, 545, 607
energy management 17, 18
Energy Research Corporation (ERC) 35, 1155
energy storage 9, 1088
  need for longer duration 16–18

energy storage (*contd.*)
  parameters for 14
  pumped hydro 18
energy storage cost (ESC) 1153
energy storage market 673
energy storage services 1091
energy storage system (ESS) 176, 1107, 1127
energy supply, electricity in 3–4
Energy Warehouse™ (EW™) 1128
Enerox 1086
equivalent circuit fitting (ECF) 246–247
ethylene tetrafluoroethylene (ETFE) 140
exothermic reactions 647
ex-situ analysis 218
ex-situ characterization methods
  ion-exchange capacity 307–310
  ion-exchange membranes, ion permeability of 311
  ionogenic groups, ion-conductivity of 310–311
  mechanical membrane properties 317
  membrane weight loss 311–313
  microscopical membrane characterization 317–318
  (degradation) of ionomers and ionomer membranes 313–314
  spectroscopic membrane characterization methods 315–316
  thermal stability 314–315
  water transfer behavior 318–319
extended Kalman filter (EKF) 637
external leakage 722
extrusion process 596
Exxon Research and Engineering (ER&E) 711
Exxon-Zn/Br-design at project
  bipolar stack 713–716
  main problems encountered 721–723
  reservoir and periphery 720–721
  stack assembly 715–720
Eyring equation 358

f

failure mode and effects analysis (FMEA) 164, 177
Faradaic reaction 232
Faraday constant 56, 107, 345
Faraday's law 470
Fe-Cr flow battery (ICFB) 32, 34
  characteristics 742–743
  components in 745
    carbon-based electrolyte 747–754
    electrolyte 745–747
  operation for 743–745
  research results 756–758
Fenton's reagent 567
Fermi-Dirac distribution 99
Fermi energy 100
ferrocenes 926, 1009
ferrocyanide/ferricyanide $(Fe(CN)_6^{4-}/Fe(CN)_6^{3-})$ 929
Fiat Panda 1193–1196
Fick's law 598
Fick's second law 235
figures of merit (FOM) 71, 90, 92
finger prints 492
finite element method (FEM) 263
finite volume method (FVM) 369
flowable "slurry" electrode 805
flow and electrolyte distribution 267–268
flow batteries (FBs) 20, 21, 71, 74, 741, 791, 875, 1025, 1128. *see* cell-scale modelling
  advantages and environmental benefits 77–79
  ancillary systems 21–22
  application field 84–85
  bipolar plates 75
  cell-scale modelling
    influence of the state of charge 385
  cell stacks 75
  centrifugal pumps 76
  characteristics of 204–205
  chemical stability 129–130

chemistries   44–46
continuum scale   381
current collectors   75
definition of   156–159
deliver   1085
dual-flow circuit   101–105
early developments   29–31
efficiency of   205
energy efficiency   76
and energy storage   22–24
engineering aspects of   86–87
FBMS   77
Fe/Cr   32, 34
figures of merit (FOM)   90–92
flexibility and safety of   77
flow frames   76
fluid flow aspects of   87–90
fundamentals of
 non-equilibrium thermodynamics
   399
 open-circuit voltage (OCV)   391–393
heat exchangers   76
ideal characteristics   85–86
international standards for   155, 159
ion exchange capacity (IEC)   124–125
ionic conductivity   126–128
ion permeable separators   76
membrane characteristics   123–124
operation   311
organic   46–48
permselectivity of chemical species
   128–129
piping   76
PMS   76
porous electrodes   75
redox   33
regenesys polysulphide/bromide
   42–44
reliability and safety   87
science and scientific disciplines
   380
scientific domains   380
steady-state   381

swelling ratio (SR)   125–126
tanks   76
thermal and mechanical stability
   130
types of   79–84, 202
vanadium   37–42
vanadium flow batteries (VFBs)
   175–177
water transport   125–126
water uptake (WU)   125–126
zinc/bromine   34–37
flow battery cycle data   1167
flow battery energy systems (FBESs)
   1055
 history   1055–1056
 typical vanadium 1990   1058
 typical vanadium 2000   1059
flow battery management system (FBMS)
   73, 77
flow battery systems   675, 1041
 modularity/flexibility   1043
 power conditioning system   1041–1042
 reliability   1042–1043
 shunt currents   1042
 size   1043–1044
 SoC band limitation   1043
 topologies   1044
  evaluation   1050
  high voltage system with multiple
   tank pair   1048–1049
  high voltage system with one tank
   pair   1046–1048
  low voltage parallel connection
   1044–1045
  low voltage system with several
   DC/DC inverters   1046
  low voltage system with several
   inverters   1045–1046
  mixed parallel-series high voltage
   system   1049–1050
flow cell   838
flow field geometry   614
flow frames   76

flowing electrolyte battery, components   712
flow-rate control   529–530
fluorenone (FL)   985
Fourier transformation   254, 297
Fourier Transform Infrared (FTIR) spectrometer   314, 315
free energy calculations   344
free enthalpy   56
front-of-the-meter (FTM)   84
fuel cells   20, 54
fundamentals of

g

gas chromatography   128
Gaussian/Lorentzian   294
G band   284
gel permeation chromatography (GPC)   312
generation 2 (G2)   677
generation 2 vanadium bromide flow battery   678–680
generation 4 (G4) vanadium flow battery   677
generation 4 vanadium oxygen fuel cell (VOFC)   682–684
generation 1 (G1) VFB   675–677
generation 3 VFB   681–682
Gibbs energy   56
Gibbs free energy of activation   358
Gibbs–Helmholtz equation   56
glassy carbon electrode   572
goal definition   464–466
good impedance   251, 252
Gouy–Chapman–Stern theory   364
GPC/SEC (gel permeation chromatography/size exclusion chromatography)   313
graft polymerization   597
Grand Canonical Monte Carlo (GCMC)   362–364
graphene oxide (GO)   286
graphite felt (GF)   37, 525, 608, 749
grid interfacing techniques   1030

h

half-cell electrolytes   590
hazard operational process (HAZOP)   164
HCl mixed acid electrolyte   681–682
heat exchangers   76
hexadecyltrimethylammonium hydroxide   194, 195
highest occupied molecular orbital (HOMO)   100
highly ordered pyrolytic graphite (HOPG)   295
high voltage system with one tank pair (HV-1T)   1046
hot air welding   724
Hotzenblitz Commuter Car   1192
Hull cell   841
hybrid inorganic-organic IEMs   141–142
hybrid polymer electrolyte membranes (HMs)   140
  hybrid inorganic-organic IEMs   141–142
  organic polymer blends IEMs   142–143
hydrochloric acid   541
hydrodynamic voltammetry   253
hydrogen-based flow batteries   875
  cell architecture   877–880
  chemistry and cell performance   884–887
  considerations and governing phenomena   880–882
  durability   884
  H_2/halogen   889
  H_2/metal-ion   888–889
  H_2/O_2 890–891
  kinetics   882–883
  thermodynamics   882
  transport processes   883–884
hydrogen-bromine flow battery system
  application
    field tests   1150
    permittability   1150–1151
  energy domain
    electrolyte   1145–1146
    hydrogen   1146

power domain
   balance of system   1149–1150
   membrane electrolyte assembly   1147–1148
   stack   1148–1149
hydrogen bubble formation   274
hydrogen-cerium   828
hydrogen cycle   20
hydrogen electrode   57
hydrogen evolution/oxidation electrocatalysis   698
hydrogen evolution reaction (HER)   102, 268, 550, 644
hydrogen gas   60
hydrogen–halogen flow battery   1145
hydrogen infrastructure   1151
hydrophobicity/hydrophilicity   271
hydrothermal pretreatments   567
hyperbranched polymers   1017
hypothetical equivalent circuit   249

i

IEC 60050   160
IEC 60079-10-1   165
IEC 61427-1   162
IEC 61427-2   160, 162
IEC 62485-1   162, 163
IEC 62485-2   163
IEC 62932-1   160
IEC 62932-2-1   160, 163
IEC 62932-2-2   160
IEC 62932   177
IEC 62933-2-1   163, 164
IEC 62933-5-2   164
IEC 61427 series   162
IEC 62933 series   164
IEC TS 62933-3-1   164
IEC TS 62933-4-1   164
IEC TS 62933-5-1   164
IEC TS 62933-5-2   164
IEEE 1491   168
IEEE 1578   167
IEEE 1635   168
IEEE 1679-2020   167
IEEE 2030.2.1   166

imaging techniques   297
Imergy Power Systems, Inc.   747
impedance spectroscopy   310
implicit solvation models   345
independent power producers (IPP)   5
inductively coupled plasma atomic emission spectroscopy   128
infrared spectroscopy   296–297
in-situ analysis   218
in-situ approach   218
in-situ characterization methods
   charge/discharge cycling tests   319–320
   current efficiency   320–321
   discharge capacity and capacity retention   324–326
   electrochemical impedance spectroscopy   327–328
   energy efficiency   323–324
   membrane permeability estimation   328
   open-circuit voltage   326–327
   voltage efficiency   321–323
Institute of Electrical and Electronics Engineers   166, 168
interconnectors   8
interdigitated flow field (IFF)   749
interferometer   297
internal leakage   722
International Electrotechnical Commission (IEC)   160, 165
intrinsic activation energy barrier   358
inverse modelling   499
inverter
   DC link model   1031
   energy storage applications   1035–1037
   fault response   1037
   flow battery energy storage   1038–1039
   grid feeding   1033–1035
   single-phase   1037–1038
   six-switch, three-phase inverter circuit   1027–1031
   three-phase and pulse-width modulation   1031–1033

ion-conductivity, ionogenic groups 310–311
ion diffusion 642
ion exchange capacity (IEC) 124–125, 307, 310, 597
ion-exchange membrane (IEMs)
 cost of 130–131
 ion permeability of 311
ionic conductivity 126–128
ionomer membranes 307–310
ion permeability measurement 598
ion permeable separators 76
ion-selective membranes 906
iron-chromium 203
iron–chromium flow batteries (ICFB) 741
iron-chromium redox flow batteries (ICRFBs) 311
iron-flow battery (IFB) 1127
 design benefits 1129
 initial discharged state 1129
 optimization 1138–1142
 power module 1134–1138
 proton pump 1131–1134

k

Ketjenblack (KB) 954
kinetic Monte Carlo (kMC) method 360, 361, 363
Kohn–Sham partitioning 341
Kohn–Sham theory 341
Kramers–Kronig (KK) relations 251–252

l

lab-scale prototype 479
Lattice Boltzmann method (LBM) 263
 methodology of 367–369
 RFBs 369–371
laws of quantum mechanics 336
layer-by-layer (LbL) approach 140
layer to layer self-assembled method 597
lead acid 12
lead acid batteries 674
lead-acid flow battery 196
lead-cerium 828

lead deposition 837
lead flow battery 200–201
levelized cost of energy (LCOE) 473
 Annual Usable KWh 1142
 CapEx cost 1139
 equation 1138
 installation cost 1139
 O&M cost 1142
levelized cost of storage (LCOS) 71, 92, 473
Levich equation 253, 895
Lichtenegg battery 1097
life cycle assessment (LCA)
 definition 661
 methodology 660–661
lifetime models
limiting current 253
linear combination of atomic orbitals (LCAO) 343
linear-sweep voltammetry (LSV) 231
linear sweep voltammograms (LSV) 570
liquid side electrode 1148
liquid spillage 179
lithium-ion-based flow battery (LFB) 951
 challenges 968
 redox-targeting
  development 961–965
  lithium-oxygen flow batteries 966–968
  lithium-sulfur flow batteries 965
  principle 960–961
 semi-solid electroactive materials 951
  Li^+ insertion-extraction chemistry 953–955
  multiple redox reactions 959
  precipitation-dissolution chemistry 955–958
lithium-ion batteries (LIBs) 13, 71, 951
 cost for 674
 energy density of 675
 energy storage market 674
lithium-ion cells 23
lithium-oxygen flow batteries 966
lithium-sulfur (Li-S) batteries 955

local density approximation (LDA)   341
low-cost hydrocarbon membranes   693–694
lowest impact scores   1141
low voltage parallel connection (LV-P)   1044
low voltage system with several inverters (LV-DC/AC)   1045

m

machine learning techniques   493–494
Marcus equation   358
Marcus-Hush-Chidsey theory (MHC theory)   359
Marcus-Hush kinetic model   577
Marcus-Hush theory   357
market estimates   17, 18
Markovian process   360
mass spectrometer data   274
mass transfer functions   236, 238
mass transport problems   385–386
mean time between failures (MTBF)   92
mechanical membrane properties   317
mechanical stability   130
mediator-based FB   47
membrane electrode assembly (MEA)   682
membrane preparation methods
   casting method   596
   extrusion process   596
   graft polymerization   597
   layer to layer self-assembled method   597
membrane weight loss   311–313
mercurous-sulphate electrode (MSE)   563
mesoscale modeling   355–356
   discrete element method   366–367
   of dissipative particles dynamics   364–366
   of electrochemical kinetics   356
   electron transfer process   356–360
   Lattice Boltzmann Method (LBM)   367–371
   Monte Carlo methods   360–362
metal coordination complexes (MCC)   924–925
   anolyte chemistries
      Cr-based complexes   933
      Fe-based complexes   931–932
   aqueous solvents   926
      catholyte chemistries   926
      cobalt complexes   931
      ferrocene complexes   926–929
      ferrocyanide/ferricyanide complexes   929–930
      oligo-aminocarboxylate complexes   930–931
metal-free and aqueous FB   46
metal-free flow batteries   1115
meta-polybenzimidazole (m-PBI)   694
methanesulfonic acid (MSA)   541
methyl ethyl pyrrolidinium (MEP)   677, 1165
methyl viologen   1118
micro-computed tomography (CT)   269
microgrids   1074
microporous plastic-sheet   715
microporous polyethylene separators (MPS)   321
microporous separator (MPS)   76, 326
Micro Réseau Intégré Seraing (MiRIS)   1072
microscopical membrane characterization   317–318
microstructure characterization
   reconstruction of experimentally gained images   416–418
   stochastic microstructure generation   418–419
   structured electrodes   419
mixed acid electrolyte   681, 689
mixed parallel-series high voltage system (HV-MIX)   1049
mobile/transportable battery   13
modular electrochemical flow reactors   73
Mohr method   309
molecular orbital (MO)   340

molecular weight, ionomers 313–314
molecular wiring 107
monochromatic X-rays 216, 286
Monte Carlo methods 360–362
    Grand Canonical Monte Carlo (GCMC) 362–364
    kinetic Monte Carlo (kMC) method 360–363
multi-level modelling of RFB 468
    component level 470–471
    electrochemical level 469–470
    system level 471–472
multiple redox semi-solid-liquid (MRSSL) flow battery 959
multiscale modelling of porous electrodes
    electrochemical interface models 400–401
    porous electrode modelling 401–403
Murdoch University (MU) 1155
MW-plus battery 1088

n

Nafion 591, 592
    composite membrane 592
    membranes 756
Nafion/PBI 140
nanopores 609
naphthoquinone (NQ) 909
NAP-XPS 215
Navier Stokes equation 367, 387, 415, 420, 425
near-ambient pressure 215
near edge X-ray absorption fine structure (NEXAFS) 294
Nernst diffusion layer 386
Nernst equation 59, 469, 621
Nernstian diffusion layer 64
Nernstian relationship 109
Nernst layer 64
Nernst–Planck equation 599
neutron imaging 274
nickel plate 194
nitrobenzene (NB) 984
nitroxide radicals 1115, 1117
noise amplitude 365

non-aqueous metal coordination complex chalcogen chemistries
    acetylacetonate complexes 933–937
    amino-alcohol complexes 937
    tunable oxo complexes 937
    pyridyl chemistries 938
non-aqueous organic flow batteries (NAOFBs) 975
    amines 982
    anolytes 982
        benzophenone 984
        benzothiadiazole 984
        fluorenone 985
        nitrobenzene 984–985
        phthalimide 982–984
        pyridine derivatives 985–986
        quinones 987
        quinoxaline 987
        viologen 986–987
    applications 976
    bipolar redox organic molecules 987–989
    catholytes 979
        1,4-dialkoxybenzene derivatives 980
        nitroxyl radicals 979
    challenges 989
    cyclopropenium derivatives 981
    electrolyte conductivity 992–993
    electrolyte cost 993–995
    electrolyte stability 989–992
    membrane incompatibility 993
    phenothiazine and phenazine 980–981
    solubility and energy density 995
    transitional developments 996–997
non-charged metals 100
non-equilibrium thermodynamics
    concentrated electrolyte solutions 397–398
    constitutive relations 397
    general balance law 396–397
    generalized Nernst-Planck models 398–399

transport through Membranes 399–400
non-flammable electrolytes 463
non-flammable rating, for VFB electrolyte 177
non-ionic porous membrane 595–596
non-linear compression effect 266
non-perfluorinated membranes 134
normalised rate constants 573
normalization method 248–249
novel electrode materials 566–567
nuclear magnetic resonance (NMR) spectroscopy 216, 315
numerical inverse Laplace transform (NILT) techniques 237
Nyquist diagram 246
Nyquist plots 572

o

Ohmic losses 204, 617
Ohmic overvoltage 65
Ohmic resistances 231, 247–248, 385
Ohm's law 203, 246
oligo-aminocarboxylate (OAC) ligands 930
one-electron approximation 340
open-circuit voltage (OCV) 326, 327, 361, 391–393
open cycle gas turbines (OCGTs) 19
operando conditions 215, 218
operational expenditure (OPEX) 465
organic flow batteries (OFBs) 46–48, 897, 1007
organic polymer blends, IEMs 142–143
organic radical battery 1117
oxidative imbalance 644
oxygen evolution reaction (OER) 102
oxygen evolution (OER) side reaction 550

p

Paasch model 248
paper electrode 266
parallel plate flow 194
partial remixing 648

peak current 232
peak-to-peak separation 542
perfluorinated membranes 132–134
personnel protective equipment (PPE) 178
photoelectron 216
photovoltaics 8, 12
physico-chemical methods 215, 216
plane waves (PWs) 342
poly(arylene ether) (PAE) 138
poly(arylene sulfide) (PAS) 138
polyacrylonitrile(PAN) 218
polyacrylonitrile-based GF (PANGF) 751
polyaniline films 524
polyaromatic hydrocarbons (PAHs) 908
polybenzimidazole (PBI) 136, 139, 310, 600, 609
poly(tetrafluoroethylene) (PTFE) chains 132
polyethersulfone/sulfonated polyetheretherketone (PES/SPEEK) 595
polyimide-based membranes 135–136
polymer flow battery (PFB)
  basic organic redox moieties 1008–1010
  oligomers and polymers 1010–1014
polysulfide-bromine 203
polysulfide/bromine flow battery
  components 766
  electrochemical couple 770–771
  electrolytes 766–773
  module 779–780
    electrode 782–783
    frame 783–785
    membrane 781–782
    plant 785–786
  self-discharge reaction 771–773
  technology drivers 765
pore network modeling (PNM) 263
pore networks 266
pore-scale FB modeling
  boundary conditions
    for charge transport 423–424

pore-scale FB modeling (contd.)
　computational details   424
　　for fluid transport   423
　　for species transport   423
　charge transport   422
　electrochemical reaction   421–422
　finite volume method   424–425
　fluid transport   420
　Lattice Boltzmann method   425
　mass transfer coefficient   436–437
　microstructure characterization
　　415–419
　multi-phase reactive transport model
　　435–436
　pore network models   434–435
　reconstructed microstructure   430–434
　species transport   420–421
　structured electrodes   426–427
　　different initial concentrations   427
　　electrode active surface area
　　　427–429
　Taylor-Galerkin finite element method
　　425
pore-scale model (PSM)   263
porous electrodes   75, 361–362
porous membranes   143–144, 595
potential energy surface (PES)   338
potentiostat   232
powder pressing methods   88
Power Conditioning System (PCS)   1041
power domain
　membrane electrolyte assembly
　　1147–1148
　stack   1148–1149
power flow control   458–459
power management system (PMS)   73, 76
power networks   8, 10, 11
power rating   193
power systems
　development of DC and AC   4
　early use of energy storage on   4
　infrastructure   5–7
　operate and control   8–10
pressure drop   87
primary batteries   54

product life cycle   465
Proton Exchange Membrane Fuel Cell
　(PEMFC)   361
Proton Pump   1131
pumped-hydro
　alternatives storage to   19
　energy storage   18
pumping energy   623
pyridine   1118
pyridine (Py) derivatives   985

q

quantified risk analysis (QRA)   1150
quantitative crossover analysis   620
quantum level model   355
quantum-mechanical modeling   347
quantum mechanics   336
　computational electrochemistry
　　343–348
　concepts of   337–339
　density functional theory   339–343
　FB materials application   348–351
quasi-reversible   234
quinones   908, 987, 1008
quinoxaline   987

r

radiography   263
　electrode materials, morphology of
　　263–266
　flow and electrolyte distribution
　　267–268
Raman spectroscopy   216, 283
Randles circuit   247
Randles–Ševčík relations   233–235
random cycling (RC)   1175
raw polymer felts   564
rayon-based GF (RGF)   751
reaction kinetics   232
real-space simulation approach   240–242
rebalancing processes
　chemical regeneration   649–650
　electrochemical regeneration   650–652
　physical regeneration   647–649

receding time horizon optimisation control   459
rechargeable flow battery
　chemicals   194
　construction materials   194–195
　current density, effect of   198–199
　electrochemical reactions   196–197
　experimental procedure   201–202
　hazards assessment   199
　learning outcomes   199–200
　preparation of   195–196
　supplementary materials   200
　teaching assessment   199–200
redox-active anthraquinones   915
redox-active fluorenone derivatives   909
redox flow batteries (RFBs)   33, 215, 229, 231, 236, 237, 247, 308, 361, 363, 366
　data basis and quality   474–476
　economic modelling methods   472–474
　electrochemical characterization   221–224
　general observations   224
　goal definition   464–466
　multi-level modelling   468–472
　physico-chemical characterization   218–221
　scope and definition   463–464
　scope definition   466–468
redox flow Li-O_2 battery (RFLOB)   966
redox-mediated process   99, 101
　dual-flow circuit flow battery   101–105
　solid boosters   106–115
Redox Solid Energy Boosters   109
redox targeting system   108
reduction/oxidation (redox) reactions   336
reductive imbalance   644
ReFlex   1108
regeneration cell electrolyser   683
regenerative solid-oxide fuel cell (RSOFC)   876
regenesys polysulphide/bromide flow battery   42, 44

renewable energy generation   12
renewable energy sources   689
representation and features   491–493
reversible electrochemical flow reactors   72
reversible fuel cells   689–699
reversible VOFC system   683
Reynolds number   368
rotating ring-disc electrode (RRDE)   253
rotating ring-disc voltammetry (RRDE)   252
round trip efficiency (RTE)   91, 1142

S

scanning electron microscopy (SEM)   216, 297
Scherrer constant   287
Scherrer's equation   216
Schrödinger equation   338
scope definition
　economic system definition   468
　technical system definition   466–467
secondary batteries   54, 689
segmented felt   88
Selemion   593
semi-infinite diffusion domain   233
semi-solid suspensions   959
short-circuit testing   182
shunt current protection   723
shunt currents   87, 623, 1158
silent period   711
simulation study   452–456
single electrochemical reaction   231
single flow circuit   203
single-relaxation-time (SRT)   368
size exclusion effect   144
size exclusion membranes   1012, 1015, 1017, 1025
size-selective membranes   906
Slater determinant   340
slotted flow field   88
slurry   107
slurry iron flow battery concept and performance   809

smart energy cell stack   1168
SOC monitoring based on electrolyte properties
   Coulomb counting   636–637
   electrolyte conductivity   632–633
   electrolyte density and viscosity   634
   half-cell potential   630–632
   open circuit potential   628–630
   spectroscopic methods   634–635
   state observer based estimation   637
   ultrasonic methods   635–636
solid boosters   106–115
   kinetics of   113
   system design   113–115
   thermodynamics of   109–113
solid energy booster   47
solvation free energy   345
spectroscopic membrane characterization methods   315–316
sp^2-hybridised carbon   283
sp^3-hybridised carbon   283
spinning reserve   7
ssealing   718
stack voltage   75
standard cell potential   676, 690
standard cell voltage   90
standard hydrogen electrode (SHE)   563
state-of-charge (SoC)   74, 469, 539, 578, 1176
static water transport   598
stationary storage applications   14, 15
steam turbines   3
step rate-determining (rds)   62
stoichiometric coefficients   60
stoichiometric flow rate   529
Stripping after every cycle (SAEC)   1175
sulfonated poly (ether) ether ketone (SPEEK)   134, 470, 595
sulfonated polyimide (SPI)   136
sulfonated Radel R membranes (sRadelR)   315
sulfur balancing   777
sulfuric acid   541
supercapacitors   54
surface area determination and porosity   298–299
Sustainable Development Goals (SDGs)   659
swarm control   13
swelling ratio (SR)   125–126, 597
synchrotron X-ray radiography   266
system design analysis   463
system efficiency modelling
system level, RFB   471–472
system operator   8

t

tapered tunnel shunt current protection   717
technical system definition   466, 467
techno-economic assessment   463, 464
techno-economic FOM   92
techno-economic modelling
   classification of   477
   and data in literature   478–480
   goal and scope definitions   477–478
temperature-programmed desorption (TPD)   291–292
temperature-programmed oxidation (TPO)   292
temperature-programmed reduction (TPR)   292
2,3,5,6-tetrakis((dimethylamino)methyl) hydroquinone (FQH_2)   908
2,2,6,6-tetramethylpiperidine-N-oxyl (TEMPO)
   battery-cycling behavior   1115
   fundamental research   1120–1122
   redox reactions   1115–1120
   scale up   1122–1123
thermal activation   271
thermal degradation   314–315
thermal desorption spectroscopy   291
thermal hazards, VFBs   177–178
thermally regeneratable FB   47
thermal management   87, 620
thermal method, carbon materials   290–291
thermal reduction   286

thermal stability   130, 314, 315
thermodynamic equilibrium potential   469
thermodynamics   55, 61
thermogravimetric analysis (TGA)   292, 293, 314
thermogravimetry (TG)   292–293
thermoneutral cell voltage   57
thin-film composite membrane (TFCM)   595
three-dimensional pore-scale Lattice Boltzmann model   437
time-independent Schrödinger equation   338
time shifting electricity   10
titanium-cerium   829
titanium/iron system   30
TOMCAT beamline   263
total cost of storage   473–474
TPower   1108
transformers   4
transmission system operator (TSO)   15
triarylamines (TAA)   982
tunnel-protection complex   718
turbostratic carbon   282
two-way power converter   13

u

UFC 3-520-05   171
ultra-high vacuum (UHV) conditions   218
undivided copper-lead dioxide flow battery
  charge-discharge experiments   847–852
  experimental details   838
  flow cell and electrolyte circulation   838–841
  hull cell deposition   841–846
  methodology   841–842
  results   842–852
undivided zinc-cerium   826
uniform electron gas (UEG)   342
unit power/energy cost   71

UV-visible absorption spectroscopy   528
UV-vis spectroscopy   128, 311

v

vanadium-based FBs   37–42
vanadium batteries   1117
vanadium bromide system (V/Br)   678
vanadium-bromine   203
vanadium-cerium   827
vanadium electrolytes   659
vanadium FBES
  microgrid operation   1072–1074
  with momentary voltage drop compensation function   1059–1061
  off-grid area   1074–1076
  optical energy   1065
  since 2010   1065
  with underground tanks   1059
  wholesale market   1070–1072
  in wind farm   1061–1064
vanadium flow batteries (VFB)   74, 79, 83, 122, 335, 741, 1007
  amphoteric ion exchange membranes (AIEM)   594–595
  anion exchange membrane (AEM)   593–594
  applications   1099
  area specific resistance   598
  battery control and management system
    electrolyte flowrate control   457–458
    power flow control   458–459
  battery management   529–530
  bipolar electrode development   525–527
  bipolar plate   1101
  bipolar plates   567
  capital cost of   480–481
  carbon corrosion   579–580
  carbon or graphite felts   564–565
  carbon paper   565–566
  cation exchange membrane (CEM)   590–593

vanadium flow batteries (VFB) (contd.)
  chemical hazards 178–180
  chemical stability 599
  classification and working principle 591
  components of 664
  composition and physico-chemical properties 540
  cost structure of 480
  definition of 661–662
  degradation of electrode performance 580–581
  dynamic stack mass balance model 444–445
    mass balance during normal operation 445–448
    self-discharge during standby 448–450
  dynamic stack thermal model 450–452
  early field trials and demonstrations 513–516
  electrical hazards 180–187
  electrocatalysis of graphite electrodes 524–525
  electrochemical reactions in 547–550
  electrode kinetics and mechanism 568–577
  electrode material screening and development 523–524
  electrode requirements and materials 563–564
  electrode substrate materials 525–527
  electrode, thermal and chemical pretreatments of 567
  electrolyte 1100
    characterisation and optimisation 519–523
    composition 539–542
    degradation and mitigation strategies 550–554
  degradation processes 551
  physico-chemical properties of 544–547
  production 554–558
  evaluation and modification 527–529
  first licence 511–512
  flow cell research 518–519
  flow-rate control 529–530
  impact of 666
  ion conducting membrane 1102
  ion exchange capacity (IEC) 597
  ion permeability measurement 598
  LCA of 663
  LCOS of 481–482
  long-term performance of 578–581
  low cost electrolyte process breakthrough 512–513
  membrane characteristics and measurement 597
  membrane preparation methods 596–597
  membranes 590
  membrane screening 527–529
  nomenclatures 590
  non-ionic porous membrane 595–596
  novel electrode materials 566–567
  with other batteries 664–668
  patents 516–518
  performance in 599
  regulatory framework 175–177
  side reactions at 578–579
  speciation 542–544
  stack and energy storage system 1105
  state-of-charge (SoC) 539
  thermal hazards 177–178
  water transport measurement 598
  water uptake (WU) and swelling ratio (SR) 597
vanadium oxygen hybrid fuel cell (VOFC) 677
vanadium pentoxide 554, 646

vanadium redox flow battery (VRFB)
253, 311, 923, 1080, 1083. *see* SOC
monitoring based on electrolyte
properties
  capacity retention 619–620
  cycling of 621
  discharge polarization 617
  electrochemical reactor design for
    610–615
  electrochemical reactor for 616
  energy efficiency 615
  engineering design 614
  enhanced electrolyte utilization
    620–622
  high performance 616–619
  lower parasitic losses 622–623
  reactor design, principles of
    615–616
  zero-gap architecture 610
vanadium species 269
vanadium waste 513
vanadyl sulfate 554–555
V-Ce FB 700
V-Fe electrolyte 691–693
V-Fe flow battery 690–691
V-H_2 flow battery 694–698
  advantages of 698–699
  electrode for 698
VII/VIII and VIV/VV Redox Reactions
  568
viologen 986, 1118
viologens 910
virtual power plant 9
voltage efficiency (VE) 137, 321–323,
  544, 599
voltaic pile battery 193
voltammetry 842
volumetric energy density 90
VPower 1107

w

water transfer behavior 318–319
water transfer cell 598
water transfer test 528
water transport 125–126
water transport measurement 598
water uptake (WU) 125–126, 597
wedge welding process 728
white beam 216
wind power 12
wind-turbine 12
working electrode 231

x

X-ray absorption spectroscopy (XAS)
  216
X-ray and neutron radiography 264
X-ray beam 295
X-ray computed tomography (X-CT) 369
X-ray diffraction (XRD) 216
X-ray imaging 270
X-ray photoelectron spectroscopy (XPS)
  40, 215, 216, 293, 296, 315, 575
X-ray powder diffraction (XRD) 216, 286
X-ray probe 216
X-ray radiation 216
X-ray tomograms 264

z

zero-current or rest potential 391
zero-gap architecture 610
zero-gap configuration 613
Zero Waste Movement 662
zinc-bromine 203
zinc bromine battery (ZBB) 711, 1154,
  1172
  electric vehicles 1185
  EV-application 1173
    crash test 1181–1183
    electrical short 1184
    hydrogen recombination 1184
    measures against 1178–1181
    operation 1175–1177
    performance 1174–175
    safety 1178
    safety relevant investigations 1184

zinc bromine battery (ZBB) (*contd.*)
   shunt currents   1176
   state of charge   1176
   thermal management   1177
   vibration test   1183
zinc-bromine flow battery (ZBFB)
     34–37, 1154
   auxiliary power   1167–1168
   efficiency   1163–1166
   Gould Corporation   1155
   shunt currents   1158–1160
   solid bed   1166–1167
   stack   1155
     circulation system   1157–1158
     electrolyte system   1157
zinc-cerium   203, 823
zinc-nickel   203
zinc reservoir   736